Oxford's Sedleian Professors of Natural Philosophy

OXFORD'S SEDLEIAN PROFESSORS OF NATURAL PHILOSOPHY

The First 400 Years

Edited by

CHRISTOPHER D. HOLLINGS
AND
MARK MCCARTNEY

OXFORD
UNIVERSITY PRESS

OXFORD
UNIVERSITY PRESS

Great Clarendon Street, Oxford, OX2 6DP,
United Kingdom

Oxford University Press is a department of the University of Oxford.
It furthers the University's objective of excellence in research, scholarship,
and education by publishing worldwide. Oxford is a registered trade mark of
Oxford University Press in the UK and in certain other countries

Published in the United States of America by Oxford University Press
198 Madison Avenue, New York, NY 10016, United States of America

British Library Cataloguing in Publication Data
Data available

Library of Congress Control Number: 2023945982

ISBN 9780192843210

DOI: 10.1093/oso/9780192843210.001.0001

Printed and bound by
CPI Group (UK) Ltd, Croydon, CR0 4YY

FOREWORD

The 400th anniversary of the Sedleian Professorship of Natural Philosophy, the oldest scientific chair at the University of Oxford, provides a welcome opportunity to reflect on the individuals who have held this position in the past, from Edward Lapworth, who was appointed in 1621, to my immediate predecessor, Sir John Ball, who retired from the chair in 2019.

This volume weaves together the personal stories of the Sedleian Professors with descriptions of their academic achievements. It has to be said that they are a disparate group; if one were to imagine a class reunion, it isn't clear they would have much in common with one another. The seventeenth-century incumbents were primarily physicians. In the eighteenth century the Chair became something of a sinecure: post holders appear to have focused more on their ecclesiastical interests than on scientific teaching or research; many seem to have had no discernible credentials in natural philosophy. From the end of the eighteenth and through the nineteenth centuries there was a slow and somewhat uneven transition to what we would now recognize as a chair of natural philosophy, beginning with the election of the astronomer Thomas Hornsby in 1782, the first scientifically active professor of natural philosophy for almost a hundred years. Hornsby oversaw the establishment of the Radcliffe Observatory in the final decades of the eighteenth century.

Since the beginning of the twentieth century, the Sedleian Professorship has been held by some of the leading figures in applied mathematics and mathematical physics, including Augustus Love, who made important contributions to the theory of elasticity and to wave propagation, especially in the context of the Earth sciences; Sydney Chapman, who is renowned for his contributions to solar–terrestrial physics; George Temple, who did significant work in mathematical physics; Albert Green, who was a leading researcher in continuum mechanics; Brooke Benjamin, who did seminal work on mathematical analysis and fluid dynamics; and Sir John Ball, who is distinguished for his foundational research into the mathematical theories of elasticity and material science, and their connections to the calculus of variations.

Over the past 400 years the Sedleian Chair has been held by some remarkable characters, and it is illuminating to have their stories brought together in this volume. Reading them in chronological order gives a fascinating insight into the development of science in universities during this period.

It is understandable that the early Sedleian Professors would be physicians: in the seventeenth century, medicine was one of the most active and rapidly developing areas of scientific research. This group includes Thomas Willis, who was an important figure in the development of the understanding of the brain and nervous system.

What is, perhaps, surprising is that the important ideas introduced elsewhere in natural philosophy, by Galileo and Newton for example, took a relatively long time to influence teaching and research in the subject in Oxford. The early statutes required the Sedleian Professor to lecture specifically on Aristotle's *Physics*, and it was not until 1810 that these were changed to stipulate lectures 'in Natural Philosophy as grounded on Mathematical Principles, and particularly in the Principia [...] of Sir Isaac Newton'. Nevertheless, it is striking how, once it had decided to act, the university then made a conspicuous success of the Chair, appointing a succession of leading applied mathematicians through the twentieth century. It could be argued that the history of the first 400 years of the Chair throws as much light on the evolution of decision-making procedures within universities as it does on the development of science.

This volume follows on from a meeting held in June 2022 at the Bodleian Library in Oxford and supported by the British Society for the History of Mathematics and the London Mathematical Society. Christopher Hollings and Mark McCartney organized that meeting and have edited this book. They are to be congratulated for having drawn together contributors who have produced such vivid and insightful biographies.

Almost certainly Sir William Sedley, who made the original bequest that funded the Chair, would not have anticipated the intellectual trajectory it has taken over the past 400 years. That is as it should be in a dynamic and evolving field of study, but one might hope that he would be pleased to see the state it has achieved since the start of the twentieth century. It is their ability and mindset to plan over such long timeframes that makes universities such singular institutions.

Since the late nineteenth century, the Sedleian Professors have, in addition to their university appointments, also been fellows of The Queen's College in Oxford. By convention, only the Sedleian Professor or the Patroness of the College may operate the orrery in the Upper Library there. I suspect that this tradition is one that Sedley would have appreciated.

Jon Keating
Sedleian Professor of Natural Philosophy, University of Oxford;
Fellow of The Queen's College, Oxford.

CONTENTS

THE SEDLEIAN PROFESSORS OF NATURAL PHILOSOPHY[1]

1621 **Edward Lapworth** (1574–1636)
1636 **John Edwards** (1600–late 1650s)
1648 **Joshua Crosse** (*c.*1615–1676)
1660 **Thomas Willis** (1621–1675)
1675 **Thomas Millington** (1628–1704)
1704 **James Fayrer** (*c.*1655–1720)
1720 **Charles Bertie** (*c.*1679–1741)
1741 **Joseph Browne** (1700–1767)
1767 **Benjamin Wheeler** (*c.*1733–1783)
1782 **Thomas Hornsby** (1733–1810)
1810 **George Leigh Cooke** (1779–1853)
1853 **Bartholomew Price** (1818–1898)
1899 **Augustus Love** (1863–1940)
1946 **Sydney Chapman** (1888–1970)
1953 **George Temple** (1901–1992)
1968 **Albert Edward Green** (1912–1999)
1979 **Brooke Benjamin** (1929–1995)
1996 **John Macleod Ball** (1948–)
2019 **Jonathan Keating** (1963–)

Notes

[1] For simplicity, modern dating conventions have been adopted in this table.

CHAPTER I

Four Centuries of Sedleian Professors

CHRISTOPHER D. HOLLINGS

The Sedleian Professorship of Natural Philosophy is one of Oxford's oldest chairs, founded in the early part of the seventeenth century, alongside a number of other such posts, most notably the Savilian Professorships of Geometry and Astronomy, with which it has always been closely associated.[1] Originally a university position whose sole duty was to lecture on natural philosophy, the Sedleian Chair has developed over four centuries into a role that encompasses both teaching and research.[2] Along the way its focus has been narrowed from natural philosophy to applied mathematics. Indeed, the Sedleian Chair arguably has had the most interesting and varied development of any such professorship, with several twists and turns on its trajectory. In this first chapter, we describe the foundation and general trends in the development of the chair by way of providing background for the stories of the individual professors that follow in the remaining chapters.

By the beginning of the seventeenth century, the collegiate system in Oxford was already well established, with the power of the colleges having come to dominate over that of the central university. Most teaching took place at this time within a student's own college. The traditional mediaeval liberal arts curriculum formed the core of college instruction, and there was little room for newer or higher-level ideas. In particular, advanced mathematics and natural philosophy, comprising the study of nature and the physical universe, received little attention in the standard Oxford education leading to the BA.[3] That being said, a small number of colleges did employ lecturers to teach these subjects, and instruction was also provided within the central university by recent MAs known as the 'regent masters', but the quality of this education was not always particularly high.[4] It was precisely this inadequacy that the various professorial foundations of the early seventeenth century sought to address by providing for centralized university teaching in subjects outside the usual scope of college instruction. The sciences, which, with the notable exception of medicine, had barely been represented in teaching posts, benefitted in particular, and came gradually to be seen as valid subjects of university study.[5] New chairs and lectureships were created on the basis of benefactions to the university, and a

Christopher D. Hollings, *Four Centuries of Sedleian Professors*. In: *Oxford's Sedleian Professors of Natural Philosophy*.
Edited by: Christopher Hollings and Mark McCartney, Oxford University Press. © Christopher D. Hollings (2023).
DOI: 10.1093/oso/9780192843210.003.0001

Figure 1.1 Sir Henry Savile, founder of Oxford's Savilian Professorships of Geometry and Astronomy, probably also influenced Sir William Sedley in the establishment of the Sedleian Professorship of Natural Philosophy.

range of motives drove the donors, from a simple concern for scholarship, to a desire to leave a visible and lasting legacy: most of the founders had no sons.[6]

One figure who emerges as particularly prominent among the university's various benefactors at this time is Sir Henry Savile (1549–1622), Warden of Merton College, Oxford (from 1585), and Provost of Eton College (from 1596).[7] Besides founding the professorships that bear his name, Savile was also heavily involved in the establishment of the new Oxford library by Sir Thomas Bodley (1545–1613) in 1602, and may indeed have prompted Bodley in this direction in the first place.[8] The question of legacy was perhaps uppermost in Savile's mind (he had no male heir), but, having himself taught mathematics and astronomy in Oxford, he also had an insider's view of the teaching of these subjects, and held firm opinions about how they ought to be handled. In particular, he sought to establish the teaching of mathematics on humanist principles: it should stem from the works of classical antiquity, and scholars ought not only to have a concern for the history of their subject, but should actively seek to restore and edit surviving texts.[9] In asserting these principles, Savile was especially hands-on in laying down the rules for his foundation, and personally appointed his first two professors. He also advised and encouraged others in their bids to create similar posts, and almost certainly exerted influence in this way over the founder of the Sedleian Chair, Sir William Sedley.

Sir William Sedley

The Sedley (occasionally Sidley) family appears to have had its origins in the area along the Kent–Sussex border around Romney Marsh.[10] By the mid-fourteenth century, the branch of the family into which the founder of the Sedleian Chair would be born had established itself a little further north in the parish of Southfleet in Kent, where it occupied Scadbury Manor.[11] The family moved in high circles: one John Sedley of Southfleet (d. 1520) was an auditor of the exchequer of Henry VIII, and both his son William and grandson John went on to be High Sheriffs of Kent.[12] This last-named John Sedley (d. 1581) married Anne Colepepper of Aylesford near Maidstone, and together they had six children, among whom were sons John, William, and Richard.[13] It was William, born c.1558, who would go on to found the Sedleian Chair.

Upon their father's death, John Sedley, brother of the Sedleian founder, inherited the family's lands in Southfleet; he already held Aylesford Priory, around 10 miles away, having been granted the manor by Elizabeth I.[14] This John Sedley, however, ran up considerable debts, apparently through lavish charitable works, but also perhaps by spending the substantial sum of £4,000 in converting the ancient Carmelite priory at Aylesford into a suitable manor house.[15] He died childless in 1605 at the age of 44, whereupon the lands and properties in both Southfleet and Aylesford passed to his brother William.[16]

It is with this latest William Sedley that we come to the first definite Sedley connection to Oxford.[17] William Sedley matriculated at Hart Hall (now Hertford College) in 1574 at the age of 16, but he soon migrated to Balliol College, where he obtained his BA in 1574/5, with an MA following in 1577.[18] A year earlier, he had entered Lincoln's Inn, and was eventually admitted to the bar in 1584.[19] By the 1590s, he appears repeatedly as a Justice of the Peace in the rolls of the Kent Quarter Sessions.[20] As a local landowner, his name may also be found in records relating to the upkeep of the Kent Militia in the years following the defeat of the Spanish Armada.[21] Sedley was knighted in 1605 and was subsequently created one of the first baronets in 1611. In these later years, he appears from time to time in the *Calendar of State Papers*, particularly in connection with the governance of Kent.[22]

Figure 1.2 The Sedley family arms (Cambridge, Trinity College, MS R.17.8, p.103).

Early accounts of Sedley paint him as a charitable figure. The antiquarian William Camden (1551–1623), for example, noted his establishment of an almshouse and the building of a bridge in Aylesford, referring to Sedley as

> a learned knight painefully, and expensfully studious of the common good of his countrie as both his endowed house for the poore, and the bridge heere with the common voice doe plentifully testifie.[23]

In fact, the creation of the almshouse may have been the doing of Sedley's brother John in the final year of his life.[24] William Sedley's generosity also extended into the arts: he was patron of a Latin poet, John Owen (1563/4–1622?), who dedicated a book and several verses therein to Sedley, including the following:

> *To* William Sidley, *Knight and Baronet.*
> Thy noble worth doth Gold, doth Honour merit;
> Yet both flag far beneath thy gen'rous spirit.[25]

Charity appears to have run in the Sedley family: William's son, another John (c.1597–1638), would later establish a school in Southfleet that exists to this day.[26]

That William Sedley should go on to endow an Oxford professorship is at least consistent with his evidently charitable nature. But why he should make such a bequest, rather than, say, making further provision for the Kentish poor is at first glance somewhat mysterious. Sedley does not appear particularly to have been a scholar, nor did he have any obvious connection to the study of natural philosophy. Moreover, he differed significantly from the various other early-seventeenth-century founders of Oxford chairs in that he had a son. Sometime after 1587, Sedley had married Elizabeth, Dowager Baroness Abergavenny, and together they had a single son, the John Sedley mentioned earlier. Thus, Sedley did not need to found a chair in order to continue the family name. It may simply have been through his son, and his own acquaintanceship with Henry Savile, that the Oxford endowment came about.

Although there is nothing to document their acquaintance before 1613, it is possible that Savile and Sedley had moved in some of the same official circles for some time.[27] Savile had twice served as a Member of Parliament; as both Warden of Merton and Provost of Eton, he was particularly visible at the state level. Savile was knighted by James I the year before Sedley. His elder brother John (1546–1607) was, like Sedley, a barrister. The first time that we find Savile and Sedley mentioned together in state papers is in February 1613, when the negotiations began for Savile's daughter Elizabeth (1599–1660?) to marry Sedley's son John.[28] Savile evidently declined an offer subsequently made by another aristocratic suitor, for John Sedley and Elizabeth Savile were married later that same year.[29] It has been suggested that the price of that marriage, for the father of the groom, was a promise to endow an Oxford professorship.[30] Whatever the truth of the matter, when William Sedley came to write his will four years later, he included provision for 'a Lecture of naturall philosophie to be read and exercised in the universitie of Oxford'.[31] The university did not have to wait long for the bequest: Sedley died on 27 February 1618/19 and was buried at Southfleet among his Sedley ancestors.[32]

Figure 1.3 St Nicholas's Church, Southfleet.

Figure 1.4 The Sedley family tomb in St Nicholas's Church, Southfleet. The tomb contains the remains of Sir William Sedley's parents and great-grandparents, but his own final resting place is uncertain.

Sedley's will

William Sedley's will, dated 29 October 1617/18, is in many respects a standard example of such a document. It begins with the usual profession of faith and declaration of soundness of mind. A request to be buried at Southfleet follows, after which Sedley's various lands and riches are dispersed among his extended family. Naturally, Sedley's son and daughter-in-law are the main beneficiaries, but lump sums and annuities are also bequeathed to Sedley's brother Richard, his nephew and godson William, and also numerous cousins in the extended Sedley family. Sedley's servants are also well provided for, as are the servants of his late brother John, who it seems had not been receiving the sums promised to them in John's will. Money is left for the poor of several parishes, including an annual payment of £3 6s 8d for the poor of Aylesford, whose church also received £10 for repairs.[33]

Some university-related bequests appear early in the will. Sedley left £100 each for 'the newe Librarye of the universitie of Oxford' (namely, the Bodleian Library) and the library of Trinity College, Cambridge. Though the former is not so surprising—Sedley may have been acquainted with Thomas Bodley via Henry Savile—the latter is more puzzling, as the nature of Sedley's connection to Cambridge generally, let alone to Trinity specifically, is unclear. Having already taken his Oxford MA, he later received the Cambridge equivalent in 1586, but otherwise he does not appear to have had any links to that university.[34] Even the author of a seventeenth-century manuscript memorializing the donors to the Trinity College Library seems to have been mystified by Sedley's bequest, and suggested, with not a little conceit, that Sedley must simply have heard tell of the college's fame.[35] The same manuscript notes the titles of some of the books that were purchased with Sedley's money, but we have no such record for the bequest to the Bodleian.[36]

Our main point of interest in Sedley's will—his provision for the Sedleian Professorship—appears about halfway through, and is substantially different in style from his other bequests:

> I will and Devise that my Executor shall within twoe yeres next ensewing my decease Dispose and ymploye twoe thousand poundes of the money which I shall leave unto in the purchasing and buying of mannors Landes and Tenements so farre as the saied somme will extend: And I will that the Rentes and profitts of the mannors Landes and Tenem^ts so purchased by the advise of Councell learned in the Lawe shalbe for ever assured and Conveyed for the mayntennce of a Lecture of naturall philosophie to be read and exercised in the universitie of Oxford and to be Directed ordered and governed by the vice channcello^r of the saied universitie for the tyme beyng: the President of Magdalen Colledge in Oxford for the tyme beyng and the warden of All Sowles Colledge in Oxford for the tyme beyng or by any twoe of them [. . .][37]

Unlike many of the other founders of Oxford chairs—Savile in particular—Sedley imposed no further requirements on his endowment. It would be for the university to decide upon the conditions and duties of its new lecturer.

Foundation

Further evidence that William Sedley and his family were not intimately connected with Oxford may perhaps be seen in the fact that the university did not become aware of Sedley's bequest until January 1620, when they were informed of the legacy by Sedley's son John. The latter also offered to pay the professor's stipend until the appropriate sum had been handed over to the university and the necessary lands purchased. Once it learnt of the legacy, the university tasked a small panel of legally trained men with the securing of suitable lands, in accordance with Sedley's terms.[38] Before very long, they had settled on the manor of Waddesdon in Buckinghamshire, which they purchased for the university in 1622 for the sum of £1,600. Within a couple of years, the manor was returning rents of £120 p.a., and these funds would be paid to the professor until well into the nineteenth century. The main house on the Waddesdon site came to be known as 'Philosophy Farm'.[39]

Whilst the search for a suitable property went on, the university set about defining the role of its new professor. Its first set of basic conditions was set down by Convocation, the university's governing body, very shortly after it learnt of Sedley's bequest, and these are concerned very much with the *form* of the lectures, rather than the content: the professor would be required to lecture on natural philosophy at 8am every Wednesday and Saturday during term time to an audience of Bachelors of Arts, who would be fined one groat (4*d*) for absence; the professor himself would be fined 10*s* if he failed to deliver his lectures.[40] The precise content of the lectures was left for the professor to decide, which was by no means an unusual position for the university to take at this time, when the codifying of statutes was done largely to regulate religious and disciplinary matters.[41] Where further details of the expected teaching were enshrined in statute, they tended simply to suggest the fundamental texts that might be used, and nothing more. As we will see, such a suggestion was eventually built into the statutes governing the Sedleian Chair, but we will also see that the incumbent professor had the freedom to ignore this.

The first Sedleian Professor, taking up post in 1621, was the physician Edward Lapworth (1574–1636).[42] Like most of his successors up to the end of the nineteenth century, he was an Oxford man, and so his receipt of the chair has the nature of an internal appointment. The influence of Henry Savile may again be relevant, for Lapworth had links to Merton College: he had obtained a BA in 1592 at St Alban Hall, one of the university's mediaeval academic halls, which by the middle of the sixteenth century was owned by Merton but operated as a separate institution.[43] Between 1598 and 1610, Lapworth was Master of Magdalen College School, and from 1605 was licenced (by Magdalen College) to practice medicine. Following the award of his medical degrees in 1611, he left Oxford for private medical practice, possibly in Kent, but then returned to Oxford in 1619, when he was appointed to a Senior Linacre Lectureship, a medical teaching post controlled by Merton College.[44] Thereafter, he appears to have split his time between Oxford and private practice in Bath. No direct evidence survives of Lapworth's Linacre or Sedleian teaching, but the latter certainly took place on at least some occasions, since the diarist Thomas Crosfield (1602–63) makes passing references to Lapworth's lectures on natural philosophy during the 1620s.[45] It has also been argued that this teaching must indeed have taken place more extensively; otherwise, Lapworth would have had little reason to live part of the year in Oxford.[46] Some slight doubt is cast on the energy that he could have brought to his teaching activities though by the physical description given of Lapworth by the Bath physician Thomas Guidott (1638–1705): 'being in body not tall, fat and corpulent, which inclin'd him the more to take his ease'.[47] Guidott's description cannot have been based on first-hand knowledge, however, as Lapworth died in Bath in May 1636.

By coincidence, the year in which Lapworth died was also the year in which the statutes governing the Sedleian Professorship received the first of many modifications. By the early decades of the seventeenth century, it was increasingly recognized that the university statutes, having evolved over many centuries without a central guiding hand, had fallen into a chaotic state, with no single authoritative version. Upon being elected Chancellor of the university in 1630, the Archbishop of Canterbury, William Laud (1573–1645), set about the organization of the statutes. The resulting 'Laudian Code', approved by Charles I in 1636, would provide the basis for the government of the university for the next 200 years.[48] Most relevant for our purposes is the statute concerning the Sedleian Chair. No changes were made to the specifications set down in the early 1620s, but a further condition was added, expanding upon the expected content of the lectures:

> The lecturer in natural philosophy is to lecture in Aristotle's Physics, or the books concerning the heavens and the world, or concerning meteoric bodies, or the small Natural Phenomena of the same author, or the books which treat of the soul, and also those on generation and corruption [. . .][49]

Thus, in one respect, the lecturer was required to cover a broad range of topics under the heading of 'natural philosophy', and yet was limited by their all being rooted in the works of Aristotle. The requirement that the Sedleian Professor teach from Aristotle seems rather conservative in a world bursting with new ideas, but we must keep in mind that there were no other comprehensive texts that could be cited at that time, and moreover to have attempted to discuss natural philosophy without any mention at all of Aristotle would have been unthinkable. It has been argued that although the *letter* of the statutes may have seemed rigid and old-fashioned, the *spirit* of the statutes, widely understood, was that the professors (not just the Sedleian, but also the others whose statutes were being rewritten) were free to cover more recent ideas.[50] And, in those few places where evidence exists, this is what we see in the teaching of the subsequent Sedleian Professors.

Physicians

Lapworth's incumbency of the Sedleian Chair may perhaps have set a trend for the occupants of this new post, for most of his successors over the remainder of the seventeenth century were physicians.[51] As a result, the instruction given by the early Sedleian Professors (insofar as evidence of it survives) tended to focus on the biological sciences and natural history. Other parts of natural philosophy, such as cosmology and the physical sciences, were quickly taken over by the Savilian Professors, whose relative eminence, compared to their Sedleian colleagues, may have given them the power to encroach on other areas.[52]

First among Lapworth's Sedleian successors was John Edwards (1600–late 1650s).[53] Educated at the Merchant Taylors' School in London, Edwards entered St John's College, Oxford in 1618, where he fell under the influence of the college's President, William Laud. As a fellow of St John's during the 1620s, Edwards appears to have been active in a variety of college roles, and had at least some contact with new ideas in natural philosophy: around 1621, he presented the St John's College Library with a copy of Johannes Kepler's *Astronomia Nova* (1609).[54] In 1632, through Laud's influence, Edwards returned to the Merchant Taylors' School as headmaster, though he resigned the post after only two years and settled in Oxford once more. It was probably also under Laud's patronage that Edwards succeeded to the Sedleian Chair upon Lapworth's death in 1636. Edwards obtained his medical degrees at the end of the 1630s, and thereafter practiced medicine in Oxford. Of the Sedleian teaching that he carried out during this time (if any), there survives no trace. At the outbreak of the Civil War in 1642, Edwards was involved in arranging the accommodation of Royalist troops within the city, and in the securing of arms to defend the university against Parliamentarian forces. His loyalty to the king, however, led subsequently to a loss of status: in 1648, the Parliamentary Visitors, charged with purging Royalist sympathizers from the university, expelled Edwards both from his fellowship at St John's and from the Sedleian Chair, in light of his 'manifold misdemeanours'—though he still received his professorial stipend until into the following year.[55] Thereafter, he continued to live in Oxford and practice medicine. Upon his death in February 1657/8, Edwards was buried in London at the church of All Hallows-by-the-Tower, the last resting place of his mentor Laud.

The candidate whom the Visitors 'intruded' in Edwards's place as Sedleian Professor was Joshua Crosse (c.1615–76), then a fellow of Lincoln College.[56] Born in Lincolnshire, Crosse had studied at Magdalen Hall from 1632, and appears to have had interests in natural philosophy, since his name is linked to the Oxford Experimental Philosophy Club that met in the city during the 1640s (see Chapter 2).[57] It was his Presbyterian sympathies that caused the Visitors to appoint him simultaneously Sedleian Professor, Senior Proctor, and a fellow of Magdalen College in 1648.[58] Crosse had already been active in various roles within the university, and since 1644 had also been Professor of Law at Gresham College in London, though his 'affairs at Oxford had prevented his residence, and the due performance of his lectures'.[59] His wider Oxford duties may also have hindered his Sedleian teaching, for here again we have a professor who has left no trace of his lecturing activities.

Crosse moved in high Parliamentarian circles: he was awarded a DCL by the university in 1649 on the personal recommendations of Oliver Cromwell and Thomas Fairfax, and was one of the delegates sent to inform Cromwell of his election to the university chancellorship the following year.[60] However, as for Edwards before him, Crosse's political affiliations eventually came back to bite him. At the Restoration in 1660, Crosse was expelled from all his university and college positions, and even charged with embezzlement.[61] Towards the end of his life, he was appointed an official of the Archdeaconry of Norwich, but otherwise lived out his days in Oxford; he died there in May 1676 and was buried in the church of St Peter-in-the-East.[62] Crosse's tenure as Sedleian Professor appears to have been an episode that the university wished to forget: in appointing his replacement, the university declined even to acknowledge Crosse's existence, instead deeming the new Sedleian Professor to be the successor to John Edwards.[63]

With Crosse's successor, we come to the best documented of the seventeenth-century holders of the Sedleian Chair, the physician Thomas Willis (1621–75), a pioneer in the study of the anatomy of the human brain, nervous system, and musculature, and a founder member of the Royal Society.[64] At the time of his election to the Sedleian Professorship, Willis was already famed as a physician, but the fact that he was also a staunch Anglican, and that his brother-in-law, John Fell (1625–86), had recently been appointed Dean of Christ Church, cannot have hurt his position.[65] For the first time in our survey of the early Sedleian Professors, we have a record of the lectures that were delivered: notes taken by students Richard Lower (1631–91) and John Locke (1632–1704) in the first half of the 1660s are preserved in the British Library and in the archives of the Royal Society.[66] Willis's life and work are discussed in greater detail by Alastair Compston in Chapter 2. For the moment, we note simply that far from confining himself to the letter of the statute that would have had him teach from Aristotle, Willis instead lectured on topics from the cutting edge of his own medical researches.

Figure 1.5 Thomas Willis, seventeenth-century Sedleian Professor.

Upon his death in 1675, Willis was succeeded in the Sedleian Chair by another prominent physician, his sometime collaborator in the Experimental Club, Thomas Millington (1628–1704).[67] Though educated at Trinity College, Cambridge, Millington had been made a fellow of All Souls College, Oxford by the Parliamentary Visitors in 1649, and had managed to retain this position even after the Restoration.[68] He went on to build a high-profile career, as a fellow and later president of the Royal College of Physicians. Millington was knighted in 1680, attended Charles II in his final illness in 1685, and was later physician both to William III and to Queen Anne.

Millington probably owed his election to the Sedleian Chair to the influence of Ralph Bathurst (1620–1704), a fellow physician from the circle around Willis, who by 1675 was President of Trinity College, Oxford and the university's Vice Chancellor.[69] In addition, Millington

INTRODUCTIO

A D

VERAM PHYSICAM.

SEU

LECTIONES PHYSICÆ.

Habitæ in Schola Naturalis Philofophiæ Aca-
demiæ Oxoniensis,

Quibus accedunt Chriftiani Hugenii *Theoremata de Vi
Centrifuga & Motu Circulari demonftrata,*

Per Jo. KEILL è Coll. Ball. A. M. & Reg. Soc. Socium.

MB. f. U. D.

OXONIÆ,

E THEATRO SHELDONIANO,

Impenfis *Thomæ Bennet,* ad Infigne Lunæ Falcatæ in Cœ-
meterio S. *Pauli* LONDINI, An. Dom. MDCCII.

Figure 1.6 Title page of John Keill's *Introductio ad veram physicam* (1701), concerning natural philosophy as taught in Oxford at the turn of the eighteenth century.

was deemed to have a sufficiently broad scientific background for the post, at least in contrast to the botanist Robert Morison (1620–83), whose application for the chair failed because he was regarded as being too narrow in his expertise.[70] On the whole, however, Millington does not appear to have followed Willis in engaging actively with his professorial teaching duties. By the time that he delivered his inaugural lecture as Sedleian Professor in 1676, he had probably already left Oxford for lucrative private medical practice in London.[71] There is a little evidence that he gave some lectures relating to chemistry early in his tenure, but otherwise Millington appointed deputies to carry out most of his Sedleian teaching.[72] Most prominent of these was the Scottish natural philosopher John Keill (1671–1721), who was responsible for the introduction of Newtonian ideas to Oxford via the lectures on experimental philosophy that he delivered at Hart Hall from the mid-1690s.[73] The version of these lectures that he subsequently published as *Introductio ad veram physicam* (1701) suggests that not all of his demonstrations of Newtonian natural philosophy were practical, but that some were mathematical in nature.[74] The study of natural philosophy in Oxford was thus nudged slightly in the mathematical direction that would go on to affect the nature of the Sedleian Chair.[75] Keill began deputizing for Millington in 1699, and here again used the opportunity to promote Newtonian physics, as well as to apply Newtonian concepts to physiology and chemistry.[76] This arrangement probably continued until the end of Millington's life in 1704.

The eighteenth century

Upon Millington's death, Keill might have hoped to have been elected to the Sedleian Chair, but it was not to be, for the reasons discussed by Nigel Aston in Chapter 3. Instead, he appears to have continued with his lectures on experimental philosophy, before finally securing the Savilian Professorship of Astronomy (at the second attempt) in 1712.[77] In the meantime, the Sedleian Chair took a rather different turn, far from natural philosophy.

The Sedleian Professors of the first part of the eighteenth century are discussed in greater detail in Chapter 3, where they are rescued from the obscurity into which they have often faded, particularly in writings on science and mathematics in Oxford. None of these professors had any credentials to speak of in natural philosophy, and probably did not deliver their statutory lectures. Instead, they were appointed successively to the Sedleian Chair under the patronage system that governed public life in eighteenth-century Britain, where the biases and manoeuvrings of the electors outweighed any consideration of scientific credentials. Election to a chair was a recognition of status, and often a reward for prior service. Only rarely did the specific title of the professorship have any bearing on the appointment. This is why, for example, we find a physician, John Smith (1721–97), occupying the Savilian Professorship of Geometry during the later decades of the eighteenth century. A lack of expertise in a particular field could easily be remedied by the appointment of a competent deputy to deliver the necessary lectures—Smith's deputy, for instance, was Abraham Robertson (1751–1826), who would eventually succeed him in the Geometry Chair.[78]

The first Sedleian Professor to be elected in this way, perhaps because he shared politics with the electors, was James Fayrer (or Farrer) (*c*.1655–1719/20).[79] Little evidence survives of Fayrer's scholarly interests, if any, but we gain a glimpse of the man—though not an entirely unbiased one—from the diaries of the antiquarian Thomas Hearne (1678–1735). Fayrer's lack of qualification for the Sedleian Chair appears to have won him Hearne's enmity: he was referred to by Hearne as 'a Fellow all Gutts without Brains',

> who to the great prejudise [sic] and Dishonour of yᵉ University, by yᵉ Interest of a few Corrupt Electors got to be Natural Philosophy Readʳ [...][80]

Hearne appears to have been particularly offended by the academic contrast between Fayrer and the then Savilian Professor of Geometry, Edmond Halley (1656–1741/2): in mentioning a lecture by Halley concerning observations of a recent meteor in 1719, Hearne noted that Halley had described the motion of the meteor

> but left the Causes to be explain'd by the Natural Philosophy Reader, Dr. Fayrer [...], who is by no means qualify'd.[81]

Fayrer died in 1719/20 and was buried in the chapel of Magdalen College, of which he had long been an active fellow.[82] Hearne, however, was equally unimpressed by Fayrer's successor as Sedleian Professor, Charles Bertie (c.1679–1746), whose election, according to Hearne, was 'not upon account of any skill (for he hath none) in Natural Philosophy'.[83] Instead, Bertie's expertise lay in the law, and he was appointed to the Sedleian Chair thanks to the influence of his friend and colleague Bernard Gardiner (1668–1726), Warden of All Souls.[84] Like Fayrer before him, Bertie did not deliver any lectures as Sedleian Professor, and in fact spent the last 15 years of his tenure away from Oxford as rector of Kenn in Devon.[85]

Similarly absent for much of his incumbency of the Sedleian Chair was Bertie's successor, Joseph Browne (1700–67). Appointed to the professorship in 1741, perhaps because he was viewed as a sound academic administrator (see Chapter 3), Browne left Oxford for the rectorship of Bramshott in Hampshire in 1746, but returned a decade later to take up the provostship of his alma mater, The Queen's College.[86] His stature within the university means that Browne appears frequently in many of the surviving records and reminiscences of the time, but always as an official, never as a scholar.[87]

For the fourth and final Sedleian Professor in this sequence of figures remote from natural philosophy, scholarship was, however, a matter of concern. Benjamin Wheeler (c.1733–83) was elected to the chair in 1767 following Browne's death, by which time he had already been an active tutor at Magdalen for some years.[88] He would go on to fill a number of roles within the university, and does indeed appear to have delivered lectures on experimental philosophy on at least one occasion.[89] But the greater part of Wheeler's teaching energies were directed towards his duties as Regius Professor of Divinity (from 1776). Materials relating to Wheeler and his divinity lectures were preserved by the young Martin Joseph Routh (1755–1854), future President of Magdalen, who would go on to exert an influence on the Sedleian Chair in the nineteenth century (see Chapters 5 and 6).[90] Wheeler held the divinity chair until the end of his life in 1783, but for reasons that are lost to us, he resigned the Sedleian Chair the year before. At this point, a change of attitude within the university meant that the time had come for the professorship to be steered back towards its founder's intentions.

A return to natural philosophy

Although the Sedleian Professors of the period 1704 to 1782 gave little attention to natural philosophy, this did not mean that the subject was absent from the university, but rather it persisted in the hands of others.[91] We have already seen, for example, that in the years following his Sedleian disappointment, John Keill nevertheless continued to deliver lectures in experimental philosophy that were akin to those that he had given as Millington's deputy. There is no evidence of Millington's successors having appointed sub-lecturers, but teaching in natural philosophy was carried out by others. At the beginning of the century, these included David Gregory (1659–1708), the Savilian Professor of Astronomy; John Whiteside (1679–1729), Keeper of the Ashmolean Museum; and John Theophilus Desaguliers (1683–1744), who continued Keill's lectures at Hart Hall during the latter's absence from Oxford between 1709 and 1712.[92] In particular, they promoted the Newtonian philosophy that had been introduced to Oxford by Keill. During the following decades, James Bradley (1692–1762), Keill's successor as Savilian Professor of Astronomy, and Astronomer Royal from 1742, maintained this tradition, lecturing in experimental philosophy, and also communicating Newtonian ideas to non-specialist audiences in largely non-mathematical terms.[93]

As Jim Bennett discusses in Chapter 4, the conception of 'natural philosophy' within the university underwent a shift during the eighteenth century, from its traditional basis in the works of Aristotle, to the embracing of newer forms of knowledge of the natural world. Thus, the nomenclature of the subject caught up with what had already been the reality for some time. No longer would the teaching of natural philosophy be founded on learned authority, however nominal that might have been in the case of prior Sedleian teaching, but on experiment and demonstration. This new emphasis was institutionalized in 1749 with the creation of a Readership of Experimental Philosophy, which over the following century went on to become one of the major scientific teaching posts within the university.[94] The creation of such a position, whose duties might reasonably have come under the purview of the Sedleian Chair, indicates how thoroughly the latter had drifted away from its founding purpose. To begin with, the stipend associated with the readership was too low for the post to be an independent one—it was not until 1839 with the appointment of Robert Walker (1801–65),[95] whom we shall meet again in Chapters 5 and 6, that the readership was able financially to stand on its own—and so its early incumbents necessarily held the position in conjunction with other appointments. The first Reader of Experimental Philosophy was the then Savilian Professor of Astronomy, James Bradley, and when in 1763 Thomas Hornsby (1733–1810) succeeded Bradley in the Savilian Chair, he also took on the Readership. During the following decades, Hornsby would accumulate other positions within the university, including those of Radcliffe Observer (from 1772) and Radcliffe Librarian (from 1783).[96] And upon the resignation of Benjamin Wheeler from the Sedleian Chair in 1782, it seemed entirely natural that this professorship should pass into the hands of someone who had already been teaching natural philosophy, in its latest incarnation, within the university for nearly twenty years. The life and career of Thomas Hornsby, and the circumstances surrounding his election to the Sedleian Chair, are described in greater detail by Jim Bennett in Chapter 4.

A mathematical turn

Beginning in the lectures of John Keill, and continuing in those of Edmond Halley, David Gregory, James Bradley, and others, a Newtonian mathematical conception of natural philosophy had been introduced to Oxford, and remained present to varying degrees throughout the eighteenth century, alongside the experimental parts of the subject.[97] Thus, natural philosophy was shifted from its qualitative Aristotelean origins towards a more quantitative Newtonian discipline.[98] At the same time, the medical instruction that had been carried out by the Sedleian Professors of the seventeenth century was now more naturally carried out elsewhere within the university, far from the chair in natural philosophy,[99] and the teaching of chemistry, which had occasionally fallen to the Sedleian Professors, was similarly regularized in the hands of others, with the more clearly defined role of Professor of Chemistry emerging in the late eighteenth century.[100] Regardless of the letter of the statutes governing the Sedleian Chair, by the beginning of the nineteenth century its domain of responsibility had shrunk to those subjects that we would now place under the heading of 'physics', both theoretical and practical.

A further break occurred in 1810 with the death of Thomas Hornsby. As we have seen, Hornsby accumulated a number of positions during his career, which were now to be distributed among a new generation of Oxford academics, although some would go to the same person. In particular, the link between the Savilian Professorship of Astronomy and the Readership of Experimental Philosophy that had been forged and strengthened over the course of six decades, first under Bradley and then under Hornsby, remained in place.[101] Abraham Robertson, Savilian Professor of Geometry, took the sideways step into the Astronomy Chair, and at the same time assumed the Readership of Experimental Philosophy, with responsibility for practical demonstrations in natural philosophy.[102] The Sedleian Chair, however, went elsewhere, taking with it the theoretical (i.e., mathematical) aspects of the subject. This was reflected in the bond that Hornsby's Sedleian successor, George Leigh Cooke (1779–1853), signed when taking up the chair: in it, he agreed to lecture 'in natural Philosophy as grounded on Mathematical Principles, and particularly in the Principia Mathematica Naturalis Philosophiae of Sir Isaac Newton'.[103] Somewhat later during Cooke's tenure, in 1839, a modified statute would decree more generally that the Sedleian Professor should 'expound the authors of best repute in physics'.[104]

Cooke's life and career are discussed more fully in Chapter 5, but we note here that he was in many ways a transitional figure. The first of the Sedleian Professors to take (applied) mathematics as the focus of his teaching, he seems to have been diligent in delivering his lectures during his early days in the chair, but he soon followed many of his eighteenth-century predecessors into academic inactivity, preferring instead to focus on his duties within the Church. As in the previous century, instruction in applied mathematics was picked up by others, but by the 1850s, such neglect of professorial duties was just one of several interlinked problems within the university that were leading to calls for reform.

A period of reform

Changes to the Oxford system had been underway, gradually and with much argument, since the beginning of the nineteenth century. The inadequacy of examination arrangements, for example, had long been recognized: candidates for the BA were expected to do little more than attend ritualized oral exams in which they provided pre-prepared answers to standard questions.[105] It was in response to external criticism of this system that three leading figures within Oxford—John Parsons (1761–1819), Master of Balliol; John Eveleigh (1748–1814), Provost of Oriel; and Cyril Jackson (1746–1819), Dean of Christ Church—pushed a new examination statute through the university in 1800.[106] Inspired by the success of fellowship examinations within their respective colleges (Oriel in particular),[107] the trio argued for a new system of public oral examinations in which candidates would be tested rigorously on aspects of Oxford's traditional curriculum: grammar, rhetoric, logic, moral philosophy, mathematics, and the articles of the Church of England—but not natural philosophy. The first examinations under the new system took place in December 1801. Over the following decades, the examination statute was frequently debated and modified, and not always to the benefit of science and mathematics; in 1807, for instance, mathematics was dropped as one of the compulsory subjects for the BA and was examined instead under its own optional honour school, with the result that very few candidates thereafter elected to sit for honours in mathematics.[108]

During the early part of the nineteenth century, the low status of mathematics within the university became a point of contention, particularly in light of the central position that it was given in a Cambridge education.[109] Oxford clung to the idea of a university as a place for training the clergy, for whom 'classics constituted the only effective mind-sharpener'.[110] Mathematics, and science more generally, was seen as less valuable in training young minds because its study was deemed to require less exercise of judgement.[111] Nevertheless, mathematics had its advocates, most notably Baden Powell (1796–1860), who was elected Savilian Professor of Geometry in 1827.[112] Warned against delivering an inaugural lecture, since it was unlikely to draw an audience, Powell instead wrote a polemical pamphlet, *The Present State and Future Prospects of Mathematical and Physical Studies at the University of Oxford*, in which he argued for mathematics and the physical sciences as part of a liberal education, and condemned those who would neglect them.[113] Although the tone of Powell's pamphlet made it slightly counterproductive,[114] his efforts elsewhere, in combination with those of like-minded colleagues, had a strong long-term effect on the study of mathematics in Oxford. He was involved, for example, in the establishment of the university's mathematical scholarships in 1831, from which many of Oxford's subsequent advocates of mathematics (including the future Sedleian Professor Bartholomew Price) would eventually benefit.[115]

One of Powell's greatest concerns was the fact that he could rarely muster the student numbers to run his lectures. As Savilian Professor, he was obliged to lecture on topics in higher mathematics (for example, calculus, on which he wrote an undergraduate textbook),[116] and yet these were subjects that were not examined. Thus, although attendance was supposedly mandatory, this regulation was not enforced, and undergraduates otherwise had little motivation to attend these lectures.[117] The situation was broadly similar across other scientific disciplines, or indeed arguably worse, for in general the sciences were taught only within the central university, and rarely in individual colleges. An Oxford undergraduate might therefore complete the supposedly liberal education leading to the BA without encountering any substantial amount of science. As we shall see in Chapter 5, there is little evidence of any teaching delivered by George Leigh Cooke, at least during the later years of his tenure as Sedleian Professor, but this may have had as much to do with the difficulty of attracting students as it had with any disinclination on his part to lecture.

(a)   (b)

THE PRESENT STATE

AND

FUTURE PROSPECTS

OF

MATHEMATICAL AND PHYSICAL

STUDIES

IN THE UNIVERSITY OF OXFORD,

CONSIDERED

IN A

PUBLIC LECTURE,

INTRODUCTORY TO HIS USUAL COURSE,

IN EASTER TERM, MDCCCXXXII,

BY THE

REV. BADEN POWELL, M.A. F.R.S.

OF ORIEL COLLEGE,
SAVILIAN PROFESSOR OF GEOMETRY.

OXFORD,
PRINTED BY W. BAXTER FOR THE AUTHOR.
SOLD BY J. H. PARKER;
AND BY J. G. AND F. RIVINGTON, LONDON.
1832.

(5)

Figure 1.7 Baden Powell, advocate for mathematics in nineteenth-century Oxford.

By the beginning of the 1850s, matters were coming to a head in relation to the various issues alluded to above. The value of a scientific education in an increasingly industrialized country was being recognized at high levels, as was the fact that—despite Cambridge's acknowledged strength in mathematics—neither of the ancient English universities provided effective instruction in this direction. University scientific education in particular suffered from the mismatch of teaching and examination. More broadly, the eighteenth-century habits of pluralism and absenteeism persisted among the professoriates of both universities. On top of all this, the finances behind many of the professorships had reached a perilous state: centuries-old endowments were no longer providing the returns necessary to pay an adequate stipend. A Royal Commission, established by Lord John Russell's Whig government in 1850, was tasked with investigating and resolving these matters.[118]

Reform and the Sedleian Chair

The story of the changes made within the university in the mid-nineteenth century, and their effects on the teaching of science in particular, is a complicated one that we will not address in full here.[119] Instead, we will confine the discussion largely to the impact of the reforms on the Sedleian Chair.

The enquiries that the Commissioners began to make into university and college teaching and finances in the early 1850s were not widely welcomed in Oxford. In an age when universities received no government funding, and were essentially private institutions (from a legal-technical point of view, they were corporations), the Commission was seen as an unwarranted intrusion, and most senior figures simply refused to answer its questions. The result was that the Commission's report, when it appeared in print in 1852, consisted largely of evidence given by people with an axe to grind.[120] The Commission laid bare some of the long-standing divisions within the university, and was seen (rightly) as an attack on the supremacy of the colleges over the central university. Heads of colleges and senior college fellows defended their vested interests, as well as the traditional Oxford education, by dismissing the Commission out of hand, whereas many other figures, such as junior fellows or college lecturers, who normally had little voice in the running of the university, rushed to give evidence, and to share their opinions on how reforms might be effected. This being said, not all of the established figures within the university were opposed to reform: Francis Jeune (1806–68), Master of Pembroke College, was an instrumental figure in the establishment of the Commission in the first place, for which action he incurred 'the odium (with social consequences) of his fellow heads of house [i.e., heads of colleges]'.[121] Jeune subsequently served as one of the Commissioners, along with Baden Powell. These two at least would have been receptive to the evidence given by Jeune's close colleague from Pembroke, Bartholomew Price (1818–98), who was already a well-established teacher of mathematics in the college setting, and would very soon take up the Sedleian Chair (see Chapter 6).

Unsurprisingly, Price's evidence to the Commission centred upon the arrangement of mathematics teaching within the university.[122] In his view, sufficient provision was available, but it needed to be better organized. Pure mathematics instruction should remain the responsibility of the Savilian Professor of Geometry, while mixed mathematics (i.e., mathematics applied to the physical world, including its applications in astronomy) should be divided between the Savilian Professor of Astronomy and the Sedleian Professor of Natural Philosophy. The former should live up to his title by teaching astronomy, which would leave such topics as mechanics, geometrical optics, and the theories of light, heat, and electricity ('Mixed Mathematics, short of Astronomy')[123] for the latter. Price reinforced the theoretical nature of the Sedleian Professor's domain of responsibility by explicitly placing all practical instruction into the hands of the Reader of Experimental Philosophy, where indeed it had already been for several decades. The evidence given by other interested parties, such as the Savilian Professor of Astronomy William Fishburn Donkin (1814–69), followed broadly similar lines. One point that emerged quite clearly was that some kind of revival of the Sedleian Chair should feature in any reform of university mathematics teaching. There is no suggestion that the elderly and absent Cooke played any part in these discussions. In his evidence to the Commissioners, the geologist Nevil Story Maskelyne (1823–1911) described the Sedleian Professorship as 'practically obsolete'.[124]

With regard to mathematical instruction within the university, the Commissioners' 1852 report broadly adopted the suggestions that had been made by Price and others.[125] The corresponding Act of Parliament that followed in 1854, however, was somewhat toothless: it did not compel change, and its recommendations were largely ignored within the university.[126] Nonetheless, the grounds for reform had been prepared: major changes—such as the dropping of the requirement for college fellows to be ordained priests, which was already being discussed in the 1850s—would unfold gradually over the remainder of the century

Figure 1.8 Francis Jeune, reforming Master of Pembroke College.

as successive Commissions sat and further bills passed through Parliament.[127] Mathematics instruction would slowly take the form advocated by the witnesses at the first Commission.[128]

Hidebound figures within the university of the 1850s were, however, not blind to the fact that reform might yet be forced upon them. From the beginning of the decade, a number of pre-emptive changes were introduced, in the hopes of staving off enforced reform from without. One such change involved the acknowledgement of additional fields of study via the creation, in 1853, of two new honour schools to sit alongside the existing schools of classics (*literae humaniores*) and mathematics: one in natural sciences, and another in law and modern history. A corresponding change to the examination statutes now compelled an undergraduate to sit final examinations in one other honour school besides the still-obligatory classics. In practice, however, the dominance of the latter, which remained compulsory until 1864, diluted the attention that students were able to give to these other subjects.[129]

By the mid-1850s, the university was also giving attention to the other concern of the Commissioners: the state of the professoriate. Discussions were underway as to how to shore up the finances underpinning the various professorships, and organizational changes were being suggested at the same time.[130] In the case of the Sedleian Professorship, it was proposed to enlarge the board of electors to include both Savilian Professors, and perhaps also the Senior Examiners within the schools of mathematics and natural sciences.[131] Such a move would give greater emphasis to the candidates' scientific credentials in any future elections, although it was criticized in an anonymous pamphlet for being a departure from Sedley's original stipulations.[132] Nevertheless, after much debate within the university, a new professorial statute received the necessary royal approval in April 1858.[133] Apart from specifying an enlarged board of electors, the statute set out few conditions for the Sedleian Professorship. The professor was to reside in Oxford for at least six months of the year, and was now free to deliver his lecture courses in whatever manner and at whatever times he saw fit. He could now also be more easily removed from his position if he neglected his duties, but at the same time procedures were put in place for appointing a deputy in the event of the professor's illness. Perhaps with the physicians of the seventeenth century in mind (but also mirroring conditions that Henry Savile had earlier placed on his professors), the university barred the Sedleian Professor from pursuing any other profession whilst holding the chair. A previously variable stipend was now guaranteed at £300 p.a., to be made up from the University Chest, should the income from the Waddesdon estate be insufficient. And at this stage, it was also made explicit that the Sedleian Professor should lecture in 'Physical Mathematics'.[134]

Securing the finances

The enlarged board of electors for the Sedleian Chair that appeared in the approved version of the new professorial statute was not quite the same as that which had been discussed at first: both Savilian Professors were to be electors, but neither of the proposed examiners subsequently found their way onto the list. Instead, the board was further expanded by the addition of both the Astronomer Royal and the President of the Royal Society, perhaps to secure the scientific underpinnings of the chair more fully, and to lend it a greater visibility in the British scientific community at large. The Vice Chancellor remained an elector, but the two other original members of the board, the President of Magdalen and the Warden of All Souls, would alternate in their participation.[135] A further new and permanent appointment to the board was—a little unexpectedly, given the lack of any prior involvement—the Provost of The Queen's College.[136] This point was, however, addressed in the anonymous pamphlet cited earlier, which informs us that rumours had been circulating within the university that Queen's might soon make 'certain augmentations' to the Sedleian Professorship.[137]

Entrance to Queen's Coll: Oxford.
14. Jan 1868

Figure 1.9 The Queen's College, which has part-funded the Sedleian Professorship since the late 1850s.

A suggestion that had come out of the Commission of the early 1850s, and which would eventually be enforced by an Act of Parliament of 1877,[138] was that funds could be diverted from colleges into the central university in order to pay the stipends of professors. The proposal was that when a fellowship next fell vacant at any given college, it would not be refilled and instead the monies saved would be handed over to the university. As an idea, this was rather radical: the financing of the professorships had previously been a matter for the central university alone, with no input from the individual colleges, each of which fiercely maintained its financial independence. Perhaps more than any other aspect of the Commission's enquires,

Figure 1.10 William Thomson, Provost of The Queen's College, and supporter of scientific education.

its insistence on investigating the closely guarded financial arrangements of the colleges was seen as an impertinence and an imposition.

The mechanism by which Queen's became involved with the Sedleian Chair is rather mysterious, and may simply have come about via personal connections not recorded in any documents of the time.[139] The proposed connection between the college and the chair first appears, without comment, in university records in November 1856.[140] At this stage, there was no compulsion to enter into such an arrangement, so this was perhaps a pre-emptive move by the college, similar to the university's attempts to forestall the external imposition of less favourable measures; the college was probably painfully aware that its wealth had already been noted by the Commissioners in connection with possible funding for professorships. College records, however, are tantalizingly vague on this point. The College Register of the time, which records in outline the decisions made by the college's Governing Body, shows the college gradually shifting its position over the course of the 1850s: in June 1851, for example, it was agreed 'that it is not expedient to answer certain questions which have been proposed by Commissioners appointed by Her Majesty'.[142] In a letter to the Vice Chancellor of January 1854, the college was exceedingly cagey about the matter of funding professorships, pleading that 'great sacrifices of income have been made and are under consideration, for purposes more immediately within our scope'.[143] By June the following year, however, the college had finally agreed to provide details of its properties to the members of a further Commission concerned with the revision of college statutes. Cooperation with that Commission probably became smoother still with the election later that year of William Thomson (1819–90), a figure known for his support of university reform, and indeed of scientific education, as Provost of Queen's.[145] In 1857, when the new professorial statute was being debated within the university, the College Register indicates that there was much correspondence back and forth between Thomson and the Commissioners, but gives no indication as to its content. What is evident from later records, however, is that somewhere within this process, Queen's had agreed to contribute to the funding of the Sedleian Professorship. The college contribution was to be £270 p.a. in the first instance, but this would increase in later years of the century (and beyond).[146] In the meantime, the rents returned by Philosophy Farm were at last deemed to be insufficient to sustain the professorship in the long term, and so the estate was sold to Baron Rothschild in 1875 for £10,000.[147] This latter sum was invested, and the returns earmarked for the university's portion of the professor's salary.[148]

The twentieth century

At the end of the nineteenth century, the Sedleian Chair remained explicitly a teaching position, although its occupant during the second half of the century, Bartholomew Price, did occasionally turn his attention to questions of mathematical research (see Chapter 6). Original scientific investigations would be even more central to the activities of his successor, Augustus Love (1863–1940), whose election to the chair following Price's death in 1898 was also the first occasion upon which an external candidate was appointed, for Love had previously had no Oxford connection; at the time of his election, he was a fellow of St John's College, Cambridge.[149]

The appointment of an external candidate to the chair also raised an issue that was deemed of great importance by those within the university, but which might reasonably be viewed by observers elsewhere as a tedious detail: to which college would Love belong? Until this point, the various professorships within the university had not been attached to particular colleges: they had always been internal university appointments, given to people who already had college associations. To have moved college upon appointment to a chair would have been unthinkable—or even infeasible, in the cases of figures such as Hornsby who held more than one such post; in general, it was not permitted to hold an ordinary fellowship at more than one college simultaneously. Indeed, the resolution of this point in the case of the Sedleian Chair had been one of the issues that had gravely concerned the anonymous author of the 1857 pamphlet on the new professorial statute that we cited in the section 'Reform and the Sedleian Chair'.[150] But the answer was simple, for financial contributions to stipends created natural connections. Bartholomew Price, while remaining a fellow of Pembroke, had been elected to an honorary fellowship of Queen's in 1868.[151] And when Love arrived from Cambridge in 1899 without any other Oxford association, it was natural for him to become a member of Queen's.[152] The connection between the professorship and the college was thus cemented, and remains in place to this day.[153] Love held the Sedleian Chair and lectured on applied mathematics until his death in 1940; his life and career are discussed in more detail by June Barrow-Green in Chapter 7.

With Love's death occurring during wartime, the university deferred any attempts to refill the professorship, but in the meantime discussions were opened up as to the position of the chair within the teaching of mathematics and theoretical physics in Oxford.[154] As early as 1933, concerns had been raised as to whether there would be any suitable British candidates the next time the chair fell vacant, and by the early 1940s there were further worries about the difficulties of post-war recruitment, and the need to act quickly before the best candidates were snapped up for similar positions in Cambridge.[155] What is striking about the various surviving university memoranda is the weight that was now given to the research activities of any prospective professor: 'The primary need is to elect a man of distinction in the field of Mathematical Physics'.[156] Nevertheless, teaching activities remained at the heart of the discussions. In particular, a shortfall in the provision of university teaching in theoretical physics raised the question of on which side of the rather nebulous dividing line between applied mathematics and theoretical physics the Sedleian Chair ought to be placed—more concretely, whether it should sit within the Sub-Faculty of Mathematics or that of Physical Sciences. The answer hinged somewhat on how the term 'natural philosophy' ought to be interpreted in the mid-twentieth century. Asked for his opinion on this point, the mathematician and astrophysicist E. A. Milne (1896–1950), Rouse Ball Professor of Mathematics in Oxford, listed the following subjects as coming under the heading of 'natural philosophy':

> It includes statics, dynamics, hydrodynamics and aerodynamics, elasticity, gravitation, electricity and magnetism, relativity, kinetic theory of gases, thermodynamics, radiation and optics, as well as atomic physics and quantum mechanics.[157]

He thus listed many subjects that would have been entirely familiar to the professors of the previous century (such as elasticity, which had been Love's main interest), but also several that certainly would not. If anything, Milne's list threw the mathematics versus physics question into further confusion, but later in the same letter he made his own opinion abundantly clear:

> I should view with dismay any movement towards separating this Chair from its general participation in the work of the Sub-Faculty of Mathematics, with which it properly belongs and to which it has belonged in the past.[158]

Figure 1.11 The Upper Library of The Queen's College, featuring an orrery that, according to recent tradition, may only be operated by the Sedleian Professor or the college's Patroness.

By 1944, the university was at last preparing to advertise the Sedleian vacancy once again, but deliberately held back until some of the issues mentioned above could be resolved. In particular, it awaited a report on the future of physics in Oxford.[159] This report, which was being prepared by a committee chaired by the mathematician W. L. Ferrar (1893–1990), related to an ongoing discussion concerning the balance and teaching provision of theoretical and experimental physics in Oxford.[160] These deliberations were themselves a part of the broader discussions of the post-war needs for scientific education and fundamental research in science.[161] As things then stood, the university had two departments of experimental physics (the Clarendon Laboratory and the Electrical Laboratory), but no academic staff specifically linked to the theoretical parts of the subject. Ferrar's committee eventually made two proposals: either that the two departments be amalgamated into a single School of Physics, with the existing Wykeham Professorship of Physics (vacant at that time) being earmarked specifically for theoretical physics, or that the two departments remain separate, and a theoretical physicist be appointed to the Sedleian Chair. In the end, it was the first of the two proposals that was adopted, perhaps because the second would have had too detrimental an effect on mathematical teaching, but this did not deter interested parties from encouraging theoretical physicists to apply when the Sedleian Chair was eventually advertised in 1945.[162]

The appointment of a successor to Love proved to be a rather protracted process, with several candidates either withdrawing or declining the chair when it was offered.[163] The circumstances under which the geophysicist Sydney Chapman (1888–1970) was elected Sedleian Professor in 1946 are detailed by Peter Cargill in Chapter 8, as is his wider career. Upon Chapman's retirement in 1953, there were again attempts to manoeuvre a physicist into the Sedleian Chair,[164] but it was perhaps by now too firmly embedded into the Oxford mathematical community, for all the remaining Sedleian Professors of the twentieth century, George Temple (1901–92), Albert Green (1912–99), Brooke Benjamin (1929–95), and John M. Ball (1948–), were more closely aligned to applied mathematics than to theoretical physics. The lives of Temple, Green, and Benjamin are considered in Chapters 9 and 10. John Ball, the last Sedleian Professor of the twentieth century, is the subject of an interview which makes up the final chapter, Chapter 11, of the book.

In a curious way, the main university statute that now governs the Sedleian Professorship mirrors the simplicity of Sedley's original specification in that it gives more space to listing the electors than to detailing the professor's duties. The latter are set out very plainly and very generally: 'The Sedleian Professor of Natural Philosophy shall lecture and give instruction in Mathematics and its applications'.[165] And when the Sedleian Professorship most recently fell vacant in 2019, the job advertisement described it in the following terms:

> The Sedleian Professorship of Natural Philosophy (Applied Mathematics) is one of the most senior positions in Oxford's Mathematical Institute and is a cornerstone of Oxford's commitment to research in applied mathematics.
>
> The successful candidate will engage in original research of the highest calibre in any field of applied mathematics, will teach at undergraduate and graduate level and will provide research leadership in applied mathematics and contribute to leadership and research strategy in the Mathematical Institute more broadly.[166]

A far cry indeed from weekly instruction in the works of Aristotle.

CHAPTER 2

Thomas Willis

ALASTAIR COMPSTON

A s fourth incumbent of the Sedleian Professorship of Natural Philosophy in Oxford and a distinguished physician, Thomas Willis replaced classical dogma in medicine with empirical knowledge of the human body in health and disease. His contributions to medical science in early modern England are set down in fourteen treatises published between 1659 and 1691; the evidence from one of three surviving manuscript casebooks; and through notes taken during delivery of his lectures in Oxford.

Willis was imaginative and prepared to speculate beyond the immediate evidence; firm in his opinions whilst observant of other people's work; remote but respected by his contemporaries. John Aubrey (1626–97) describes Willis as being of middle height, with dark red brindle hair like a red pig, and a stammer; and Anthony à Wood (1632–95) added, 'a plain man, of no carriage, little discourse or complaisance and with no powers for appearing with advantage or brilliance in society' (Figure 2.1).[1] In the days after his death from pneumonia on 11 November 1675, Willis's brother-in-law, John Fell (1625–86), added a postscript to the preliminary pages of the posthumously published second part of *Pharmaceutice rationalis* (1675). Fell laments a person who showed courage in adversity, modesty, temperance, a forgiving nature, loyalty to the Crown, devotion to the Church, ingenuity in his work, diligence in his studies, and precision in his speech.[2]

Thomas Willis in Oxford and London

Willis was born on 27 January 1621 at Great Bedwyn, Wiltshire. He attended the school of Edward Sylvester (*c.*1586–1653) at 135 High Street, Oxford. He matriculated at Christ Church on 3 March 1637 (or 1636 before adoption of the Gregorian calendar), graduating Bachelor of Arts on 19 June 1639 and Master of Arts on 18 June 1642. Life changed for Willis when Oxford became the Royalist headquarters during the Civil War. Fell wrote that Willis '[r]eturned to the Tents of the King as well of the Muses [...] listing himself a Souldier in the University Legions, he received Pay for some years; until the Cause of the Best Prince being overcome, Cromwell's Tyranny afforded to this wretched Nation a Peace more cruel than any War'.

Alastair Compston, *Thomas Willis*. In: *Oxford's Sedleian Professors of Natural Philosophy*. Edited by: Christopher Hollings and Mark McCartney, Oxford University Press. © Oxford University Press (2023). DOI: 10.1093/oso/9780192843210.003.0002

Figure 2.1 Portrait of Thomas Willis engraved by George Vertue, 1742, commissioned by Browne Willis for £21 and published in Thomas Birch, *Heads of various illustrious persons of Great Britain* (1743–51).

Willis had studied under the patronage of Dr Thomas Iles (d.1649) where 'His genius lay to Mathematiques and Chymistry'; but Iles's wife knew about physick and discussed this interest with Willis. After the Civil War, Willis avoided the obligation for ordination on members of the university. He graduated Bachelor of Medicine on 8 December 1646. Although the teaching was faithful to ancient doctrines, there was an undercurrent of new learning influenced by William Harvey (1578–1657), amongst others, who had been attached to the Royal household in Oxford when King Charles I (1600–49) took up residence at Christ Church. Willis referred to 'our most famous Harvey [who] hath laid the Circulation of the Blood as a new Foundation in medicine'.

During the Commonwealth period Willis developed his medical practice and taught in the university. From 1657, he lived at Beam Hall opposite Merton College (Figure 2.2).[3] Despite all Anglican rituals being prohibited, 'even to the danger of his life', Willis used 'part of his House for Holy uses [holding] Assemblies, and Publick Prayers, and other Offices of Piety [...] according to the Rites of the Church of England'. Those present included John Dolben (1625–86), Richard Allestry (1619–81), and John Fell. As his reputation increased, Willis became wealthy and, by 1659, his income was the highest for any physician in Oxford.

On 7 April 1657 Willis married Mary Fell (*c*.1630–70), daughter of Samuel Fell (1584–1649). Only four of their nine children survived to adult life. Thomas (born 26 January 1658) 'falling into consumption, he [the father] sent him to Montpellier in France, for the recovery of his health, and it proved successful, Thomas returning to his studies at Westminster School'. Anne (born 30 June 1665) and Jane (born 8 September 1666) were baptized in Oxford, whereas Rachel (born in 1667) was not registered in the city because Willis had moved to London. That decision may have been sudden since Willis had recently taken a forty-year lease with Dr Peter Eliot (1618–82) on the Angel, a coaching inn on the High Street, and Bostar Hall, also in the High Street, as premises convenient for seeing patients, including those on their way

Figure 2.2 Numbers 3 and 4, Merton Street, with the plaque in memory of Thomas Willis identifying Beam Hall, formerly Biham Hall, placed on number 4.

to the medicinal waters at Bath. The reason for moving from Oxford may have been pressure from Gilbert Sheldon (1598–1677), Archbishop of Canterbury, and Oxford friends who had themselves moved; and because London offered a better place for Willis to enhance his clinical reputation.

Willis became even more prosperous in London, acquiring a great many properties but gave away his wealth. He donated fees earned each Sunday (estimated at one-tenth of his income) to the poor of the parish. He arranged for the schoolmaster who taught at St Martin's Church to read prayers so that the ordinary working people could attend (Figure 2.3). He endowed the salary of a priest to provide 'sacred offices of the Church' for those in the neighbourhood otherwise detained in commerce and trade. His faith was repeatedly declared in Willis's treatises. The arterial circle at the base of the brain was evidence of the 'Finger and Divine Workmanship of the Deity, a most strong and invincible Argument may be opposed to the most perverse Atheist'; the earthworm made him speculate that this 'Little living Creature, tho it be esteemed Vile and Contemptible, hath allotted to it vital organs, as also other Viscera and Members, made most admirable by a Divine Workmanship'; and 'The Physician, however Skilful he be, ought always to implore the help of the Heavenly Power, to be assisting to him'.

Willis had little time for leisure. After becoming interested in medicine, he could 'hardly intend or admit thoughts of any other thing' and did not suffer his 'Faculties although small (as the Talent entrusted with me by God Almighty) to perish through sloth'.[4] He was abstemious and considered illness to be caused by 'guzzling of strong wine'. He allowed himself some pleasures: 'musick does not only affect the Phantasie with a certain delight, but besides chears

Figure 2.3 The west prospect of St Martin's Church in the Fields, Westminster, by George Vertue, 1744.

a sad and sorrowful Heart; yea allays all turbulent Passions'; coffee 'wonderfully clears and enlightens each part of the Soul, and disperses all the clouds of every Function'; 'the use of Tobacco is not only good, but almost necessary for Souldiers or Seamen [...] [although] most men take too much of it'.

Mary died on 31 October 1670 from pulmonary tuberculosis and was buried in Westminster Abbey. On 1 September 1672, Willis married a widow, Elizabeth Calley (1644–1709). His surviving family and descendants evidently disapproved. Browne Willis (1682–1760) preferred not to acknowledge the second Mrs Willis 'for I heard she was for getting all she could from his Family'.[5] Thomas Willis died at his house in St Martin's Lane from pneumonia on 11 November 1675, St Martin's Day. He was buried in the north transept of Westminster Abbey near his first wife Mary and one child, Katherine (1663–7). His memorial stone became worn and its location in the Abbey subsequently lost until identified and replaced in the 1960s.

Medical practice in Oxford and London

Lacking medical experience, Willis 'fell to the practice of it and every Monday kept Abingdon market', sharing a horse with Dr Richard Lydall (1621–1704). Willis formulated his ideas on disease by studying patients:

> Sitting oftentimes by the Sick, I was wont carefully to search out their Cases, to weigh all the Symptoms, and to put them, with exact Diaries of the Diseases, into writing; then diligently to meditate on these, and to compare some with others; and then began to adapt general Notions from particular Events.[6]

Surviving letters describe individual cases.[7] Rather more is learned from the manuscript of the casebook for 1650–2, one of at least three known to have been kept in his lifetime. The calligraphy suggests that an amanuensis took notes while Willis assessed patients, although he did scribble on the end-papers.[8] These fifty cases demonstrate the transition to an approach based on empirical observation and interpretation, no longer shackled by ancient doctrines, and a standard method for clinical description.[9]

Most patients are named, often with a brief demographic statement. Many come from wealthy families or are relatives of Willis's. There are adults and children. Symptoms relate to respiration, appetite, the bowel and bladder and appearance of their effluents, pain, weakness, and mental function. Signs reflect what can easily be observed: fever, patterns of respiration and the pulse, frequency of excretion, material emitted from the throat and other orifices, appearance of the external genitalia, the presence of swellings, and skin eruptions. Fluids—urine, sweat, and semen—might be tasted. Not every patient is seen personally. Some are treated simply on the basis of examining a sample of urine.

Willis records past medical history and contributing factors. For women: virginity, menstrual activity, childbirth, and any tendency to hysteria. For men: excessive use of alcohol. Willis's formulations on mechanisms mainly involve the blood and its circulation. He is thinking in terms of five chemical constituents: spirit, sulphur, and salt, with water and earth binding these together. With heat produced from their fermentation, these elements compete in his analyses with the four Galenic humours: blood, yellow bile, black bile, and phlegm. At fault are usually the intestines, liver, spleen, kidneys, uterus, and lungs. Willis's treatments are based on correcting altered chemistry of the blood, removing ill-digested material by purging, vomiting, sweating, or bleeding using remedies made up by apothecaries. When the outlook seems ominous, Willis is not afraid to confirm the prognosis. When, as is often the case, the patient dies, post-mortem examination may be carried out.

The Oxford casebook is important as a record of medicine as practised in the 1650s, and because it illustrates the convergence of classical humoral doctrines with Willis's ideas on fermentation as the basis of human vitality and, when perturbed, a generic mechanism of disease. Largely missing from these cases is the interest in disorders of the nervous system which later dominated Willis's thinking.

The 1636 charter issued by Charles I required that a 'sounde body' of a person executed after the Lent Assizes in Oxford be dissected annually by a surgeon who demonstrated the different parts and their 'site, nature, use and office'. On 14 December 1650, William Petty (1649–52) secured for anatomical demonstration the body of Anne Greene (1628–[65]), who had been hanged by Jack Ketch (d.1686) in the Oxford Parks for the murder of her child following conviction by Serjeant Upton Croke (1593–1670). Anne had been seduced by Jeffery Read, the seventeen-year-old grandson of Sir Thomas Read (1575–1650), 'a youth of forward Growth and Stature', and was delivered of 'a Man-child: Which being never made knowne, and the Infant found dead in the House of Office, caused a suspicion, that she, being the Mother had Murthered it' (Figure 2.4). Contemporary accounts of the episode originate from number 126 of Petty's medical papers, 'History of the Magdalen (or the raising of Anne Greene)'.[10]

Petty describes attempts to accelerate Anne's death on the gallows by dragging on her legs, beating her breasts, 'and moreover a souldier did the same severall tymes with the butt end of his musquett'. But, on arrival at the place for dissection, signs of life were detected which led to stamping on her chest and stomach by a lusty fellow: 'presently after myselfe and Mr. Willis came in, and found her yet to rattle a little as before notwithstanding the said latter strokes'. Realizing that she was alive, those present now tried to revive Anne. She was bled: 'As soon as her arme was bound [. . .] shee suddenly bent it, as if it had been contracted by a convulsion fit. The vayne being cut, she bled so freely that having bled 5 ounces, when we desired to stop, wee could not easily'.

Petty provides details of the medications that he and Willis administered over the next few days. These included cordials (rainwater and cinnamon in equal parts one ounce and Acetosella, a preparation of sorrel, four ounces); anointing the neck wound with turpentine and stimulating Anne's nasal mucosa with oil of cloves; poultices applied to her breasts; and a healing and odiferous enema which 'came away' after some hours with the addition of a candle end by way of a suppository. Starting to eat bread soaked in water and drinking 'small beere', after two days Anne 'made some water but had no stoole' until the following day.

After a few days, Anne was asked what she could recall. The answer was nothing beyond taking off her bodice and giving this to her mother on the morning of the execution. Later she did remember seeing the executioner. By 17 December, Anne had a slow and irregular pulse but could stand in order to have her bed made. The tip of her tongue was numb. By 25 December, she had residual bruising but was otherwise well recovered.

Many came to gaze at Anne. They were charged for the pleasure which earned her several pounds 'whereby the Apothecaries Bill, and other Necessaries for her Diet and Lodging, were discharged, but some Overplus remained towards the sueing out her Pardon'. Petty submitted Anne's petition to the Privy Council explaining that, on the testimony of the midwife, she had

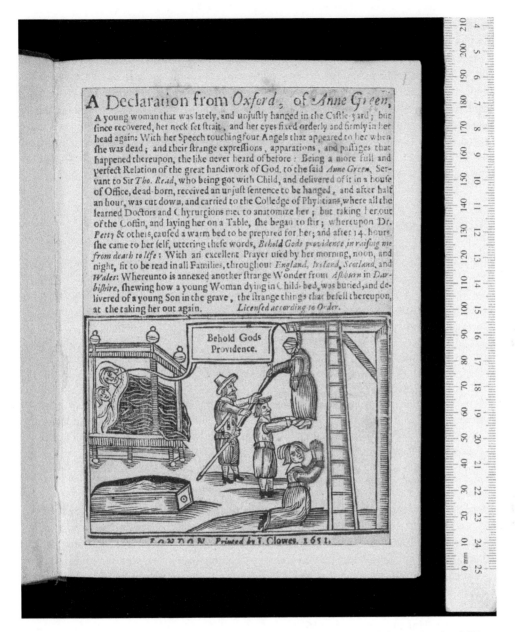

Figure 2.4 Contemporary depiction of the story of Anne Greene. Printed by T. Clowes, London, 1651.

suffered a spontaneous miscarriage at 18 weeks of a foetus dead at birth, and so deformed and stunted as be of uncertain sex and never capable of life: 'more a lump of flesh than a mature and onely formed child'. Anne was duly pardoned and became something of a circus act, being exhibited laid in her coffin in the room where she was to have been dissected and requiring 'guards to control the crowd' when 'multitudes flooded daily to see her'. Her father continued to collect the fees. Eventually she was sent home with her coffin as a 'Trophey of her wonderful preservation'. Anne married and had three children during the 15 years she survived after this unhappy but profitable episode.

The extensive experience gained early in his career left Willis confident in his clinical abilities. By the time he moved to London, Willis was a fashionable physician, although occasionally outspoken. Suspecting his wife of infidelity with the Duke of York (1633–1701: later King James II), the Earl of Southesk (c.1649–88) acquired venereal disease and deliberately infected Anna, Countess of Southesk (dates unknown), from whom the condition eventually reached the Duchess of York accounting for ill-health in several of her children. Consulted about one of the Duke of York's sons, Willis gave as his opinion 'mala stamina vitæ' (threads of the evils of life) which caused such offence that he was never consulted thereafter.[11]

Clubs and societies for natural philosophy

Willis was part of a group committed to advancing knowledge of natural philosophy which first met in the 1640s. This became the Oxford Experimental Philosophical Club.[12] Others, interacting at much the same time in London, later founded the Royal Society. Some individuals were active in both communities. In the first full account of the Royal Society, Thomas Sprat (1635–1713) explained:

> It was therefore, some space after the end of the Civil Wars at Oxford, in Dr Wilkins his lodgings, in Wadham College, which was then the place of Resort for Vertuous and Learned Men, that the first meetings were made, which laid the foundation of all this that follow'd [. . .] The principal, and most consistent of them, were Doctor Seth Ward, the present Lord Bishop of Exeter, Mr. Boyl, Dr. Wilkins, Sir William Petty, Mr. Mathew Wren, Dr. Wallis, Dr. Goddard, Dr. Willis, Dr. Bathurst, Dr. Christopher Wren, Mr. Rook: besides several others, who joyn'd themselves to them, upon occasions [. . .] For such a candid, and unpassionate company, as that was, and for such a gloomy season, what could have been a fitter subject to pitch upon, than Natural Philosophy.[13]

Four of the Oxford group were physicians: John Ward (1629–81), Ralph Bathurst (1620–1704), Petty, and Willis.[14] As their activities became more formalized, the club met in Petty's rooms at Buckley Hall (107 High Street) where discussions were held and simple experiments carried out. Later, Wadham College offered access to scientific instruments for 'making chymicall experiments'. By 1651, the club was meeting every Thursday at 2pm with those who failed to attend for a period of six weeks deemed to have forfeited their membership. One person appointed for the day was expected to carry out an experiment. Willis was responsible for the Thursday experiment in November 1654:

> Dr. Wellis [sic] of Dr Wilkins acquaintance a very experimenting ingenious gentleman communicating every weeke some experiment or other to Mr. Boyles chymical servant, who is a kind of cozen to him. Hee is a great Verulamian philosopher, from him all may bee had by means of Mr. Boyle.[15]

Willis recorded his expenses for this and other occasions: £21 18s 10d for previous expenditure and new costs for glassware and 'carriage [. . .] of thinges from my Chamber' (£10 4s 8d), 'drugges from London' (£1 19s 6d), carpentry, masonry, and other tradesmen (£6 7s 6d), and 'to old Watt for stilling fires' (£0 1s 6d).[16] He provided continuity when the founder group started to drift away but rarely attended meetings in the 1660s before also moving to London. Thereafter, the Oxford Experimental Philosophical Club started to disaggregate, whereas the Royal Society flourished.

'A Colledge for the promoting of Physico-Mathematicall Experimental Learning' was founded on 28 November 1660 by twelve individuals who met after a lecture by Christopher Wren (1632–1723) at Gresham College in London. The founders agreed to meet weekly and 'put downe ten shillings [. . .] and one shilling weekly whether present or absent'. A further forty-one names were proposed of which Willis was number 30. When it was granted a Royal Charter on 15 July 1662, the college became the Royal Society for improving Natural Knowledge. Willis received the invitation to participate on 30 January 1661; but despite confirmation of his fellowship at the meeting on 18 November 1663, he was not admitted until 24 October 1667:

> Dr Willis was chosen, having been forgot to be chosen again at the time when, upon the renewall of the charter, the Council, according to the power granted them therein, did receive and admitt into the Society such persons as had been elected Fellows afore upon the first charter.[17]

Contemporary accounts place the origins of the Royal Society much earlier than 1660. John Wallis (1616–1703) wrote that, in about 1645, whilst living in London and at a time when academic life was disrupted in 'both Universities' (Oxford and Cambridge), he became involved with a group interested in experimental philosophy. Mention of divinity, state affairs, and current news was prohibited. Legitimate topics included physic and anatomy: 'we there discoursed the Circulation of the Blood, the Valves in the Veins. Scientific discussion combined with friendship and conviviality'.[18]

Thomas Sprat provided a narrative of what happened in the first few years. After 1658 when some of the group were 'called away to several parts of the Nation and the greatest number of them coming to London', they met at Gresham College after the Wednesday and Thursday lectures of Wren and Laurence Rooke (1622–62). The mix of professions which avoided any one subject in particular overweighing another, 'or making the Oracle onely speak their private fence', made for success. Sprat singled out the role of physicians and their commitment to the Royal Society even though they had their own long-established 'Colledge peculiar to their Profession'. He lists Willis amongst the twelve 'principal, and most constant' men who in their printed works in both Latin and the vernacular 'have much conduc'd to the Fame of our Nation *abroad*, and to the spreading of profitable Light, at *home*'. Although not attentive to activities of the Society, Willis's involvement in chemistry applied to medicine (iatrochemistry) identified him as a leading scientific physician of that period. In line with the Society's motto, 'Nullius in verba' ('Take nobody's word for it'), he was comfortable with the rejection of scholasticism and authority of empirical observation as the most reliable route to knowledge.

A few Fellows had 'Club suppers' after meetings for ordinary business. This became the Royal Society (Dining) Club.[19] Although there is no mention of Thomas Willis having been a member, at the dinner on 21 November 1667 John Wilkins (1614–72) mentioned an experiment about to be performed by Gresham College philosophers in which the blood of a sheep was to be transfused 'into the body of a poor debauched man'. Willis may have been one of the 'brave men' who attended this (and other) Club dinners. In discussions with Wilkins and Robert Boyle (1627–91), Wren suggested a method for injecting poisons into the blood stream of a dog with access using a large vein in the hind leg. When the discussion turned to the principle of using this technique on patients in London hospitals, Willis suggested that the experiment 'might be proper to make use of in rotten sheep'. It is unclear whether this was intended as disparaging or a cautionary next step in moving towards human transfusions.

Willis remains largely absent from records of the Royal Society. He accumulated debts. On 22 October 1673, he was among eighteen who were in arrears and asked to comply with the new form of obligation. By 28 January 1675, he still had not signed and further application was made. In the end, the Society grew weary and, on 21 October 1675, recorded that Willis would no longer be asked to sign the bond. That did not prevent it being minuted on 29 November 1675 that 'it be recommended to the care of Dr King [Edmund King: 1630–1709] to solicit the executors of the will of the late dr Willis for the payment of his arrears to the Society, amounting to twenty pounds and eleven shillings, as appeared from the treasurer's book'. No obituary of Willis was published by the Royal Society.

Willis's lectures as Sedleian Professor of Natural Philosophy

At the Restoration, Willis was appointed to the Professorship of Natural Philosophy in the University of Oxford endowed by Sir William Sedley (c.1558–1618). Following election on 30 October 1660, Willis received his doctorate of medicine. He acknowledged the patronage of Gilbert Sheldon (1598–1677: consecrated as Bishop of London on 28 October 1660) in the updated dedication to the third edition of *Diatribæ duæ medico-philosophicæ* (1662):

> It was by your means (most Noble Prelate) that I obtained the Votes in this Famous University for the place of Sidly Professor for how small soever my merits might seem they were helped by the greatness and weight of your opinion.[20]

Yet more adulatory was the dedication to 'His Grace Gilbert' in *Cerebri anatome* (1664) which preceded Willis's cleverly worded explanation of how, in placing his attempts to 'unlock the secret places on Mans Mind, and to look into the living and breathing Chapel of the Deity' at the 'most holy Altar of Your Grace', he had avoided 'the School-house of Atheism'. He went on:

> Once more your Sidley Professor and your Servant (the more happy Title) flings himself at Your feet, with this only Ambition, that he might render something of Thanks for your Kindness and benefits.

The homage to 'His Grace Gilbert' was further developed in *De anima brutorum* (1672) making clear that:

> The Academical Readings by a necessitated Duty belong to you, for that I received them from your Favours; and indeed, neither had ever seen the Light, not perhaps my self had ever been in the number of Authors, unless I had been made at first, your Sidlie Professor at Oxford.

Dedicating books to Sheldon was wise since, although Willis was unlikely to be guilty of sectarianism, the Act of Uniformity (1662) required adherence to liturgy and governance under the direction of the Archbishop of Canterbury (to which Sheldon was translated in 1663). Sheldon instructed all bishops to:

Certify me, the names sirnames, Degrees and qualities of all Practicers of Physick within their respective dioceses: In what towns villages or Places they live; whether licensed & by whom; & how they appear affected to His Ma(jesty) gouerm't & ye Doctrine and Discipline of ye church of Engl'd.[21]

Willis gave three series of lectures as Sedleian Professor. Details of the first are lost but much can be learned from notes taken down by two who attended the second and third courses: Richard Lower (1631–91), between 1661 and 1662; and John Locke (1632–1704), in 1663 or 1664 (Figure 2.5). Together, they provide details of thirty-eight lectures, from approximately fifty that Willis may have been required to deliver between 1660 and 1667.[22]

The lectures attended by Lower are instructive with respect to Willis's evolving views on fermentation, delivered shortly after he published on the topic in 1659. He considers four classes of living creatures: plants and worms; insects; animals with a primitive circulation; and warm-blooded highly developed species. The properties of circulating blood are related to heat resulting from chemical interaction of the five chemical particles. Like a lamp which requires fuel, space in which to burn, and removal of its soot, the blood is fuel, the heart is the burner, and a furnace provided by the lungs and other vents. In disease, the blood may be insufficient, lost following haemorrhage, or burned out through fever. The heart may receive too little or be swamped with blood, and fail to propel the fluid if forces that move the heart under influence of the nerves falter. The flame may be extinguished if ventilation is prevented, congested, or obstructed. Robert Plot (1640–96) records that

In Natural Philosophy, the famous Dr Willis [...] Sidleyan Professor [...] in this University, first taught us, that the Generations, Perfections, and Corruptions of Natural Bodies, whether Mineral, Vegetable, or Animal; and so like-wise of Bodies Artificial, do depend upon Fermentation, raised from the different Proportions and Motions of the Spiriti, Sulphur, Salt, water, and Earth, which he did constitute the ultimate sensible Principles of mixed Bodies.[23]

Locke's notes headed 'Willis Praelector' summarize Willis's emerging interest in the nervous system, and mental functions or 'psychologia'. Ignoring statutory instructions, Willis's emerging ideas appealed to his audience and attracted 'an unusual number of auditors'. Willis was starting to explore topics which were later expanded in his printed works on structure and function of the brain and nerves, and the relationship of mind and body in health and disease. He tried out ideas on sensation and movement, convulsions as a consequence of brain disease, the nature of emotions and their disorders, and sleep as a physiological function.

When, by movement, shape, or touch, an external object excites the spirits, sensation is conveyed to the cerebrum. But if the brain is deficient, or distracted, flow is suspended and the sensation not registered. Imagination occurs when sensation agitates spirits present in the medulla oblongata (the structure connecting the lower part of the brainstem to the spinal cord). Ideas result from permeation of this material into the cerebral 'orbits'. Memory is a retrieval involving agitation of spirits in preformed tracts, phantasy occurring when the process is disordered. Pleasure results from agreeable dancing of spirits moving from the nerves to the medulla. Joy and sadness reflect expansion and stasis of the nervous spirits, respectively. Melancholy is persistent sadness, the imagination disproportionately magnifying small objects like a microscope, due to spirits permeating newly carved pathways. Whereas the localization of these events was spurious, the concept that mind is embodied and experience shapes

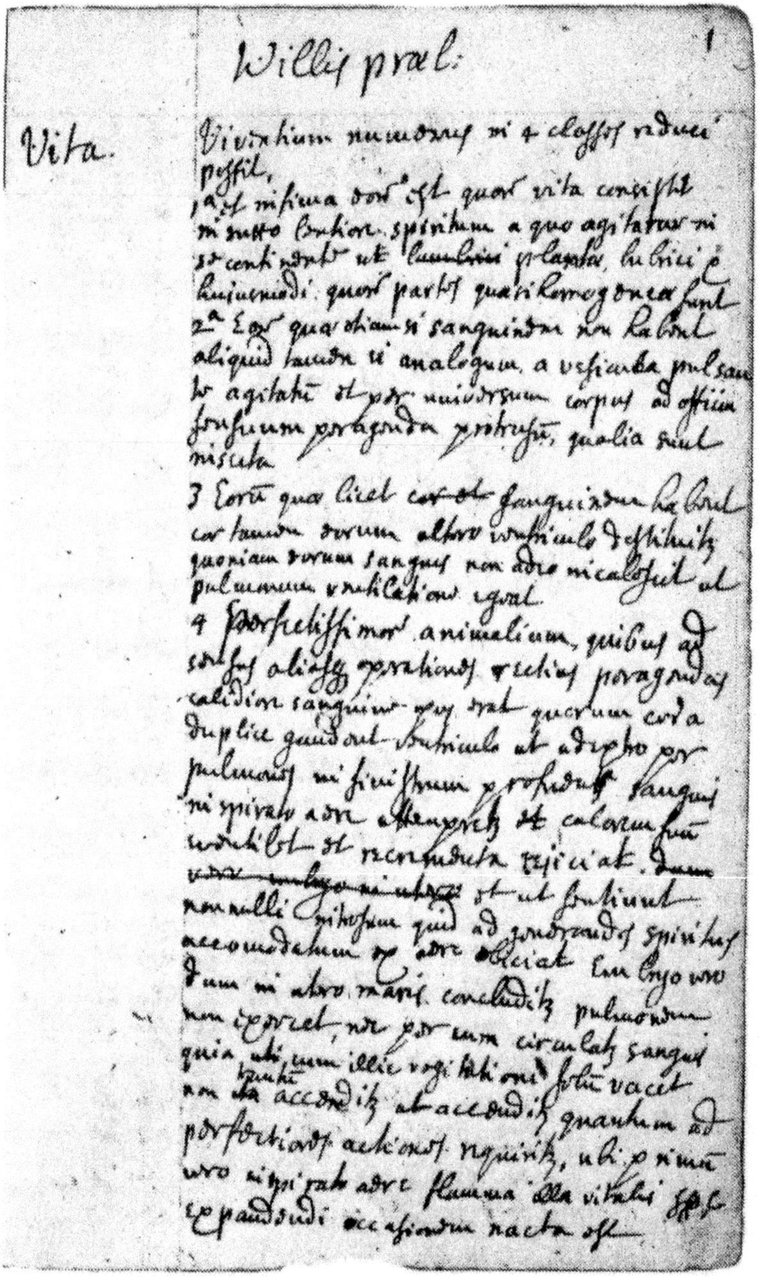
Willis prael:

Vita.

Viventium numerus in 4 classes reduci possit.

Figure 2.5 The first page of John Locke's notebook recording Willis's lecture on 'life'.

behaviour with the traces of past events etched as alterations in architecture of the brain were profound and novel insights.

Willis was already formulating novel ideas on the roles of the cerebrum and cerebellum (the structure placed at the back of the skull under the cerebral hemispheres) in movement.[24] He argued that the cerebrum governs the will and voluntary movement, whereas the cerebellum is responsible for involuntary activity, including rhythms of the heart, lungs, and abdominal organs. Willis thought that these vegetative functions are transmitted through cranial nerves which originate in the cerebellum. Reciprocal flow of spirits between the cerebrum and cerebellum explains the involuntary associated gestures of action and behaviour. Willis realized that, ultimately, all movement depends on animal spirits reaching muscles which expand and shorten, thereby drawing together the structures to which they are attached. The muscle fibres have lakes where the spirits rest until activated and mixed with material in fermented blood delivered through the arteries. The failure of spirits to flow into muscles from nerves that are 'griped, contracted and wrinkled', for any one of several reasons, leads to paralysis. Willis explained that convulsion results from turbulence of the nervous spirits resulting in failure of an opposing (antagonist) muscle to relax when the prime mover (agonist) contracts.[25] The consequence is pain and spasm with 'leapings and contractions' of the muscles. Willis's concept of convulsion, as a disorder of movement, does include epilepsy, if sensibility is lost. He suggested that characteristics of the seizure depend on whether the disturbance of spirits originates in the cerebellum or cerebrum.

Sleep replenishes the nervous spirits ready for waking.[26] Waste from brain activity is removed through nocturnal perspiration. Sleeping with the mouth open and yawning allow 'soot' to escape through the breath. Some people are phlegmatic and sleep more readily because their blood is serous and easily relaxes spirits in the brain. Sleep occurs at fixed times and cannot easily be resisted. It may be encouraged by physical and mental exercise, soothing music, and the murmur of water or the wind. But even during sleep, not all is necessarily quiet. Eruptions may occur resulting in nocturnal speech and movements, and inducing phantasies and memories in dreams as the spirits 'go over the tracks and orbits marked out in the circuits of the brain, moving around in a varied and irregular manner'. Those with spirits continually agitated and expanded experience wakefulness. Sleep deprivation can be treated with opiates but the amount tolerated differs (dogs are tolerant, whereas the same quantity given to a cat 'launches it at once into eternity') and habitual use lessens its force.

Three lectures dealt with therapeutics: specifically the use of purges, cordials, and narcotics. By the time he wrote *Pharmaceutice rationalis* (1674–5), Willis had reorganized his thoughts on these and other medicines, considerably expanding his pharmacopoeia and account of principles underlying the choice and combination of remedies.

The intellectual stimulation of being amongst the first to hear ideas articulated in these lectures must have been considerable. For all their whimsicality, spurious chemicals, and flighty hypotheses, by any analysis Willis's lectures given as Sedleian Professor of Natural Philosophy are sufficiently full of wisdom and novelty to constitute vital transitional readings in the history of ideas relating to order and disorder in the nervous system; and a taster of the monographs that followed.

The printed works of Thomas Willis

Readership on science and medicine in the seventeenth century was largely confined to scholars and physicians. Usually written in Latin, it took time for the written records on natural phenomena to replace the slavish rehearsal of scholastic dogma. But change was underway and physicians were advised that 'the sick should be the doctor's books'. By the late 1650s Willis felt that his own ideas were ready to be set down. Nutrition provides particles which are enkindled through interaction with air. The fermented blood is circulated to organs where it is further changed into vital spirits. Waste is removed. When the process goes wrong, the aberrant state of the blood and animal spirits provides a rational basis for treatment. Willis was not the first to articulate the principles of iatrochemistry but no one had previously extended particulate theory to the entirety of health and disease, and set out ideas in such a comprehensive series of printed books linked by a common intellectual thread. Thomas Willis wrote fourteen treatises, originally published under seven titles, and later as collected works in Latin and English.[27]

Despite working in Oxford, Willis turned to London printers for publication of his first book: *Diatribæ duæ medico-philosophicæ* (1659). One person closely identified with printing in late-seventeenth-century Oxford has special significance for Thomas Willis. His brother-in-law, John Fell, was instrumental in reinvigorating the university press.[28] In 1672, Fell listed the first of eight forthcoming publications. Number seven was 'A treatise concerning the Soul (in Latin) to be printed in quarto'. This is *De anima brutorum* (1672) considered to be the 'first considerable medical work for the Sheldonian Press'. Commercially the most successful of Thomas Willis's works printed in Oxford was *Pharmaceutice rationalis* (1674–5: Figure 2.6). Both *De anima brutorum* and *Pharmaceutice rationalis* were unusual amongst Oxford medical publications from that period in having engraved plates.

The works of Thomas Willis variously incorporate an engraved title-page, printers' ornaments, decorated initials, and anatomical plates.[29] The engraved title-page added to the second edition of *Diatribæ duæ medico-philosophicæ* (London, 1660) has oblique text taken from Titus Lucretius Carus (*c*.99–*c*.55 BC): Ita res accendunt lumina rebus (In this way facts throw light on facts: Figure 2.7). Medicine and science are explaining that chemistry illuminates their understanding of the natural world. In *Pharmaceutice rationalis* (part 1: Amsterdam, 1674) this imagery is extended to the principles of medical treatment (Figure 2.8). More elaborate is the allegorical engraved title-page to *Opera omnia* (Amsterdam, 1682). Medicine is turning her back on classical knowledge; seated on a pedestal of anatomical knowledge, she is dissecting and adding details whilst an observer is setting down the new knowledge and explaining to physicians how this can be used better to help the sick (Figure 2.9).

Most relevant to the community which advanced these ideas, the edition of *Cerebri anatome* published by Gerbrandt Schagen (b. 1633; Amsterdam, 1666) has an engraved title-page (dated 1665) which does not appear in any other printing. This shows seven individuals (and a dog) gathered around a cadaver undergoing dissection of contents within the skull. Those depicted are considered to be Robert Hooke (1635–1703), Edmund King, John Locke, Richard Lower, Thomas Millington (1628–1703: Willis's successor as Sedleian Professor of Natural Philosophy), Thomas Willis, and Christopher Wren. As such, it represents the first illustration of the Oxford Experimental Philosophical Club and Fellows of the Royal Society at work (Figure 2.10).

PHARMACEUTICE RATIONALIS.
SIVE
DIATRIBA
DE MEDICAMENTORUM
Operationibus in humano Corpore.
AUTHORE
Tho. Willis M. D. in *Univ. Oxon* Prof. Sedleiano,
Nec non *Coll. Med. Lond.* & Societ. Reg. Socio.

E THEATRO SHELDONIANO. M. DC. LXXIV.
Proſtant apud *Robertum Scot:* Bibliopolam *LONDINENSEM*

Figure 2.6 The title-page of *Pharmaceutice rationalis*, part 1, 1674, printed at the university press and with a depiction of the Sheldonian Theatre engraved by David Loggan.

Figure 2.7 Engraved allegorical title-page: *Diatribæ duæ medico-philosophicæ* (London, 1660).

Figure 2.8 Engraved allegorical title-page: *Pharmaceutice rationalis* part 1 (Amsterdam, 1674).

Figure 2.9 Engraved allegorical title-page: *Opera Omnia* (Amsterdam, 1682).

A portrait of Thomas Willis, engraved by David Loggan (1634–92), was first printed for *Pathologiae cerebri* (1667). It shows Willis aged 45 turned to his left with the eyes facing, dressed in a fine coat with white collar, trimmed moustache, and hair to the shoulders. The expression is reflective and conveys the impression of authority. Versions of this image

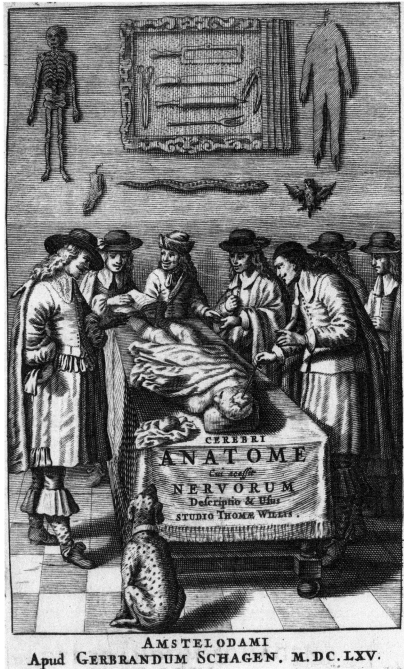

Figure 2.10 Engraved title-page thought to depict Robert Hooke, Edmund King, John Locke, Richard Lower, Thomas Millington, Thomas Willis, and Christopher Wren: *Cerebri anatome* (Amsterdam, 1666).

were reproduced in many printings of Willis's treatises, and other works. The most elaborate, from 1742, is by George Vertue (1684–1756), commissioned by Browne Willis for £21 (see Figure 2.1). There are two surviving oil paintings of Thomas Willis. One, attributed to John Wollaston (1710–75), is now in the Lower Reading Room of the Bodleian Library. The other, in the Royal College of Physicians of London, is by an unknown artist. It is inscribed 'THOMAS WILLIS MD, FRS Morum Suavitate insignis. Summo Omnium, ac imprimis merito' (Impressive by the sweetness of his manners. At the summit of all, and especially as deserved).

The first illustrated work is *Cerebri anatome* (1664); this was followed by *De motu musculari* (1670), *De anima brutorum* (1672), and *Pharmaceutice rationalis* (1674, 1675). The evidence on who was responsible for these anatomical illustrations is ambiguous. Wren was drafting images for *Cerebri anatome* in the spring of 1663. Richard Lower wrote that 'Dr Wren hath drawn most excellent schemes of the brain, and the several parts of it, according to the doctor's design, and the next week he will have finished the scheme of the eighth pair of nerves, and then all the work is at an end'. This is confirmed by Hooke who comments favourably in a letter to Robert Boyle on the careful and detailed drawings made by Wren for the group of 'neurologists'.[30] More details are provided in a clear statement from Willis in the preface that, in addition to Wren, Lower was responsible for images depicting the nerves (Tabulæ IX–XIII):

> For having prosecuted, with a most exact search all the divarications, wandring on every side of the Nerve, how minute or small soever, and immersed, and variously unfolded within other Bodies, and so turning over the Labyrinths of the Branches, and shoots of every pair [. . .] he [Lower] drew out with his own hand the Schemes, Images or Draughts of them, also of many passages of the Blood, as they appear in this Tract.

Annotation in a copy of *De anima brutorum* owned by Theophilus Metcalfe (fl. 1635) reads:

> Figuras in hoc Libro contentas (quae mea quidem sententia paene aequiparent Cowperianas) ipse manu sua designavit Dno. Christ. Wren Eques; & postea exsculp. curavit: ubi etiam eas, quas in tractatu Loweriano de Corde invenias. (The figures contained in this book (which at least in my opinion are almost as good as Cowper's) Mr. Christopher Wren Esq. drew by hand and later engraved: he also provided those that one can find in Lower's tract on the heart.)[31]

One interpretation is that in the early 1660s, Willis and Wren were working together in Oxford and Wren agreed to record their dissections as images that Loggan would engrave. Lower, inferior as an artist but close to the work and engaged with Willis in medical practice, was party to this arrangement. As the need to complete the work and have the plates ready for binding with the text sheets became increasingly urgent, and Wren still had not completed the task, Lower stepped in and drew several images, allowing the book to be published. After September 1666, Wren was busy as an architect but did complete four drawings illustrating comparative anatomy of the brain and nerves in various animal species for *De anima brutorum* (1672). Comparison of the artistry suggests that Wren drew Figures I–VIII of *Cerebri anatome* (1664), and was responsible for Tabulae V–VIII of *De anima brutorum* (1672). There is very close similarity between Figura VII in *Cerebri anatome* and Tabula V in *De anima brutorum*. The other eleven plates, Tabulae IX–XIII and the two unnumbered plates from *Cerebri anatome*, and Tabulae I–IV from *De anima brutorum*, are the work of Richard Lower. Christopher Wren can have had nothing to do with the four figures of muscle in *De motu musculari* (1670), or the fourteen plates in *Pharmaceutice rationalis* (parts 1 and 2), depicting the stomach, blood vessels, and lungs, which are relatively crude and uninformative. Although Willis writes in his preface to the reader of help from John Masters (fl. 1672–80) and Edmund King in 'illustrating the pharmaceutical Doctrin', this may imply elucidating the concepts rather than depicting the work.

Willis on medical chemistry

Those working in the second half of the seventeenth century who committed to a better understanding of vitality needed to understand how the material substrate of animal life is sustained. By the end of their productive lives, these scientists and physicians had resolved that a component of air interacts with venous blood in capillaries of the lungs; red blood circulates allowing recipient structures to function; darkened blood is then returned to the lungs where it is again enriched. In this context, Willis was primarily interested in the chemical activity on nutrients in blood.[32] Included as topics in his lectures as Sedleian Professor, Willis wrote on fermentation in *Diatribæ duæ medico-philosophicæ* (1659), with further references in *Cerebri anatome* (1664), *Pathologiæ cerebri* and *De scorbuto* (1667), and a last summary in *De sanguinis accensione* (1670).

Willis did not concern himself with the substance in air on which fermentation of the blood depends. But others were giving this matter much thought. The experiments of Boyle and Hooke using the vacuum pump showed that animals breathe in a confined space but, despite maintaining respiratory movements, do not survive if air is removed.[33] John Locke argued that during inspiration particles of aerial nitre volatize and mix with blood in the lungs, changing the colour from dusky blue to bright red only for the hue to be lost as blood moves throughout the body. Hooke showed that life is maintained for as long as air passes through the lungs, even without the mechanical excursions that many had argued were necessary.

The issue that concerned Willis was the next stage in the process. He accepted that fermentation requires exposure to air and a supply of sulphurous food. Blood could be described in terms of the state and balance of spirit, sulphur, and salt with water and earth, the five particles acting independently or as 'copula'. Critical to their function was the tendency to interact and 'fly away'. Motion 'inflamed' the blood, generating heat. Fermentation mainly took place in the 'Chimny of the Heart'. Willis never made entirely clear whether his concept of flame was a reality or metaphor.

As a physician, Willis applied his chemical theories. The natural place to start was diseases characterized by fever, classified by periodicity and severity: intermittent or continuous; putrid, malignant, or pestilential. He understood that contagion spreads through contact between one infected person and another; the causative matter might survive between epidemics, reappear with altered virulence, and preferentially infect vulnerable hosts. The symptoms of fever are heat, weariness, anorexia, diarrhoea, burning pains, thirst, coating of the mouth and tongue, headache, disordered sleep (the 'Watchings'), swooning, convulsions, stupefaction, and phrensie. The pulse is nature's 'Thermometer or Weather-Glass', its state the harbinger of 'Death or Health'. Fever might resolve by 'fit' or crisis in which perspiration, compartmentalization, discharge through the nostrils, formation of boils and pustules, or excretion in the urine, stools, or vomit remove the over-fermenting principles. Willis's medicines aimed to induce purging, vomiting, and sweating or enable the unruly particles to be shaken apart.

Willis advanced ideas that resonate with therapeutic medicine down the ages: early treatment is more likely to succeed than rescuing advanced disease; combinations of remedies are often required; and medicines may do more harm than good.

Willis's case histories are rich in conveying the domestic and often foetid atmosphere of the sickbed and infected environment. His account of intermittent fever (malaria) is conspicuous for the use of quinine in the form of Peruvian bark. His recurrent epidemics are presumed to include influenza. 'Putrid fever' is an account of typhoid. The 'Flux of the Belly' is cholera. Willis gives a brief account of the plague in Oxford during 1645. The attendant physician, Dr Sayer, protected himself by taking a large draught of sack before perambulating 'about the Borders of Death and the very Jaws of the Grave' and another slug once his work was done. But his luck did not last because 'being so bold, as to lie in the same Bed with a certain Captain (his intimate Companion), who was taken with the Plague', Dr Sayer became infected and 'with great sorrow of the Inhabitants, nor without great loss to the Medical Science, he died of that Disease'.[34] Willis wrote *A plain and easie method etc.* (1691) after the 1665 London plague. His advised that 'wine and confidence are a good preservation against the plague'. Willis's description of the pestilential or camp fever describes typhus among the Oxford troops during the Civil War, as a result of which each army was affected so that both sides fought 'not with the Enemy, but with the Disease'. Willis makes the prescient point that 'it happens for every man only and once to be distempered with the Small Pox or Measles'. He gives a description of the chin-cough (pertussis, or whooping cough). Willis discusses puerperal fever: 'Vulgar Experience abundantly testifies, that the Feavers of Women lying-in are very dangerous'.

Given his main interest, it is not surprising that Willis described fevers accompanied by neurological symptoms. Those affected lapsed into 'deep stupidity or Insensibility', their 'knowledge and Speech failed' and, with incontinence, they remained ill for several weeks during which they were reduced 'to the highest *Atrophie*, or wasting of all parts'. When older people were affected the outlook for survival was poor: 'those who were before obnoxious to cephalic distempers, as the *Lethargie*, *Apoplexie*, or Convulsion, it oftentimes kill'd them in a short space'.

Because excretion was fundamental to Willis's concept of fermentation in health and disease, his treatise on urines was published with the *Diatribæ duæ medico-philosophicæ* on fermentation and fevers. It is a pioneering work on biomarkers of disease. 'Uromancy' (divination by studying urine) made sense since the liquor had washed all 'the parts and taken from many some little parcels'. Urine could reveal the many 'dyscrasies of our Bodies and their habitudes'. Willis's method was to note the volume and observe the sediment, colour, and smell. However, he was clear that generalizations based on the urine usually denote the consequences not causation of disease. Willis concluded that practitioners of uromancy 'deserve rather the name of a jugling Quack, than of a Physician'.

Willis on anatomy of the brain

Willis understood that the brain is a system for sensing the external and internal environment, integrating responses, and learning from these experiences to adapt physically and interact socially. In the conclusion to *Cerebri anatome* (1664) Willis explained that he planned three books on the brain. The second, *Pathologiæ cerebri* (1667), was to describe the nature of disease. The 'Crown of the Work', *De anima brutorum* (1672), would set out Willis's ideas on the relationship between mind and body together with accounts of neurological and neuropsychiatric disease. Taken with *De motu musculari* (1670), these printed works form the basis for his lasting reputation in medicine.

'Addicted to the opening of heads', Willis's method of dissection allowed him to remove the brain from the skull, whole and inverted. His observations were mainly macroscopic but he picked away at structures with a knife and occasionally used a microscope. Although most widely known for description of the arterial arrangement at the base of the brain, arguably his most significant contribution was the concept of reflex function in which the nervous system senses and responds to its external and internal environments: 'Motus [. . .] est reflexus qui scilicet a sensione prævia [...] dependens illico retorquetur' (a movement is reflex if it depends from a previous sensation, and is turned back at once).[35]

Willis realized that function depends on the substance of the brain, not its cavities: 'almost all anatomists of more recent times have given [. . .] the vile duty of a Jakes or sink to [these] more inward chamber[s] of the Brain'. The convolutions of the brain surface increase the amount of available tissue whilst leaving the underlying structures in continuity thereby allowing preservation of function when one part is diseased. Connection between the two halves of the brain ensures that 'every impression coming this or that way, becomes still one and the same'. He thought that the centre for receiving sensation and sending motor signals was a collection of ganglia known as the striatum (because it is striped on cross section) placed deep within the cerebral hemispheres: 'as it were the King's High-Way [which] leads from the Brain, as the Metropolis, into many Provinces of the nervous stock, by private recesses and cross-ways'.

Nervorumque descriptio et usus is the treatise in which Willis describes 'neurologie', his 'doctrine of the nerves', by which he meant the structure and function of cranial nerves. Now the term refers more generally to the study of disease affecting the nervous system. Not original to Willis, the word combined -λογία as the body of knowledge on a given subject prefaced by νεῦρον (neuron), the Greek word for sinew, tendon, or bowstring. The legend for Willis's depiction of blood vessels at the base of the brain shows his classification of the 'ten' cranial nerves. This was a transitional view of the twelve that are now accepted.[36] In a much expanded version of his Sedleian lecture, Willis describes the two nerves which innervate and orchestrate workings of structures in the chest and abdomen. Willis considered that the 'wandring' and 'intercostal' nerves originate from the cerebellum and orchestrate 'the beating of the Heart, easie Respiration, the concoction of the Aliment, the protrusion of the Chyle, and many others' (Figure 2.11). Although anatomically separate, these two nerves form part of one physiological system bringing internal organs into 'sympathy'. Now designated the parasympathetic and sympathetic nervous systems, respectively, this is the original account of the autonomic nervous system.

It was natural that Willis wanted to understand how fermented blood reaches the nervous system. He was not the first to describe the arterial network at the base of the brain, nor did he claim priority.[37] But no one before him had combined an account of the anatomy and physiology, or explained the clinical significance of a grid-like arrangement for maintaining perfusion of the whole following partial failure of the circulation. In a famous passage, Willis described a patient who had no neurological symptoms in life but in whom one of the four main vessels feeding blood to the brain (right carotid artery) was blocked. Nature had provided a remedy

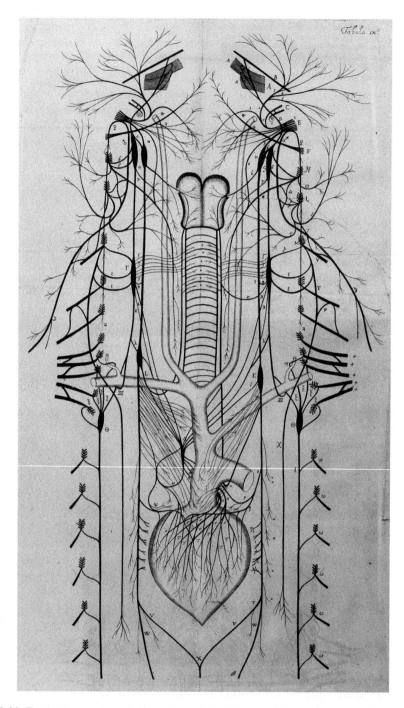

Figure 2.11 The distribution of Willis's fifth, sixth, seventh, ninth, and tenth cranial nerves, and the roots of the intercostal and 'wandring' nerves distributed to the chest and abdomen as found in man. Tabula IX from *Cerebri anatome* (London, 1664).

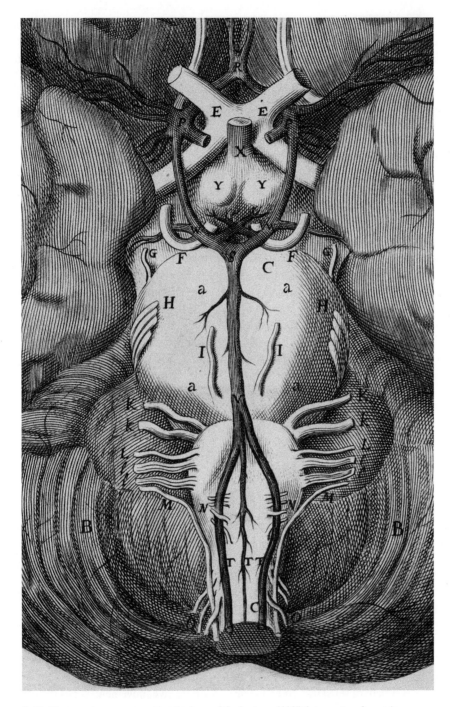

Figure 2.12 The vascular arrangement at the base of the brain and Willis's ten pairs of cranial nerves. Figura 1 (detail) from *Cerebri anatome*, London, 1664.

by enlarging one of the other (the vertebral artery) and so ensuring adequate blood supply to the otherwise compromised right side of the brain.[38]

Willis made several points in describing this 'fourfold Chariot' of vessels. In rich language, he explains that, after entering through the skull base, the arteries at the front and back on each side are connected to form a circle at the base of the brain through which blood can be redistributed before the feeding vessels divide into further branches which 'creep through and, like vine shoots, tightly bind not only the external circumference of this sphere, but also its inner parts and recesses'. Blood eventually reaches sinuses within the membrane covering the surface of the brain (meninges) which feed into veins draining the brain (Figure 2.12). His emphasis is on the arteries at the front of the neck (carotids) but no less precise is his description of those at the back (vertebral arteries) which, after joining to form the basilar artery, 'meeting as it were in a threefold way, are inoculated one with another'. Willis's also describes blood vessels supplying the spinal cord.

Willis had outlined ideas on muscle contraction in his Sedleian lectures. These were expanded in *De motu musculari* (1670). He argues that volatile nitro-aerial-enriched animal spirits in the nerves and saline-sulphurous particles in arterial blood combine as particles which are capable of exploding. This increases the breadth and reduces length of the muscle belly, bringing together the 'hanging' parts and thereby moving the joints. Willis also considered muscle mechanics: 'a Collection of moving Fibres, so framed together, that the middle flesh constitute an oblique angular Parallelopipedum (a three dimensional figure formed by six parallelograms) but the opposite Tendons compose two quadrangular Prisms or Figures'. Contraction starts at either end of the muscle belly and proceeds towards the centre, increasing the angle between the insertion of the tendon and the greatest width of the muscle belly.[39]

The discourse of the soul

Willis agreed with others that there are two souls.[40] The corporeal soul of brutes aligns with Willis's ideas on reflected (reflex) responses: 'nothing is in the brain or heart which has not been in the senses first'. Man has this corporeal soul but also a rational soul gifted as 'a particle of divine breath' responsible for decision and reasoning. The rational soul can abstract 'universal things from singulars'; contemplate 'God, Angels, It Self, Infinity, Eternity, and many other Notions remote from Sense and Imagination'; resist slavish adherence to behaviour and habits; exercise judgement and restraint; be aware of what it has not necessarily sensed; devise the methods of algebra, geometry, and astronomy; resolve natural philosophy from its causes; and contemplate metaphysics and the supernatural. This argues that the 'Substance or Nature of the Rational Soul is Immaterial and Immortal'.

But Willis did not always find it easy to maintain this position. The corporeal soul is more than a mute recipient of sensation and a machine for reflective movement. It has the capacity for perception and intention, properties more easily assigned to the rational soul of man. Conversely, diseases which manifest as disturbance of functions belonging to the rational soul of man have an organic basis. Willis was a man of faith and conformed to doctrines of the church; but, as a scientist, too often he allows reasoning in brutes and a material basis for the rational soul of man.

Discussion of the passions represents Willis's first direct excursion into human psychology and a prelude to his contributions on psychiatry. The extremes are pleasure and grief within which boundaries all other emotions can be accommodated. The passions are easily assuaged, familiarity tending to quieten movement of the spirits and novelty being needed again to excite them. They are stimulated by sensory experience and moderated by judgement. Willis makes an important point about cooperation of the five senses: 'nothing pleases the Palate unless the Sight, and hearing, Smell, and Touch approve it'. In an age where reading was limited, hearing could be considered 'far more Excellent than the other senses; for as much as by its help chiefly, Sciences and Learning are acquired'. But in discussing sight Willis declares that 'if there be any strife for Dignity among the senses, the Palm is given [. . .] as the most noble Power'.

Willis had discussed the physiology of sleep and waking in his Sedleian lectures. In *De anima brutorum* (1672), he adds that the individual organs do not sleep other than in the brief pauses of their rhythms: 'condemned to the stone of Sisyphus; to wit that they still lift up the same burthen. Then resting whil'st it slides down again, they presently, and so perpetually, repeat their Labour'. Nor does sleep necessarily stifle the senses and motions. Dreams and movements occur when the animal spirits 'wander without any Guide or Ruler'. Movements are repeated 'as it were Dances before learnt [. . .] and the Cogitations of things, though after a very confused manner'. The '*Incubus* or Nightmare' is experienced when the stomach is full and people sleep lying on the back: 'as if with some Animal or heavy weight lying upon the Breast, stops our breathing'. In sleep walking, 'the Animal Spirits, being agitated [. . .] leap back, into the Nervous Stock, and [. . .] produce divers sorts of local Motions'. Here, Willis has in mind complex behaviours not mere thrashing around the bed. In their most complicated manifestations these include dressing, carrying out household activities, conversing with those they meet, 'taking them by the hand and often-times striking them', moving aside obstacles, and hearing, seeing, and responding to 'Impressions made from sensible things'.

Willis describes sleep disorders. The 'Lethargie' follows delivery of morbific matter in the blood which drowns the nervous spirits. Leading to loss of memory and impairment of reasoning, sensation, and motion, it often occurs in the context of another disease, especially fever. Those affected 'in the midst of a journey, yea at dinner, or though busied about any thing, they presently fall into a drousiness'. Willis explains:

> Whilst talking, or walking, or eating, yea their mouths being full of meat, they shall nod, and unless rouzed up by others, fall fast asleep: and thus they sleep continually almost, not only some days or months, but (as it is said of *Epemenides*) many years.[41]

Treatment includes 'Coffee, or the liquor prepared of that Berry, first boiling in it, the leaves of Sage or Rosemary, till it has got a greenish Tincture'; or wearing a 'quilted thing of Cephalicks and Spices' under a cap over the shaven skull. The 'sleepy coma' and *carus* occur when morbific matter seeps throughout the brain, confining victims to their chair or bed and interfering with imagination and memory. In the 'preternatural Watchings', insomnia may be symptomatic of another illness such as 'Phrensie, Madness, the Colick, Gout or such like'; a disorder in which those affected 'fly to *Opiates*'; or due to 'drinking coffee in order to study late into the night'. Individuals affected by 'the Waking *Coma*' crave sleep, but it eludes them and leads to 'a *Delirium*' in which, rambling in their talk and the limbs restless, the 'sick can no more sleep, than those on the Rack'.

Neurological and psychiatric disease

Pathologiæ cerebri (1667) mainly comprises Willis's treatise on convulsions, a topic introduced in his lectures as Sedleian Professor. In discussing epilepsy, Willis now makes the point that the fault lies not with the part that moves but from origins in the brain. His concise description of sudden falling, the aura 'ascend[ing] [. . .] like a Cold air, and creeping towards the head'; incidental injury from earth, fire, or water (more violent that the falling of 'the Syncope and Apoplexie'); gnashing of the teeth, foaming at the mouth, shaking of the head with tetanic posturing of the trunk, thrashing of the limbs, disturbances of the bowels and breathing, with slow recovery followed by headache and a period of lassitude, cannot be bettered.

Any neurologist will recognize conditions now designated as movement disorders from Willis's descriptions of convulsion without altered awareness: distortions of the face, eyes, and mouth; jerking of the head with respiratory movements and repetitive utterance of the same word; and people tormented with 'inquietude of their members that they are forc'd to walk, till they were tyred, as also to dance, leap, and run about that by this means they might shun the grievous trouble'. Witchcraft or the devil are suspected when the contortions or gesticulations are such that 'no sound man, nor mimick, or any tumbler can imitate'.[42]

The second part of *De anima brutorum* (1672) considerably extends Willis's classification and descriptions of neurological and neuropsychiatric disease. His most famous patient with migraine was Anne, countess of Conway (1631–79).[43] Willis on apoplexy (stroke) is equally impressive. Without wavering from his theory of disease as a process determined by perturbations of the heated blood and the animal spirits, here is summarized, in early modern English terminology, all that is now known of cerebrovascular disease: bursting of vessels (haemorrhage), blockage from local degeneration (atheroma) or material from a remote site, especially the heart, lodged in the main arteries (thrombosis and embolism), involvement of the main arteries in the neck or their smaller divisions in the brain (large and small vessel disease), brief episodes with recovery as harbingers of persistent stroke (transient ischaemic attacks), and the ominous manifestations of haemorrhage compressing the brain stem and medulla. Willis's patron and Sedleian elector: 'the most Reverend Father in God the Lord *Gilbert* [Sheldon] Archbishop of Canterbury, recovered of a grievous *Apoplectical* Fit, six years ago [i.e., 1666], (God prospering our medicinal help, to whom we render eternal thanks) and from that time, though he sometimes suffer'd some light skirmishes of the Disease, yet he never fell, or became speechless or senseless'.[44]

Willis attributes palsy to disorders of the pathway for voluntary movement extending from the surface of the brain through the 'Streaked Bodies [striatum], Oblong Marrow [medulla], and also to the Nerves, and nervous Fibres'. The weakness may be partial, one-sided, or more generalized, with or without disturbance of sensation. He has observed damage to individual peripheral nerves, and conditions affecting a limited part of the spinal cord. Cases are 'accidental', by which he means acute, or 'habitual', that is chronic and reflecting the cumulative burden of previous or predisposing events. 'A prudent and honest Woman [. . .] speaks freely and readily enough, but after long, hasty or laborious speaking presently she becomes mute as a fish, and cannot bring forth a word, and does not recover the use of her Voice till after an hour or two' is thought to describe a condition in which the chemical message released from the nerve ending and causing muscle fibres to contract is blocked (myasthenia gravis).[45]

The final chapters of *De anima brutorum* (1672) deal with disorders of the rational soul. Imagination and memory are 'deformed, distracted one from another, and very confused'. The brain, 'evilly affected', represents 'monsters from a multiplying or distorted Glass'. Willis's classification includes delirium. As in other contexts, he makes the important point that transient events are tolerated, whereas those that are habitual leave permanent traces etched in the brain. In the phrensie (mania), 'Ideas of things raised up out of the Memory, the old confounded with the new, and some evilly joined, or wonderfully divided, are confounded with others'. Melancholy (bipolar disease) is 'a raving without Feavour or fury, joined with fear and sadness'. Amongst the melancholic, Willis includes people who bark or howl, thinking themselves dogs, wolves 'or some Monster'; others who imagine themselves dead, asking to be buried; and those who conclude that their bodies are made of glass and so they fear being touched 'lest they should be broke to pieces'. Again, Willis cannot agree that melancholy is a disorder of the internal organs. One of Willis's less well-recognized contributions, first explored in the Sedleian lectures, was to discard the view that hysteria and hypochondriasis are due to vapours emanating from the uterus or spleen but, rather, convulsive disorders etc. of the brain resulting in stimulation of nerves and reciprocal effects on the abdominal organs. In *Affectionum hystericæ et hypochondriacæ* (1670), Willis again emphasizes these conditions as organic disorders of the brain.

Madness is an extension of melancholy: 'These two, like smoke and flame, mutually receive and give place one to another'. The mad shun no dangers, and attempt the most improbable and difficult tasks. They are strong 'so that they can break chains and cords, break down doors or walls'. Mad men never tire 'although playing mad pranks [. . .] without sleep or eating'. They 'bear cold, heat, watching, fasting, strokes, and wounds, without any sensible hurt'. In writing on foolishness and stupidity, Willis uses terms that might be considered synonymous but, although careless in always maintaining the differences, his intention is to distinguish two quite separate conditions. Foolishness characterizes individuals with preserved comprehension and memory but defective judgement, poor reasoning, and awkward social behaviour. Stupidity comprises defective apprehension, poor memory, and simplicity in appearance and behaviour.[46]

Rational therapeutics

Willis's book on rational therapeutics is a fitting conclusion to his writings. There is no clearer account of treatments used by enlightened physicians in seventeenth-century England. The theme running through these treatises had been articulated at the outset of Willis's medical career. Vitality is dependent on healthy nutrition, respiration, fermentation, circulation of the blood, and removal of waste. When one or other process goes wrong, the focus on increasing excretion of toxic material through normal channels is entirely rational in the context of Willis's ideas.

Animal, vegetable, and mineral sources are scavenged for these medicines. Most require chemical liberation of active particles from the starting materials: leaves, flowers, seeds, or roots of plants and fruits; animal material such as horn, shell, tooth, or bone and occasionally fresh organs; and metals or minerals. Willis's pharmacists made up juleps, tinctures, liquors, elixirs, apozemes, and potions made palatable with liquorish, spices, and sweeteners. Other decoctions were used as powders, boles, pills, morsels, tablets, troches, and lozenges; linctuses, lohochs, eclegmas, and lambitives (syrups and materials to be licked); or fumigations and balsams. Many treatments are administered as enemas. Rarely are Willis's patients spared topical application of irritants or liniments soaked in astringent or soothing substances: blisters, vesicatories, scarification, cupping glasses, cataplasms, poultices, foments, and leeches applied to the affected part. Many patients are subjected to surgical procedures: blood-letting to remove aberrantly fermented circulating blood; ligatures around the limbs or incisions in the skin and underlying tissues providing restraint or outlets for the toxic material. More elaborate procedures include stenting a stricture of the oesophagus, and external drainage or washing out purulent collections within the chest. In order to make such surgery tolerable, Willis offers the patient an alcoholic julep. He recommends changes in diet and the use of iron-containing (chalybeate) and other healing waters. Almost invariably his regimen offers medicines taken in combination and sequentially so that, as the distemper continues, patients are exposed to a vast number of medicines. Their tolerance is much to be admired.

Because they provide easy access for removal of waste, Willis describes contents of the chest and abdomen, and their diseases. The anatomy and physiology of breathing form the basis for discussion of 'Cough, Phthisic (wasting or cachexia), Catarrh, Asthma, Dyspnoea, and other [symptoms from] Diseases of the Brest'. Willis describes 'Pleurise' and empyaemia, 'Dropsie in the lung' (infection and collection of fluid on the surface of the lungs), 'Phthisis or Consumption' (tuberculosis), disorders attributable to immobility of the chest wall, the 'Chin-cough' (whooping cough), 'Hæmoptyis or spitting blood', and 'peripneumony' (pneumonia). Willis's remedies increase discharge through the upper airway. Emptying the 'ventricle' (stomach) is achieved by induced vomiting; purgatives and enemas ensure that 'thorow the intestinal Pipes, as it were through a sink, the whole impurity of the Body is [. . .] purged forth' of the 'chyle and *Fæces*'. On ascites (accumulation of fluid): 'I have once seen a Tub would hold 15 gallons filled with water taken out of the Abdomen of a Woman dead of a Dropsie'. The 'tympanie' results from an abdomen full of gas: pricking the distended belly may discharge 'a stinking winde' but usually leads to death. Accumulation of water 'within the pores of the flesh and skin' (dropsie or oedema) results from imbalance between outpouring from arteries and supping by veins, the surplus swelling the flesh because fluid is not excreted in the sweat or urine.

Increasing the flow of urine is prominent in Willis's therapeutics. Some medicines release the watery humour from tenacious particles; some dissolve accretions in the blood which have 'put away the Serum [. . .] in divers places'; and others 'sup up again the Serum [. . .] and deliver it to be sent away by the Reins (kidneys)'. On diabetes: 'in our Age given to good fellowship and gusling down chiefly of unalloyed Wine', as a result of which 'those labouring with this Disease, piss a great deal more than they drink [. . .] joined with it continual thirst'. In a famous statement, reminiscent of his habits from the early 1650s at Abingdon market, Willis adds 'the Urine in all [. . .] who hath hapned to have it [. . .] was wonderfully sweet as it were imbued with Honey or Sugar'.[47]

Cardiacks 'unlock and open joynting together of the Blood' when the particles are 'complicated within themselves' and will not separate for normal excretion of waste. At the risk of making retrospective diagnoses, it is remarkable that the modern reader can identify disorders of rhythm, heart failure, and bulging of the wall (ventricular aneurysm) from muscle disease, tightening of the valves (aortic stenosis), and fatty degeneration (arteriosclerosis) of the arteries.

Willis defers to Marcello Malpighi's (1628–94) description of the skin which offers another outlet.[48] Freckles and other harmless changes in appearance encourage remedies for 'the Art of Beautifying' amongst 'curious Ladies' that open the pores or drive back 'spotty matter' into the blood. The 'Psora or Scab' (psoriasis) is characterized by scaly skin; matter is produced that becomes 'Lousie' (as in lice) in those held 'long in prison [. . .] and [. . .] of a sedentary life [. . .] used to nastiness and sluttishness'; even those who live decent lives may develop the itch if they 'lye in the same bed [. . .] with a scabby person', or use linen 'tainted by the Mange'. Wheals and red pustules of the impetigo (bacterial infection of the skin) are not contagious: 'the Miasma from the Husband doth not pass to the wife, or from her to him, though they lye together'. Waste is removed through the skin by sweating. In times past, 'it was a custom of the Irish, if any one were sick of a Fevour, to roll him up in woollen cloths wetted with cold water, by which means a plentiful sweat succeeding the disease was often broken'. Willis knew of a noble Lady 'famous [. . .] for that she does not only every night, moysten or thoroughly wet ye Linnen and bedcloaths, but besides distils into a basin put under her thighs many [. . .] pounds of mere Sweat'.

The reputation of Thomas Willis

Thomas Willis might reasonably have gone to his grave in Westminster Abbey confident that posterity would celebrate his achievements. But he soon proved victim to waspish criticism, mainly from Anthony à Wood. Efforts were made by Browne Willis to correct and perpetuate his grandfather's memory but it was not until the twentieth century that Thomas Willis's originality was unreservedly acknowledged. One issue was whether Willis was merely the popularizer of work done with others. He had always been at pains to acknowledge collaboration:

> I was not ashamed to require the help of others [. . .] *Doctor Richard Lower* [. . .] the edge of whose Knife and Wit [. . .] have been an help to me in searching out both the frame and offices of before hidden Bodies [. . .] it becomes me not to hide, how much besides I did receive from these most famous Men, *Dr Thomas Millington* Doctor in Physick [. . .] and the learned *Dr. Chr. Wren* [. . .] [was] frequently to be present at our Dissections, and to confer and reason about the uses of the Parts.[49]

Willis's work on fermentation attracted criticism. The most vitriolic came from Edmund O'Meara (1614–81) and, whilst easily dismissed as sabre rattling by a disgruntled Galenist, O'Meara found it regrettable that an Oxford professor should lead students astray with erroneous and unorthodox concepts and treatments. He needed to prevent 'the lechery of the wanton mind and the mad itch for innovation [causing] a dangerous chain of errors that will end in the ruin of the human race'. This triggered a vitriolic defence of Willis from Richard Lower who described O'Meara as 'A little frog from the swamps of Ireland'.[50] Later: 'An Irishman, Meara, takes first prize for sheer perversity and stupidity [. . .] if I had to contend with him in this matter, I should not have to go into the ring so much as into the cess-pit'.

Frederic Slare (1647–1727) had a gripe against Willis's use of medicines.[51] Bezoar stone was of no therapeutic use: it was concretion of mud 'so inert that one may take 50 or 60 Grains in a Mess of Broth [. . .] without being in any respect Philosophically or Physically benefited or hurt by it'. Better to lift the 'spirits' with a 'glass of good Ale, or Beer, or of generous Wine'. Willis had written that sugar causes disease. Slare appealed for support from women accustomed to take refined sugar in 'Tea, Coffee *and* Chocolate [. . .] at the Morning Repasts, called Breakfasts'. His only caution was that 'sugar being very high a Nourisher, may dispose them to be fatter than they desire to be, who are afraid of their shapes'. Slare agreed with Mrs Grymes (dates unknown), housekeeper to (Henry Somerset) the Duke of Beaufort (1629–1700); 'That which preserves Apples and Plumbs / Will also preserve Liver and Lungs'. Slare explained that his maternal grandfather remained fit and active until he died of apoplexy aged 100 years, despite indulging in sugar at every meal and losing all his teeth at 80 at which point, happily, another set erupted.

Nor was Willis on firmer ground with his writings on anatomy. Walter Charleton (1619–1707) was asked by the Royal Society discretely to communicate discrepancies in *Anatome cerebri* to Willis: 'what shall pass between those two doctors upon this occasion, shall not be made public without their consent'. Nicolas Steno (1638–86) was politely critical: 'The best Figures of the Brain are those of *Willis*; but even these contain a great number of important Mistakes'.[52]

Eventually, Sir Charles Symonds (1890–1978) restored the reputation of Willis's work on cerebral blood vessels.[53] And Sir Charles Sherrington (1857–1952) praised Willis's ideas on reflexes:

> Willis [. . .] gave numerous instances of automatic acts, where stimulus was promptly followed by movement without conscious participation of the will. He spoke of this as being 'reflex' [. . .] Willis put the brain and nervous system on their modern footing, so far as that could then be done.[54]

From modest origins as a country doctor in Oxfordshire, Willis became an intellectual leader of medical science in early modern England. He wanted to understand the nature of vitality especially as this affected the brain. These imaginative and influential ideas were first explicated in his lectures as Sedleian Professor of Natural Philosophy in Oxford.

CHAPTER 3

The Sedleian Professors of the Eighteenth Century

A Theological Turn?

NIGEL ASTON

Eighteenth-century science, as understood and practised at Oxford, was academically eclectic.[1] Tellingly, the Savilian Professors in Astronomy and Geometry, under their foundational statutes of 1619, were expected to lecture on both ancient and modern topics as well as practical mathematics.[2] But the four Sedleian Professors between 1704 and 1782—James Fayrer, Charles Bertie, Joseph Browne, and Benjamin Wheeler—were not mathematicians and had limited claims to being, as contemporaries would have understood the phrase, 'natural philosophers', a category older than the 'new science' with its origins in the Renaissance and its emphasis on experiment.[3] All were, nevertheless, respected figures within eighteenth-century Oxford (Browne was Vice-Chancellor between 1759 and 1765) whose appointments to the chair are explicable given the workings of the patronage system. Each one merits proportionate recovery from dismissive notice along the lines of G. L'E. Turner's comment that they were 'undistinguished as holders of the chair'.[4] This contextualized chapter is an attempt to do so.

If we are to understand the basis on which these appointments were made we need to do two things. Firstly, to appreciate the extent to which the term 'natural philosophy' could be stretched in the eighteenth century and thus allow scope for nominees to profess the broadest range of intellectual interests that might qualify them for the Sedleian Chair. In particular, students of divinity were expected to be familiar with the claims made for the truth of the Christian faith on the basis of natural religion.[5] Newton's influence (he became President of the Royal Society in 1703) was unescapable and largely embraced in Oxford during the first third of the century until the growing number of Hutchinsonians within the university (a physico-theological grouping that sought to display what they viewed as the dubious, unbiblical foundations of his physics) made their influence felt.[6] Particularly in the second edition of the *Principia* (1713), Newton identified the reality of absolute space and time with God's extension

Nigel Aston, *The Sedleian Professors of the Eighteenth Century*. In: *Oxford's Sedleian Professors of Natural Philosophy*. Edited by: Christopher Hollings and Mark McCartney, Oxford University Press. © Oxford University Press (2023). DOI: 10.1093/oso/9780192843210.003.0003

Figure 3.1 Oxford High Street, *c.*1750.

and duration and made universal gravitation a manifestation of a substantive divine presence. This was a rejection of mechanical, nominalistic views of nature that such as Robert Boyle had championed in, for example, his *A Free Inquiry into the Vulgarly Received Notion of Nature* (1696).[7]

Secondly, we should be conscious that appointments were made within the usual parameters of a patronage system everywhere operative in English society in which patron–client relations determined outcomes, taking account of the nebulous concepts of influence and interest that informed it. Patronage, as Elaine Chalus has noted, 'pervaded the eighteenth-century domestic, cultural, social, and economic domains'.[8] And a successful client could be deemed to be an individual who held several lucrative and status-giving positions concurrently. He thereby became a pluralist and absenteeism was inherent in pluralism, as the history of the Sedleian Chair in this era would demonstrate. The University of Oxford and its constituent bodies functioned as a hub that formally and informally connected a network of internal and external parties with powers to confer office and status—or to withhold them. It was no different to any other corporate body operative three centuries ago in that regard with obligation and politics potentially decisive factors when an office was conferred or withheld.[9] In other words we should not expect scholarship and merit to be academic trump cards when Sedleian Professors of Natural Philosophy were decided upon—and they were not. So much depended, as in any other appointment to office, on which individuals had it in their gift. The Savilian Professorship of Geometry had a considerable number of electors, many of them holding office in institutions outside Oxford, but that was not the case with the Sedleian Chair of Natural Philosophy. This was an internal appointment that was decided upon by three men: the Vice-

SIR THO. MILLINGTON, F.R.S.
Savilian Professor at Oxford,
President of the Royal College of Physicians.

Figure 3.2 Sir Thomas Millington, mislabelled.

Chancellor, the Warden of All Souls College, and the President of Magdalen College.[10] But, of course, they were potentially subject to external pressure from ministers and bishops pushing the candidature of a particular individual, and to disdain the preferences of such a figure would be to hazard their own hopes of ascent within the *cursus honorum*.

So much can be seen when the chair fell vacant in 1704. The deceased incumbent, Sir Thomas Millington, Kt. (1628–1704), an original member of the Royal Society and a Fellow of All Souls College, had followed Thomas Willis (whom he succeeded in 1675) both in ignoring the original statutory requirement to teach only from Aristotle and in making neurological topics his research preference. Millington, by all accounts a charming personality, was primarily a London-based physician who, at his death, had been President of the Society of Physicians since 1696 (previously its treasurer, 1686–90), and a court physician from 1680.[11] As the alleged discoverer of sexuality in plants and a respected establishment figure,

Millington's stature as a natural philosopher conferred further distinction on the Sedleian Chair following on from Willis.[12] The great Oxford benefactor and fellow physician, John Radcliffe, is reported to have remarked to Millington that the whole art of medicine could be put on a sheet of notepaper, and was met with the sharp rejoinder: 'As far as you know it, it certainly could.'[13] But distinction was conferred from a distance. He was an absentee professor by the 1690s and fulfilled the duties incumbent on him vicariously through nominated deputies. Of these, the most distinguished was John Keill (1671–1721).

On the face of it, Keill was the obvious candidate in 1704. An Edinburgh graduate incorporated at Oxford in 1694 as Scottish Exhibitioner at Balliol College, the pupil of his brilliant fellow Scot, David Gregory (1661–1708), he gave the first experimental lectures on natural philosophy at Hart Hall (the precursor of Hertford College). From 1699 he was deputy to Millington, and delivered lectures published in Latin as *Introductio ad Veram Physicam*, and had become an FRS in 1701.[14] Keill was one of the rising intellectual stars of his generation in 1690s Oxford, an enthusiastic Newtonian, and regular contributor to the *Philosophical Transactions of the Royal Society*. He explicitly denounced Cartesian mechanism and reasoning and declared that all the efforts of all the mechanical philosophers did 'not amount to the tenth part of those Things, which Sir Isaac Newton alone, through his vast Skill in Geometry, has found out by his own sagacity'.[15] But, as David B. Wilson has pointed out, Keill was an eclectic thinker for whom Newtonianism was but one source of knowledge. His was 'a theological Newtonianism, modified and strengthened by Aristotelian considerations'.[16] Keill had indeed announced in his *An Introduction to Natural Philosophy* that, though Newton's genius had accomplished more than all the rest of human history combined, he would incorporate ideas from four other groups that had made useful contributions: Plato, Aristotelian peripatetics, experimenters, and the mechanical philosophers.[17] Such eclecticism was an obvious consequence of physics being still taught within the arts or philosophy faculty. It essentially covered the material found in Aristotle's *Physics* and *De caelo* (*On the Heavens*), but studied mathematically and phenomenologically under the shadow of Newton.[18]

But Keill was not appointed in 1704, and the opportunity was lost to make the Sedleian Chair consonant with cutting edge scientific proficiency within the university.[19] Keill had the learning; he had the right sort of moderate Tory politics in fashion with the Oxford academic establishment. It was probable his character may have counted against him. He was reportedly a heavy drinker, married to a servant (daughter of a bookbinder and his former mistress), and was, according to the Tory antiquarian and diarist Thomas Hearne, 'a man of very little or no Religion'.[20] And Keill was an Oxford outsider in more senses than one: he was Scottish (and Oxford Scotophobia three centuries ago is not to be discounted), and with ties only to Hart Hall and to Balliol, he did not move in the circle of the three men tasked with election to this chair.

Millington was an All Souls man and therefore his successor, on the basis of the electoral balance, was likely to be an individual acceptable to the Magdalen interest. The President of Magdalen, Thomas Bayley, had only just taken up office after the death of his predecessor in February 1703. He was an Oxford absentee throughout the 1690s, having resigned his Magdalen fellowship in 1689 on taking up the lucrative college living of Slimbridge, Gloucestershire. Then, suddenly, he was summoned back to assume the Presidency and, dutifully, in his early sixties, he did so.[21] It should be assumed Bayley would have little sense of the field for the Sedleian Professorship. The Vice-Chancellor's interest in this selection was nominal.

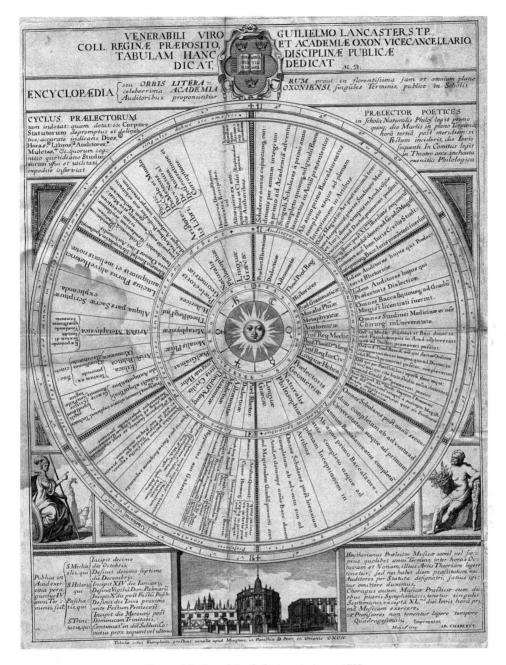

Figure 3.3 Plan of the Oxford curriculum, c.1709.

This was William Delaune (1659–1728), the President of St John's, 'a man of parts and learning' but a companionable individual with a gambling problem, and minimally connected to either Magdalen or All Souls.[22] The real power broker was Bernard Gardiner (1668–1726), Warden of All Souls, 1702–26, originally a Magdalen man but ejected from his demyship (a half-fellowship) at the college during James II's reign just as his senior, Thomas Bayley, had suffered a similar fate.[23] Gardiner was a moderate Tory, an ambitious insider, a future Vice-Chancellor, and determined to exert a personal control over the All Souls fellowship.[24] But, to that end, he needed support in high places, particularly from the Archbishop of Canterbury, the Visitor of All Souls. Since 1695 this had been the Whig Thomas Tenison, who had sanctioned the appointment of Gardiner as Warden despite their political differences and, two years later, for reasons that are unclear on either moral or political grounds, wanted to apportion the Sedleian Chair to James Fayrer or Farrer (c.1655–1720).[25]

Fayrer was known to Bayley as a fellow junior to him at Magdalen College, elected in 1683, but he was senior by some margin to Gardiner who had also come across him there in the troubled mid-1680s.[26] Electors and elected had a common Toryism that had been reinvigorated by the accession of Queen Anne two years previously but it cannot be said that this was an adventurous or imaginative choice. Far from it. In nominating Fayrer they broke the connection between the Sedleian Chair and work in the natural sciences of the kind that Willis and Millington had pioneered before him. The contrast between the lacklustre Fayer and the astonishing talents of Edmond Halley, named Savilian Professor of Geometry just the previous year—1703—was stark.[27]

Fayrer was Sedleian Professor until his death on 20 February 1720. There was never any pretence that he would seriously dabble in natural philosophy where Halley and Keill (Savilian Professor of Astronomy after 1712) dominated the Oxford field in Fayrer's time. As college divinity lecturer, he might technically qualify as a theologian:[28] he was awarded a DD in 1704 and subscribed to the first volume of transcript of the Septuagint based on the *Codex Alexandrinus* produced in 1707 by Johann Ernst Grabe, a gifted Prussian expatriate in Anglican orders resident in Oxford, holding a chaplaincy at Christ Church.[29] Although evidences of his intellectual life are elusive, that is not the case for his personal life: Fayrer, a bachelor in holy orders, was known for lechery and laziness. The rumours of his chasing after the women bedmakers in Magdalen College were persistent and, according to Hearne, Fayrer's patronage 'of these loose women' was 'severely reflected upon by some of the Fellows, who knew he laboured under a flagrant suspicion with regard to some of them'.[30] In later life he kept one Mrs West of Holywell as his mistress (wife and later widow of Richard West, formerly of Oriel College), and she was with Fayrer in his last illness.[31] All of which makes his appointment the more remarkable given that Keill's personal life was a factor in his exclusion.

Fayrer's successor had a strikingly different background. This was the Hon. Charles Bertie (c.1679–1746), the sixth and youngest son of the 1st Earl of Abingdon but, again, to label him a 'natural philosopher' would be stretching the term to breaking point. Neither was he a theologian. He was a Law Fellow at All Souls (BCL, 1706; DCL, 1711), not in holy orders, and not a bachelor but a married man with six surviving children.[32] He needed to feed his family and maintain his status as a peer's younger son. The Berties, Earls of Abingdon (1682), owned extensive estates at Rycote Park near Thame and Wytham Abbey (3 miles away from Oxford but just inside Berkshire). Charles's eldest brother, the 2nd Earl, was high steward of the borough of Oxford, and worked to ensure Tory party advantage in elections both to the mayoralty and to the city's two seats in Parliament. However, his lesser land holdings and smaller bank balance meant that the Bertie family was steadily losing influence as county grandees to the Whiggish Marlboroughs recently domiciled at Blenheim.[33]

Figure 3.4 Bernard Gardiner, sometime Warden of All Souls and Sedleian elector.

The 1720 vacancy was an All Souls turn where Bernard Gardiner remained Head of House and would expect not to be gainsaid. The Vice-Chancellor since 1718, Robert Shippen, Principal of Brasenose College, had close links to the Tory party in Parliament through his brother William, an undisguised Jacobite.[34] Vice-Chancellor Shippen was disliked by many Oxford

Tories on personal grounds but there is no indication of his falling out with Gardiner over the nomination of Bertie to the Sedleian Chair. As for Joseph Harwar, President of Magdalen since 1706, he had a consistently low profile and was not one to obtrude his opinion in these circumstances.[35] Charles Bertie had a proven record as a college administrator, often travelling away on All Souls business. He was a known quantity to Gardiner, another Law Fellow from an elite background, with the same Hanoverian Tory sympathies, and a colleague with whom he could work. Any thought of steering the vacant professorship back nearer to the founder's intentions was displaced by the desire to repay his friend and colleague. Once in post, Charles Bertie submitted a request that he should be allowed to continue in his All Souls fellowship and it was granted.[36] However, this request was indicative of his disinclination to be an active professor. For Bertie's ambitions centred on All Souls and, on Gardiner's death in 1726, he was one of the two names sent to the Archbishop of Canterbury, William Wake, for him to select the new Warden. Against the odds and in a break with custom, the archbishop named the lower-middle class Stephen Niblett, an Arts Fellow, to the Wardenship.[37] At that point, Bertie, the college's first choice, went off in a huff, taking holy orders,[38] resigning his fellowship, quitting Oxford, moving to become a country parson at Kenn in Devon (in the gift of his brother-in-law, Sir William Courtenay, 2nd Bt.),[39] but not, significantly, resigning the Sedleian Professorship until 1741, shortly after Courtenay had nominated him to another living in the same county.[40] As he had not lectured while resident in Oxford, his departure for Devonshire made no difference to the *de facto* academic dormancy of the professorship. Bertie published nothing during his tenure.

One minor divine, one civil lawyer: would the electors on Bertie's death in March 1741 try more overtly to fill this professorship with someone more closely fitting the description as a 'natural philosopher'? It was not to be. Instead their choice fell on Joseph Browne (1700–67), a Cumberland man, fellow of The Queen's College since 1728, and college bursar at the time of his election (having previously been chaplain and tutor).[41] It is hard to explain why a don from a college previously unconnected with this chair and with a *curriculum vitae* that gave minimal indication of his aptitude for the professorship was preferred, unless it be an emerging reputation for probity and competence. So who were the electors in 1741 when one might anticipate that a Magdalen man would emerge (but did not)? Two were former Vice-Chancellors likely to be familiar with all the potential candidates. The President of Magdalen was the wealthy, well-connected Edward Butler, since 1737 one of the two Members of Parliament for the university;[42] the Warden of All Souls was Stephen Niblett, who had beaten Charles Bertie to that post in 1726. And reaching the end of a three-year stint as Vice-Chancellor was Theophilus Leigh, the Master of Balliol. Leigh, like Butler, had a comfortable income and was a Justice of the Peace for Oxfordshire.[43] His family was related by marriage to the Dukes of Chandos, and was known for its loyalty to the memory of Charles I and the custom of drinking to 'Church and King' before every meal.[44] The main clue to Browne's selection lies in an early career overlap with Niblett, who was only three years older than him. They were both undergraduate contemporaries at Queen's which Niblett entered in 1713 and from which he took his MA in 1720, the same year that he was elected to All Souls.[45] In sum, these were old associates. Both men had similar talents, the one giving his friend a step up the Oxford ladder when the opportunity offered.

Browne actually quit residence in Oxford and relinquished his fellowship in 1746 to become rector of Bramshott in Hampshire (a Queen's living) and Prebend of Bartonsham, Hereford Cathedral, 1746–67, where he also served as Chancellor (1752–4).[46] The lure of a lucrative benefice and a cathedral canonry held out to Browne respectively by his college and his former pupil, Lord James Beauclerk (the grandson of Charles II), immediately on the latter's becoming Bishop of Hereford in 1746, was irresistible.[47] But the good-natured and business-like Browne was not forgotten in Oxford. He returned in 1756 to become Provost of Queen's and within three years was preferred to the Vice-Chancellorship. At no point in his progression was he willing to give up the Sedleian Chair of Natural Philosophy and its attendant stipend, but the only hint over two decades that he may have been interested in fulfilling its original job description occurs in a letter of 1741 to his predecessor as Provost, Joseph Smith, concerning the endowment of a geometry lecture at Queen's to coincide with Browne's professorial nomination.[48] Browne was not much of a scholar.[49] He owed his career success, it seems, to his genuine affability and an administrative efficiency that was recognized across the university.[50] Writing in 1759, a member of Queen's reported that 'both old and young almost adored him' and that he continued as Vice-Chancellor for the exceptional time of six years (until suffering a stroke) says a great deal.[51] It was during the early 1760s that Browne seems to have appreciated that he had to play his part—wearing his Sedleian Professor hat—to satisfy an unmistakeable appetite for the dissemination of scientific information in Oxford. He nominated an assistant, Israel Lyons (also deputizing for Humphrey Sibthorp, Professor of Botany), who could be found in 1764 lecturing to a class of over sixty.[52]

Figure 3.5 The Oxford Physic Garden, domain of the professor of botany.

The Sedleian Chair was retained by Browne for over a quarter of a century without his adding anything to enhance the intellectual lustre of the post. The last holder before Hornsby did at least manage, although his writings were more theological than scientific, to merit being the principal focus of this chapter. This was Benjamin Wheeler (c.1733–83), elected in 1767, who skilfully ran together successful careers in the university and the diocese of Oxford. Wheeler was from Magdalen, quite senior in the college fellowship (elected 1761) where, as a contemporary attested, he 'toiled with unremitted diligence and unwearied exertion' as a tutor,[53] and was known for his good nature and affability.[54] Thus was elected a fourth successive Sedleian Professor who was not primarily identifiable as a 'natural philosopher'. Wheeler turned out to be outstanding at collecting professorships of all persuasions. The Sedleian Chair was his second, for in July 1766 he had been elected to the Professorship (Praelectorship) of Poetry, trading on his popularity with the electors, that is all Oxford degree holders, for Wheeler, despite his extensive linguistic abilities, had no obvious distinction as a poet in either English or Latin. He was, however, perceived to have the capacity to judge the poetic offerings of others in Oxford and had acted as one of the selectors of suitable university verses offered to George III in August 1761 on the occasion of the king's marriage.[55] What Wheeler did have was the lifelong support and protection of the Poetry Professorship's most distinguished holder, the renowned Hebraist Robert Lowth.[56] His parents, lower middle class residents of St Giles', Oxford, had sent the young Benjamin to New College School, where Lowth recognized his latent abilities, and moved him on to Winchester College. Thence he returned to Oxford as a scholar and fellow at Trinity College, where he overlapped with Lord North (Prime Minister, 1770–82), another future patron.[57] It was no surprise that Lowth soon made Wheeler his chaplain.

This backing of Robert Lowth, who became Bishop of Oxford in the same year—1766—that Wheeler became Poetry Professor, opened most Oxford doors in the 1760s and it was probably decisive with the Sedleian electors. These were David Durell, biblical scholar, Principal of Hertford, and Vice-Chancellor 1765–8, another admirer of Bishop Lowth, and the client of the current Archbishop of Canterbury, Thomas Secker; Thomas Jenner, who had known Wheeler for many years, the President of Magdalen since 1745 (also an unexceptionable Lady Margaret Professor of Divinity since 1728), and in the penultimate year of his life; and, at All Souls, the Hon. John Tracy (1722–93), a former senior proctor, who had assumed the Wardenship just the preceding year (fellow from 1746, friend of Sir William Blackstone, and a campaigner against Oxford abuses, including Founder's Kin at his own college).[58] Wheeler's ascent was by no means over at this juncture for he was Senior Proctor in 1768, collected the degree of DD in 1770, and was made Chancellor of the Oxford diocese in 1775.[59] The following year he moved to Christ Church to take up a canonry and the Regius Professorship of Divinity thanks, in no small measure, to the favour of Lord North and Bishop Lowth.[60] This was the summit and the end of his Oxford accumulations. Like Bertie and Browne, Wheeler seems to have excelled in administration. As soon as he moved to Christ Church, he wanted to make himself master of how the college operated, and he created a volume that survives, listing allowances, costs, etc. on all aspects of Christ Church from fees for use of the library to how much wine was consumed in the Audit House annually.[61] At least Wheeler appears to have had a genuine interest in natural philosophy, probably derived from his primary vocation as a theologian. And he does appear to have given lectures himself as Sedleian Professor, part of a pattern whereby the amount of organized teaching done by holders of 'scientific' chairs increased after *c.*1760. Thus, notes taken by a student survive, almost certainly for lectures on 'experimental philosophy' given in 1772.[62]

It was as Regius Professor that Benjamin Wheeler was energetic and diligent in his teaching. In this he followed the precedent set by his immediate predecessor, Edward Bentham, who recognized that the theological formation at the English universities of the ordained ministry had to be improved, and that the Regius Professor was therefore duty bound to take his lecturing duties seriously. No more than Bentham before him, Wheeler was not attempting his own *summa* of Anglican doctrine and history, or trying in divinity to rival Blackstone in law. He was an educator not a researcher. His modest aim was to ensure that students who were intending ordinands went away with an assured, orthodox theological grasp based on scripture and the tradition of the Church. The lectures were only, in a limited sense, any sort of practical preparation for parish ministry.[63] Amid the exposition, glimpses emerge of Wheeler's essential decency and quiet enlightenment values. He was, for instance, insistent that

> Persecution [...] is so far from being a mark of the Christian Church, that on the contrary it is one sure token of the very reverse, [...] I speak here of persecution [...] existing in any form or degree whatever.[64]

The same generosity of spirit denoted his understanding of the Church as being synonymous with the body of Christ:

> In a word, the true Catholic Church is the same all over the world; and no distinction of country, or party, or mode of worship, can justify an exclusion of [...] a brother from participating in the common benefits of the Church, or the witholding of necessary assistance to him by virtue of this his spiritual connexion.[65]

With their conscientious lecturing, Bentham and Wheeler set a firm precedent so that, by *c.*1800, it was claimed that no man could be admitted to holy orders in the Oxford diocese without a certificate that he had attended them.[66] But Wheeler was, by then, long since removed from the scene. He died quite suddenly from a stroke in September 1783 aged just fifty after having been appointed as a prebendary of St Paul's Cathedral only the previous week.[67] At that point, his plans to publish his lectures as Regius Professor were nowhere near to fruition.[68] He had, unaccountably, resigned the Sedleian Chair the year before. The reasons are unclear. It may be that it was his personal choice, newly conscious of spreading himself too thinly, and an unwillingness to limit promotional opportunities for colleagues; there may have been lobbying on behalf of Thomas Hornsby's indubitable claims to the professorship; or an emerging sense within the university, given the growing public popularity of lectures and demonstrations in Oxford, that the Sedleian Chair should be allocated to an individual more identifiable as a 'natural philosopher'.

Thus ended a run of four successive men—two minor theologians, two administrators—with nominal claims to this chair on the basis of its foundational description. None of them were Fellows of the Royal Society. Yet—with the possible exception of Fayrer—they were competent figures in the life of the contemporary university and to view them as anomalous embarrassments in the history of the Sedleian Chair would be to misunderstand the workings of academic society in what a recent collection of essays has called Oxford's 'Forgotten Century'.[69] To make the point again: this was an institution in which pluralism in office holding, within limits, was unexceptionable. Scholarship mattered but research intensity was not the *summum bonum* but rather an individual don's personal choice, and none of Millington's successors went far down that route. Personal connection rather than innate talent was the overriding decider in who was awarded professorships, and academic cheerleaders and coteries were as much in evidence then as now when a vacancy occurred. There was, too, a shortage of eligible candidates, certainly those deemed genuinely eligible. That said, these nominations to the Sedleian Chair showed the inward-looking side of the university at its most unimaginative, one viewed by both appointers and appointees as primarily one in which the successful candidate was given an office and a stipend with minimal duties. It would be going too far to say that it had become a sinecure but it was not so far from that category. Where the eighteenth-century professors were culpable was in the widespread omission of nominating a deputy to discharge lecturing duties, as Millington had named Keill. Fayrer did not imitate his predecessor and the precedent appears to have been forgotten until the 1760s, one that could easily have led to an association of the Sedleian Chair with the popularization of natural knowledge that was such a feature of contemporary science and would make Hornsby's lectures so popular.[70] One of the possibilities that Sir William Sedley suggested for his lecturers was to treat 'the books which treat of the soul'. With the possible exception of Benjamin Wheeler, it was not an ambition shared by the holders of his chair between 1704 and 1782.

CHAPTER 4

Thomas Hornsby

JIM BENNETT

Thomas Hornsby is a familiar historical presence for anyone concerned with the history of astronomy in eighteenth-century England. He was a prominent Oxford figure, who was affable, cooperative, and positive. He matched a thorough command of mathematical astronomy with sustained dedication to observing and a flair for communication in experimental philosophy more generally. Despite his contemporary standing and reputation, reflected in his senior appointments, no one has been energized to write a biography but we do have valuable accounts of his life by Ruth Wallis in a journal article[1] and his entry in the *Oxford Dictionary of National Biography*.[2] His appointment to the Sedleian Professorship came relatively late in his career, after the period of his most vigorous activity, but his earlier work in Oxford is relevant to the history of the chair, creating as it did a particular significance to his accession.

Hornsby was born in Durham in 1733, into a family of some local standing in trade: his father was an apothecary and later became an alderman. He went to Corpus Christi College, Oxford, matriculating in 1749, and graduating BA in 1753 and MA in 1757. Elected a fellow of his college in 1760, he lived in Oxford for the rest of his life. He was appointed to a series of prominent positions in the university, holding them concurrently as they accumulated until his death in 1810, and for close to fifty years was a major influence on the scientific aspects of Oxford life.

Career, character, creed

The stimulus for Hornsby's passion for astronomy is not known, neither is his qualification for the senior position in the subject, Savilian Professor of Astronomy in 1763, at all evident. As an early indication of interest and aptitude for the mathematical aspect of astronomy, we can note that a letter was read at a meeting of the Royal Society in June 1762 from Hornsby to Nathaniel Bliss, Savilian Professor of Geometry, on elements of the motion of Venus derived from the observations of Jeremiah Horrocks (or Horrox) in 1639 and those of the transit of

Jim Bennett, *Thomas Hornsby*. In: *Oxford's Sedleian Professors of Natural Philosophy*. Edited by: Christopher Hollings and Mark McCartney, Oxford University Press. © Oxford University Press (2023). DOI: 10.1093/oso/9780192843210.003.0004

1761.[3] However this is hardly a substantial basis for the appointment, sound and successful though it was.

Hornsby's other astronomical talent—as an observer—quickly became evident, as he acquired a 33-inch mural quadrant[4] and a 44-inch transit instrument[5] from the London maker John Bird (Figure 4.1), and began observing from a site in the college. He also had access to refractors and reflectors, with micrometres, some at least contributed by colleagues for special astronomical events. By a letter from Bird, 10 October 1767,[6] we know that the quadrant was already in use and that Hornsby was uncertain about the accuracy of the typical Bird feature (borrowed from George Graham)—a 96-part division of the arc alongside the conventional 90°.[7] While Hornsby cited discrepancies between the two scales (after conversion) Bird staunchly defended the innovation: 'there are two to one against you, in favour of the 96'. Already we see Hornsby's meticulous attention to detail, alongside a willingness, though a novice, to challenge the leading instrument maker of the day.

Hornsby's great project was the foundation and equipping of the Radcliffe Observatory, a work of some years but generally dated to 1772.[8] It will be treated in greater detail in the section 'Savilian Professor of Astronomy, 1763–1810', but as an indication of Hornsby's character and vision, it was the outcome of a remarkable aspiration. A university observatory was not a standard or expected institutional provision of the eighteenth century. There were a few examples in continental Europe, where there were also Jesuit colleges with observatories (though they might be better grouped with the observatories of other religious orders or church authorities), and beyond this, astronomy professors would often have some accommodation for instruments. Hornsby's ambition was of a different order, based on Greenwich but going further still: 'such a Collection of Astronomical Instruments will in all human Probability be secured to the University, as will not be equalled in the whole World'.[9]

This was an impossible stretch for the university, as Hornsby knew, so he devised quite a different scheme. Having secured estimates from Bird for the instruments he wanted, Hornsby persuaded the trustees of the charitable fund established by the will of John Radcliffe to approve funding in principle. They were not, however, in a position to advance the money until the cost of the Radcliffe Infirmary had been settled. Hornsby was anxious to avoid delay: he was convinced that he had to secure the instruments from Bird but was far from confident of the state of Bird's health and his future fitness for the work.

> There is unquestionably at present but one Person living, who is capable of making them with that Precision and Accuracy, on which the Goodness of the Observations essentially depends. . . . It is much to be wished . . . that Mr. Bird might receive Orders to set about the Instruments without farther loss of time, and while there is yet a probability of his living to execute them.[10]

A printed proposal he then set before the Convocation (the governing body of the university) was for temporary funding from the Delegates of the Clarendon Press, as a form of bridging loan. This was agreed but must have required clever management on Hornsby's part. The status of the resulting 'Radcliffe Observatory' was also unusual: it operated as the observatory of the university, under the occupation and direction of a university office holder, but in fact belonged to a private trust. Hornsby demonstrated political acumen and ingenuity in devising these unusual stratagems and carrying them through successfully.

His predecessor as Savilian Professor of Astronomy had been the great James Bradley, appointed in 1721. He had lectured in experimental philosophy as well as astronomy in an

Figure 4.1 The instrument maker John Bird.

arrangement dating from 1729, itself a continuation of lectures given by the Keeper of the Ashmolean Museum, John Whiteside, teaching delivered in the 'Schola Naturalis Historiae', the room below the museum gallery in the 'Old Ashmolean' building in Broad Street.[11] (Figure 4.2) As was becoming customary in the period, these lectures were illustrated using a collection of

Figure 4.2 The Old Ashmolean Building.

demonstrational apparatus; this was accumulated by Whiteside and kept in the School. Provision for this teaching was placed on a more formal footing in 1749, when part of Lord Crewe's benefaction and some additional provision from other sources were reserved as a stipend for 'a Reader of Experimental Philosophy'.[12] The appointment was in the hands of the Vice-Chancellor, who bestowed it on Bradley and the latter, according to A. V. Simcock, moved his Savilian Lectures into the School of Natural History, adding astronomical instruments to the collection, which he augmented generally. It is worth remembering that the university already had an official locus for the teaching of natural philosophy in the form of the Sedleian Chair.

When Hornsby was appointed his successor in Oxford in 1763, Bradley had been responsible for reading and demonstrating experimental philosophy for some 30 years, though having also held the position of Astronomer Royal since 1742, the regularity of the teaching might have been altered. With his health failing, he ceased to lecture in 1760, and Nathaniel Bliss stood in for a few years before himself succeeding Bradley as Astronomer Royal in 1763.[13] The then Vice-Chancellor presumably thought it expedient to continue the long-standing arrangement with the astronomy professor, perhaps especially if the astronomy lectures were being given in the School of Natural History. Accordingly, Hornsby became 'Reader of Experimental Philosophy', began his own courses of lecture demonstrations, and continued the practice of augmenting the collection of apparatus.[14] Although no known instruments from what grew into an extensive teaching resource are known to have survived, Hornsby's inventories and—more particularly—his lecture notes and texts or transcripts have been preserved and they form a rich record of science teaching for the period in Oxford. The words actually delivered by Hornsby also offer us very valuable insights into his character and his creed.

Hornsby's opening words would be along lines such as these (there are several versions):

> The business of Natural Philosophy is to describe the Phænomena of Nature; to give some Account of their Causes so far as they depend upon the known Properties of Bodies; & to ascertain those general Laws of Nature, according to which the Motions & Actions of Bodies are regulated. A strong Curiosity has prompted Men in all times to study Nature; Every useful Art has some Connection with this Science; & the Beauty & Variety of Things makes it ever agreeable & surprising.[15]

The promise of a curiosity 'ever agreeable & surprising' seems typical of Hornsby's teaching aims and characteristic of the lectures that follow.[16] He would then explain that, since the laws of nature depend on the will of God, they cannot be found *a priori*: our knowledge of them rests on experiment: 'this is properly the Scope & Business of Natural & Experimental Philosophy'. Again, this was not the approach to natural philosophy espoused throughout the university.

It follows for Hornsby that, prior to the adoption of an experimental methodology, natural philosophy had been of very little value.

> Progress ... which the knowledge of Nature made in many former Ages is exceedingly inconsiderable when compared with the numberless improvements it has received from the Discoveries made in the Course of the last Two Centuries only: insomuch that some of the Branches of Natural Philosophy, which at present is almost complete in all its parts were utterly unknown 200 Years ago. The reason of this seems upon proper Enquiry to be justly attributed to the practice of the Philosophers of former times; who in their pursuit after Natural Knowledge disregarded Experiments & busied themselves only in framing Hypotheses to explain in what manner any particular Appearance might have been produced.[17]

Yet Aristotelian thought remained the statutory basis for natural philosophy elsewhere in the university.

Hornsby was comfortable with including cosmological generalizations and what we might call devotional asides even in his most technical of published papers. Deep within a discussion of the proper motion of Arcturus, based on observations both historic and recent, he opined that:

> The system of the world, considered in an enlarged sense, and agreeable to the idea we may entertain of an all-powerful benevolent Creator, may be taken to occupy the whole abyss of space, and to consist of an assemblage of bodies, having different magnitudes, and emitting various degrees and modifications of light.[18]

This expansive remark sits within an extraordinarily meticulous analysis of many measurements, which impressively demonstrates another of Hornsby's characteristics. He has acquired an exhaustive familiarity with Flamsteed's *Historia Cœlestis* but at the same time is not content to live with the inaccuracies of Flamsteed's instrument (the mural arc) but infers a complex pattern of errors from Flamsteed's own data, together with remarks on the instrument itself. The rest of the paper is equally remarkable and indicative of Hornsby's exhaustive concern for accuracy in ferreting out errors and unacknowledged assumptions, and we shall refer to it in the section 'Savilian Professor of Astronomy, 1763–1810'.

Hornsby's subsequent appointments in Oxford were as Radcliffe Observer in 1772, Sedleian Professor of Natural Philosophy in 1782, and Radcliffe Librarian the following year. We shall consider each of his appointments in turn, except for the last, which the *Oxford Dictionary of National Biography* tells us he treated as a sinecure.

Beyond Oxford, Hornsby was elected a Fellow of the Royal Society in 1763, but he was not very active there and held no offices. He was much more engaged as a commissioner for Longitude (an appointment entailed with the Savilian Chair),[19] attending board meetings regularly, undertaking tasks on their behalf and taking part in discussions. He put forward a case for Bird receiving a share of the longitude prize.[20]

Savilian Professor of Astronomy, 1763–1810

Hornsby opened his account as Savilian Professor by presenting six papers to the Royal Society between December 1763 and December 1772, in the period between his appointment to the chair and the beginning of the observing programme at the Radcliffe Observatory. They were published in the *Philosophical Transactions*.

The first was on the value of the parallax of the Sun.[21] Hornsby reviews the work of a wide range of earlier astronomers and summarizes the situation prior to the 1761 transit of Venus. Then he tackles Halley's proposal for 1761 that times should be noted at Hudson's Bay (Port Nelson) and the mouth of the Ganges, detecting a mistake by Halley and citing the chosen alternative stations of Toboliki (Tobolsk) in Siberia (observer, Jean-Baptiste Chappe d'Auteroches, or d'Auteroche), Bencoolen in Sumatra (Charles Mason and Jeremiah Dixon were the first choices), and Rodrigues (Rodrigue; observer Alexandre Guy Pingré), before comparing the duration of the transit at Tobolski with other stations, in Scandinavia and in India. He then considers an alternative method, requiring an accurate knowledge of longitude. Throughout, Hornsby's command of observations scattered across the globe and across the historical record is impressive and it is clear that the geographical reach of his project has engaged his imagination and enthusiasm. This is an unusual paper, where Hornsby combines an archival and geographical command with his aptitude for astronomical calculation. He concludes with an undertaking that the prospect of the 1769 transit will be 'the subject of a future paper'.[22]

A communication of 30 April 1764 (read on 10 May) reported his observation of a solar eclipse on 1 April, but on 1 December 1765 (read on 13 February 1766) he made good his promise to consider the prospect of the 1769 transit of Venus, in a paper Hornsby called a 'discourse', rather grandly 'inscribed' to the President, Council and Fellows of the Royal Society and printed over nineteen pages of the *Philosophical Transactions*.[23] Its preparation had entailed a great deal of computation, but had also involved the use of a terrestrial globe for tracing the visibility of the event around the world in a search for possible observing stations. Locations should be chosen to give the greatest difference in the duration of the transit. Tornio ('Tornea') in Lapland could be one candidate, which sets Hornsby in search of a corresponding location in the 'great South sea'[24] and in turn to the reports of early navigators, from as long ago as the late sixteenth century. Hornsby's source was the second edition of Harris's *Voyages*.[25] Hornsby draws up a table of seventeen candidates, with their latitudes and longitudes, and calculates the parallactic angles for those whose positions he considers most securely known among them, and the differences in transit durations from Tornio, which could be between twenty and twenty-four minutes. If a location in the South Sea is not practicable, Mexico could yield a difference in duration from Tornio of almost eighteen minutes. For his search of the South Seas, Hornsby has Guillaume de l'Isle's 'map of the southern hemisphere'.[26] We can think of him settled comfortably as ever in Oxford with his maps, globe, and large volume of voyages, together with other texts, deploying his computational facility in his very particular geographical, historical, and mathematical project. His conclusion is to favour stations in Lapland and the South Sea.[27] Finally he offers complementary inducements of a more worldly kind:

> How far it may be an object of attention to a commercial nation to make a settlement in the great Pacific Ocean, or to send out some ships of force with the glorious and honourable view of discovering lands towards the South pole, is not my business to enquire. Such enterprizes, if speedily undertaken, might fortunately give an advantageous position to the astronomer, and add a lustre to this nation, already so eminently distinguished both in arts and arms.[28]

Hornsby's next paper, of 14 June 1769 (read on 15 June), was a report of his own observations of the transit, made at Oxford on 3 June, and of those made at the Earl of Macclesfield's observatory at nearby Shirburn Castle, with which Hornsby maintained a cooperative interest.[29] Hornsby himself observed from the traditional station for the Savilian Professor, a room at the top of the Mathematical Tower in the schools' quad, and he reports also observations of the associated solar eclipse the following day from both observatories, together with a good many observations of the transit by others at sites across Oxford. Hornsby notes a clear record of the visual effect afterwards known as the 'black drop',[30] when observing internal contact with a 7½-ft Dollond refractor. It is a valuable description of 'live' observing by Hornsby, who

> perceived that the Planet appeared to be wholly entered upon the Sun, though the limbs of the Sun and Venus were not actually separated; that part of the Sun's edge, where the ingress happened, being very sensibly obscured by a penumbra, and the limbs appearing to be united, by a kind of ligament of a considerable breadth. This ligament became narrower and narrower, and was at length reduced to a point, and actually broken at 7^h $21'$ $57''$½ mean time, or 7^h $24'$ $13''$¼ apparent time. At 7^h $24'$ $23''$ apparent time, the thread of light between the edges of the Sun and Venus, which was, before compleated, now appeared to me of a very sensible breadth, and to equal 1/10 of the Planet's diameter. If I have estimated this breadth properly, the true internal contact must have happened considerably more than a minute sooner.[31]

Although some observers of the 1761 transit had noted the effect,[32] it seems to have come as a surprise to Hornsby in 1769, yet he remained remarkably controlled. The implications were significant: 'it will, I fear, occasion a much greater uncertainty in the quantity of the Sun's parallax deducible from these observations than was reasonably expected.'[33]

Hornsby followed this with a paper of 19 December 1771 (read the same day), in which he offers 'The Quantity of the Sun's Parallax', deduced from reported observation of the 1769 transit.[34] His travellers are not now distant historical figures but his own contemporaries, including James Cook, Charles Green, Daniel Solander, and William Wales. His method is to compare the observed values of transit duration with a computed value using an assumed solar parallax, and to find a corrected parallax from the discrepancy. This yields nine values around a mean of 8.650 seconds of arc. He then described at some length how he selected two timings in particular—Cook's value for ingress and the mean of his and Green's value for egress, both observed on 'King George's Island' (Tahiti), and taking the difference of this duration from that measured by Wales and Joseph Dymond at Prince of Wales's Fort in Hudson's Bay. Using other available data he derived values for the longitudes of both stations. He concludes a mean parallax of the sun of 8.78 seconds of arc and a distance of 93,726,900 English miles.[35] From this he calculates the distances of all the known planets.

In passing he announced that the observation made at the Cape of Good Hope in 1761 by Mason and Dixon was 'abundantly confirmed' and that of Pingré on Rodrigues was mistaken by one minute.[36] A dispute over whether Pingré or Mason and Dixon had been mistaken in this respect had hung over from the 1761 transit and had been regarded as having significance. Nevil Maskelyne, for example, wrote to Hornsby on 13 June 1769, in anticipation of his revised determination of the parallax, that if Pingré had been correct

> then Mason & Dixon must have mistook 1 whole minute, which I now somewhat suspect. If they did, they were block-heads or perhaps not quite honest in their accounts, for if they found they differed 1 minute from each other they ought to have sent their observations unaltered. The honor of our nation is concerned.[37]

Fortunately for the nation, Hornsby was able to vindicate the observers at the Cape.

Hornsby would later use an incident reported in Wales's *Journal* for the voyage, published in the *Philosophical Transactions* in 1770,[38] in the text of his Sedleian lectures delivered in 1785, and probably on other occasions.[39] His subject was 'The different Kinds of Air', and in support of the assertion that uptake of water vapour from the sea surface alters the refractive properties of the lower layers of the air, he cited Wales's direct confirmation of a seaman being unable to see from the masthead the low-lying land of Cape Churchill clearly visible from the quarter deck of the ship. The report was memorable, but so was Wales's intrepid response:

> This appeared so extraordinary to me, that I went to the main-top-mast-head myself to be satisfied of the truth thereof; and though I could see it very plain both before I went up, and after I came down, yet could I see nothing like the appearance of land when I was there.[40]

The Society were surely gratified by the adherence of their commissioned observer to their motto, 'Nullius in Verba'![41] For Hornsby as lecturer this became an engaging and instructive anecdote.

Last in the set of six is Hornsby's paper on the proper motion of Arcturus, mentioned earlier (23 December 1772; read 24 December), which uses observations of his own in 1767–8 (made with his 33-in quadrant and 44-in transit instrument, both by Bird) and observations from 1690 by Flamsteed, corrected through Hornsby's analysis of Flamsteed's mural arc.[42] Comparisons are made with observations of nearby stars, which are checked where necessary for the possibility of their having their own proper motions. The whole exercise is impressively thorough, and the results are further checked by comparisons with a set of observations made at Shirburn Castle. Hornsby concludes that the proper motion of Arcturus is 1.205 seconds of arc (per annum) in right ascension westward and 2.005 seconds in declination southward, that no other prominent star has so large a value, that Arcturus is the nearest star to our system, and that the search for an annual parallax among the stars might be concentrated there.

Hornsby moves on to the implications of this result for the question of the alleged diminution of the obliquity of the ecliptic, eventually rebutting the challenges by César-François Cassini de Thury and by Pierre Charles Le Monnier to the widespread acceptance of a decrease. A set of observations shows that the obliquity has been declining at a rate of 58 seconds of arc per century and Hornsby concludes by landing his ambitious paper, filled with observation and computation, in the arena of natural philosophy: 'a quantity which will be found nearly at a mean of the computations framed by Mr. Euler and Mr. de la Lande, upon the principles of attraction'.[43]

By now Hornsby had already ordered the foundation instruments for the Radcliffe Observatory, the project that increasingly engaged his thoughts and plans.[44] Bird was engaged to provide two mural quadrants (to face north and south, respectively) of 8-ft radius, with telescopic sights by Dollond of 3-in aperture (Figure 4.3), an 8-ft transit instrument, a 12-ft zenith sector, and an equatorial sector with a 5-ft Dollond telescope.[45] The designs followed precedents set by George Graham and the general model for the regime of instruments and observing came from Greenwich. The price agreed for the principal instruments was £1,300 but Hornsby had a long list of smaller instruments, accessories, clocks, refractors, a reflector, a sextant, barometers, thermometers, etc.,[46] which brought his total spending ambition on instruments to £2,500. To have founded and equipped a major observatory, with no clear precedent or any expectation that this should be his concern was a remarkable achievement and Hornsby's lasting legacy in Oxford.

The notes and texts for Hornsby's lectures (cited earlier) are not only for those on 'experimental philosophy': each of his teaching posts is represented. There are sixteen manuscript volumes,[47] most of them substantial, and a few printed notices announcing the lectures.[48] The

Figure 4.3 One of two 8-ft-radius mural quadrants, made by Bird for the Radcliffe Observatory.

earliest for 'A Course of Astronomical Lectures' announces one beginning in October 1764; another announced June 1776. Notes for the 1776 course survive, with an attendance list, while others show that it was given also in 'Lent 1776' and 'Lent 1778'.[49] Other notices where astronomy might arise *inter alia* referred to courses of 'Experimental Philosophy' or of 'Mathematical Lectures and Experiments', so we can be confident that the 'Astronomical Lectures' were Savilian lectures. Yet another notice identified the Savilian Professor of Astronomy as announcing a course on the transit of Venus (3–4 June 1769) beginning on 24 May 1769. Here Hornsby planned to introduce his auditors to the content of his recent research:

> the Method of computing the Places of the Sun and Planets explained and exemplified in the Case of the ensuing Transit; the Method of computing the Effect of Parallax, and of finding the Places upon the Earth's Surface, where Observations may be made with the greatest Advantage, will be pointed out; and the Manner of determining the Quantity of the Sun's Parallax from some of the principal Observations made in the Year 1761 will be shewn and illustrated by Examples.[50]

There would be serious work for the auditors, who 'must previously furnish themselves' with specified volumes of logarithms and of astronomical tables. All these lectures would take place 'at the *Museum*'. There is also a printed syllabus for a course of 'Astronomical Lectures', which details the content of an ambitious set of fourteen lectures, requiring Tobias Mayer's *Tabulæ motuum solis et lunæ* and a book of logarithms.[51]

Reader of experimental philosophy, 1763–1810

Whiteside had acquired a collection of instruments for his lectures and these must have been his own property, as they were purchased by Bradley after Whiteside's death in 1729, and were used by Bradley prior to and beyond his appointment as Reader in 1749.[52] Hornsby succeeded to their use (and probably purchased the Bradley collection) and he added to it extensively. The Danish traveller Thomas Bugge visited Hornsby in 1777 and noted 'a beautiful collection of physical instruments'.[53] We have a very precise account of the contents, from a series of inventories by Hornsby.[54] The instruments are listed under twenty-six 'days', which refer to the sequence of lectures. We know that they were kept in a set of cupboards and the sequence surely also refers to the arrangement of instruments in the cupboards.

The most extensive inventory dates from 1790.[55] Internal evidence in the archive indicates that it was prepared by Hornsby, recorded as a fair copy, and annotated with a valuation by the instrument maker Edward Nairne.[56] A very full listing mentions not only complete, self-contained instruments, but extends to the kind of loose apparatus that might be well be found in the context of lecture demonstrations. We find entries such as 'The Armillary Sphere to shew the diurnal Motion 18 Inches diameter' and others of the character, 'Large tubes terminating in a Capillary bore, capillary tubes of different bores in a piece of Cork'.

The inventory has about 330 entries, many of which refer to a number of items—'Drawer with artificial Magnets about 8 or 10', 'Four tin Machines with Threads to represent the passage of Rays thro Lenses', or a pair of globes. About 160 entries relate to coherent instruments for a particular purpose, rather than miscellaneous pieces of apparatus. It seems likely that the miscellaneous pieces did have a specific purpose, such is the structured nature of the collection, distributed across the twenty-six days, but we do not always know what this was, the functions of the odd pieces only being mentioned in a few cases, such as 'Two Tubes for the Torricellian Experiment'. Others seem less purposeful, such as 'Pieces of Copper, Lead & brass with strings'. In a number of cases, however, we can place the instruments in a very well-documented account of the teaching they supported, see where they came into play in lectures, and how they were deployed.

The 1790 inventory records many of the classic instruments of experimental philosophy. There are an electrical machine, air pumps, 'Guerricks Hemispheres', orreries, 'Two Models of the Lifting Pump', 'Two Models of the squirting Pump', 'the Pile Engine', 'The Crane', 'The Corn Mill with a Water Wheel', 'The Coal Gin (very large)', 'Machine to Illustrate the different strength of Horses', 'The large Machine to Illustrate the Wedge', and so on. There are also some unusual or special items, such as 'Wooden Table with 32 Compasses to Illustrate Dr Halley's Hypothesis' or 'Ferguson's Machine to shew that the path of [the] Moon is concave to the Sun'. Only very rarely are makers mentioned, such as a 4-ft achromatic telescope by Dollond and a pair of 18-in globes by Adams. Also listed are two 'old 12 Inch Globes, Celestial & Terrestrial'.

The use of the globes was a staple topic in this genre of lecture demonstrations and it was included by Hornsby. Most of this would have been covered by working through various operations and manipulations performed when solving problems using globes, so through performance rather than reading a text. Hornsby's notes for a lecture on revolution of the Earth, for example, contain the reminder:

Shew the Seasons upon the Orrery;
And the different Lengths of the Day on the Globes
And the reason why the Days lengthen & shorten so rapidly near the Equinoxes & so slow near the Solstices.[57]

In one text, however, he does run through the features and components of the globes, explaining their functions; more often they are simply listed as an *aide-memoire*.[58]

Most of the instruments were made for teaching but others were standard working devices brought into a teaching context, such as an azimuth compass, a variation compass, a 'Sea Compass on Gimbals', or regular items of instrument commerce like seven barometers, covering a range of types. Here they are for learning natural philosophy through illustration and demonstration.

The specific purpose for the 1790 inventory was to provide a basis for the valuation of the collection by Nairne and for a payment to be made to the Reader. The university had been given a benefaction to support and consolidate the teaching of experimental philosophy and was beginning to spend it. An element was added for the charge for the valuation and another for the cupboards and cases. Hornsby signed the document, agreeing, for example, that he will not buy any more instruments without the prior consent of the trustees of the fund, and Nairne signed in acknowledgement of his fee, which he did not receive until June the following year, 7 months later. I have found no further record of the collection itself.

What we do have is an account of how it was used in the outstandingly rich resource of Hornsby's very full lecture notes: written-out lectures in multiple, redrafted versions. They generally contain lectures written in Hornsby's hand, word for word as they were to be delivered, interrupted from time to time by reference to an experiment or a short sequence of experiments. We see how the use of instruments occurred within the lectures and the points they were intended to illustrate. Also Hornsby sometimes reflects on the history of the subjects he takes up, giving us insights into how the past of science was viewed and taught in the eighteenth century.

In Hornsby's lectures on the 'Laws of Motion', for example, he quickly establishes a pattern that is carried throughout, one of exposition interrupted by notes such as 'Proved by Expt',[59] or 'explained by the Diagonal Instrument'.[60] A few pages further on there is a more extended experimental interlude: 'The Truth of what has now been advanced may be illustrated by several Experiments',[61] whereupon he lists four:

1. By the Motion of a Brass Square in an horizontal direction . . .
2. By the Brass Diagonal Machine
3. Let three Ivory Balls of equal size be suspended from three Pins by strings of equal lengths . . .
4. By the Diagonal Jet of the Fountain.

Later still, he announces, 'I shall now proceed to illustrate & prove what has already been said by such experiments as may be proper for this purpose',[62] the list including the 'Whirling Table', of which he seems to be fond, and adds a note to himself, 'Shew Mr. Cantons Experiment of the two Whirligigs supported by an artificial Magnet'.

In a lecture on the orrery he allows himself a modest flourish in introducing the instrument: 'Let us now take a general View of the Universe'.[63] Here we learn of another of Hornsby's teaching techniques: the use of charts or large diagrams as well as instruments. He explains that the orrery can deal with relative motions but not with distances or magnitudes:

> The Proportional Distances are in the Scheme before you, their Magnitudes are also before you; but neither the proportional Distances, or Magnitudes can be conveniently express'd in an Instrument of this Sort.[64]

When dealing with the detection of parallax by double star measurement, he says simply, 'see the Scheme'. The well-known lithograph of William Buckland lecturing in the same room shows that he continued the use of schemes or wall charts into the nineteenth century.[65]

Hornsby tells the story of Pieter van Musschenbroek in Leiden in 1746, 'much attended to & frequently repeated by all Electricians, when an Accident gave birth to a very remarkable [experiment], which has since engaged the Attention of ever Lover and Promoter of this branch of Knowledge'.[66] It resulted in 'a disagreeable sensation; for which reason the Effect has been called the Electric Shock'.[67] Hornsby relates the electrical experiments of the Abbé Nollet and the discovery by Henry Gellibrand of the variation of the magnetic declination, very much in the way the stories are still told to history of science students today.

Just occasionally his accounts of the achievements of others come right up to date, as with the discovery of what he calls the Georgian Planet in 1781, and we call Uranus:

> From these considerations therefore it should seem that Mr. Herschel has had the happiness to discover a new Planet revolving round the Sun . . . & be allowed to increase the Number of those Bodies, which from the earliest period of time to the present Year have received no addition, & equalled only in Number the 7 days of the Creation.[68]

Hornsby had been involved with the momentous discovery since the beginning. The earliest letter we have from Hornsby to Herschel (and clearly the first he had sent him) is dated 22 December 1774. They had met previously and Herschel had written on 15 December with a query about eclipses of Jupiter's satellites. Hornsby replies with detailed and helpful information on their observation, along with other news.[69] He writes to the musician in Bath, and future famous astronomer and telescope builder, 'As you seem very fond of the science of Astronomy, may I presume to ask what Instruments you have'. Herschel had taken his first steps in polishing specula the previous year, in January 1774 mounted a 5½-ft reflector, and began his observation journal on 1 March.[70] He first used his famous 7-ft Newtonian in May 1776 but had already told Hornsby in December 1774 that he had 'a very good Reflector'.

Herschel first observed a 'comet' on 13 March 1781 and informed Hornsby, who struggled to find it until mid-April and then did so on the basis of further data from Herschel.[71] Whereupon he realized that he had seen it on five occasions since the discovery, but had taken it for a regular star. That Herschel had recognized its non-stellar nature at sight was a tribute to the quality of his 7-ft telescope. Thinking it was a comet seen in 1700, Hornsby promised, 'I will very soon try to construct its orbit'. On 14 October, in a letter containing his observations of the comet and orbital elements calculated by Anders Johan Lexell, he referred to 'this Comet or Planet as some are disposed to call it'.[72] It was relatively far away (the recognition of the planet Uranus would double the known size of the Solar System) but would be a telescopic object throughout its orbit. In a letter to Herschel of February 1782, Hornsby was still debating with himself whether this object should be classified as a comet or a planet, or something else entirely.[73] Yet the Copley Medal had been awarded the previous year and a consensus had formed in favour of a planet.[74] From Hornsby's text ('therefore it should seem that Mr. Herschel has had the happiness to discover a new Planet ... [in] the present Year') we see that he had chosen to acknowledge a planetary status in his Oxford lectures.

An example of how the lectures can illuminate the entries in the inventories and revise our assumptions about an instrument's use concerns an inconspicuous item in the 1790 list, 'Metal Cylinder with its Pictures', among optical bits and pieces.[75] It is obviously a cylindrical mirror with anamorphic drawings or prints, and it might be considered out of character with a teaching collection, being little more than an optical toy.

If we turn to Hornsby's lectures on optics, we find that the pattern of his practice—the use of diagrams and experiments—is as we have encountered already, with notes to himself such as: 'Shew the Scheme & prove by Expt'.[76] He first seems to agree with the idea that the cylindrical mirror is only a toy:

> Cylindrical, Conical, Prismatical & Pyramidal Mirrors are to be considered as mere Curiosities. They give irregular Images of Regular Objects presented to them; or they give a Regular Image of an Object, provided the degree of Irregularity be accommodated to the Circumstances of the Mirror.[77]

However, he goes on to give an analysis of the distortion produced by cylindrical and conical mirrors, and then to an exercise in

> the Manner of designing upon an horizontal Plane a disfigured Object, so that by placing Vertically a Cylindrical Speculum upon that Plane of a given radius, that object may appear Vertical and Regular to an Eye in a given situation.[78]

The treatment is quite mathematical for Hornsby's lectures on experimental philosophy, which do not have much mathematical content. This is a case where the lectures give an instrument an alternative use and significance.

The physical arrangement of the school reflects the method adopted in the lectures. A little sketch by Bugge in 1777 shows a large table for demonstrations surrounded by tiers of seating in the manner of an anatomy theatre.[79] This fits very well with what we read in Hornsby's lecture notes and emphasizes the close interaction between the delivery and the demonstrations.

The main lesson we should take from the record of Hornsby's lectures and the collection of instruments in the school is that he was enthusiastically committed to natural philosophy as experimental practice, the creed that was taking hold throughout Europe and beyond, with the development of similar cabinets of experimental philosophy. It was infiltrating Oxford through the activities of a composite institution whose theme was materiality—from the chemical furnaces in the basement to the collections of specimens and curiosities on the top floor. On the floor between, the school practiced and taught a manipulative, instrumental, empirical, mechanical approach to natural knowledge, with Hornsby as its latest champion. Earlier thinkers had failed because 'in their pursuit after Natural Knowledge [they] disregarded Experiments & busied themselves only in framing Hypotheses'. Drawing on his experience of teaching natural philosophy by an experimental and demonstrational method, he would carry this forward into his future lectures as the Sedleian Professor.

Radcliffe Observer, 1772–1810

Hornsby's next appointment came in 1772, when he was made Radcliffe Observer. This made perfect sense and for a time the new position went with the Savilian Professorship of Astronomy. Hornsby's case set this precedent because he had been the driving force behind the creation of the observatory (Figure 4.4). So well appointed an institution needed an observer and a programme of regular meridian observations. There was a house for the observer and his family (as well as his assistant, though no one was appointed in Hornsby's time[80]) and Hornsby filled the office very well. The arrangement then continued for subsequent Savilian Professors of Astronomy Abraham Robertson and Stephen Rigaud. The connection to the Savilian Chair was broken in 1839 with the appointment as professor of George Johnson, who was a mathematician with little inclination towards practical astronomy, and Manuel John Johnson was appointed to the separate position of Radcliffe Observer.

Bird died in 1776 and Hornsby's dealings with the trade were through other makers, notably Jesse Ramsden. Hornsby had no further major commissions to dispense, but he needed repairs and adjustments to his existing equipment. His letters to the Duke of Marlborough often complain of Ramsden's tardiness in attending to the Radcliffe's small commissions and, since Ramsden did not adopt the attitude expected from a tradesman, Hornsby jokingly referred to him as 'the Sieur Ramsden' or 'the Gentleman of Piccadilly', where Ramsden had his shop.[81] A visit from Ramsden to Oxford in April 1795 furnished Hornsby with several anecdotes in his letters to the duke regarding Ramsden's well-known absent-mindedness: 'I told him that I wondered he had not left his Head behind'.[82] Their relationship deteriorated in mid-1788 but was restored with the help of the duke and in August Hornsby entertained Ramsden and Jérôme de Lalande, travelling to Oxford together, to dinner. In 1791 Hornsby entrusted an eyepiece to Ramsden and, as had happened before, found himself pleading for its return. He had the greatest respect for Ramsden's work, even if he was 'that intolerable Villain that lives near St James Church'[83] and he found again, 'I am at his Mercy'.[84]

Radcliffe Observatory.

Figure 4.4 The Radcliffe Observatory.

When he was equipped to work efficiently, Hornsby was a good and conscientious observer (Figure 4.5). However, a considerable cloud was created for him by the observations of his predecessor James Bradley at Greenwich, which were passed to the University of Oxford in 1776 on condition that they would be published by the Clarendon Press.[85] Hornsby was appointed editor but long delays were created by his many commitments, not least to the work of the new observatory, and by increasing ill health as he developed epilepsy. By 1784 there were strong complaints from Greenwich, from the commissioners of longitude and from the Royal Society. Hornsby retained the support of the university and of the delegates of the Clarendon Press, and weathered the storm. However, the first volume did not appear until 1798 and the second was edited by Abram Robertson.

Publication of Hornsby's own observations was subject to even longer delay, since they were eventually published in 1936, following the work of a team led by Harold Knox-Shaw.[86] That the publication was still considered worthwhile was a tribute to their value. A significant step to this end had been taken by Arthur Rambaut, the then Radcliffe Observer in 1900, with the assessment of the work as a whole and, as a specimen, the reduction and publication of observations from 1774 in the *Monthly Notices of the Royal Astronomical Society*.[87] Rambaut includes an account of Hornsby's instruments and methods. Regarding the double scales on the mural quadrants, he concurred with the opinion of Bird, with which we began: 'In these,

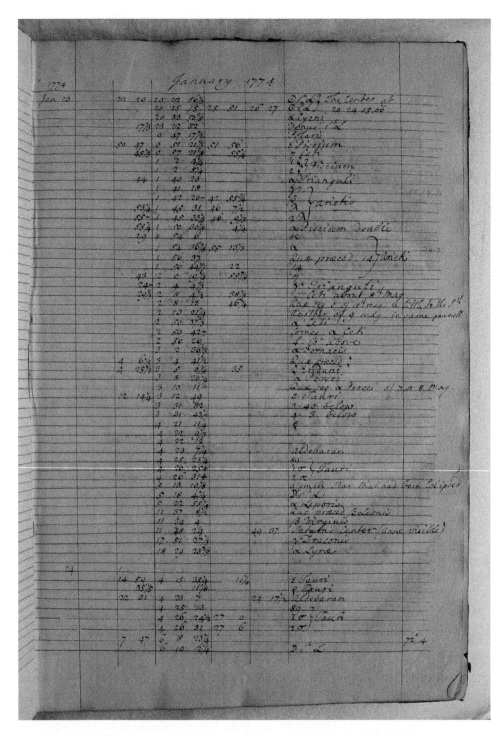

Figure 4.5 The first page of Hornsby's transit observations taken at the Radcliffe Observatory.

as with other quadrants by Bird, the divisions of the 96 arc seem to be much superior to those of the 90 arc'.[88]

The extension of this work was occasioned by a proposal from the Astronomer Royal in 1925 and a series of grants from the Royal Society.[89] Knox-Shaw and his colleagues reduced the observations made with the transit instrument and the south quadrant (the 96 arc only) between 1774 and 1798 and, even with other filters in place, they amounted to 54,000 (counting right ascension and declination separately).[90] Knox-Shaw is full of praise for Hornsby's care and dedication. As 'an example of Hornsby's enthusiasm', he drew particular attention to observations made on Christmas morning, 1798; the outside temperature was −2°F.[91]

Sedleian Professor of Natural Philosophy, 1782–1810

Hornsby's appointment as Sedleian Professor of Natural Philosophy came in 1782, after he had already been the most active teacher of natural philosophy in the university for many years, even if this teaching had not conformed to the discipline so-called in the statutes. But to have followed the Laudian statutes of 1636 in this respect would have seemed bizarre in 1782, calling as they did for lectures on the writings of Aristotle.[92]

We have met with three terms to refer to the study and teaching of an understanding of the physical world. The Sedleian Chair was devoted to 'natural philosophy', a traditional subject in the medieval university curriculum, devoted mainly to the study of Aristotle's account of the physical world and the causal explanation of its behaviour. This was a discipline much older than what might be called—though it goes by other names—the new science of the seventeenth century. Since the Sedleian Professorship became effective in 1621, and references Aristotle in its regulations, the traditional discipline may be assumed. Natural philosophy also came to act as a general, inclusive term, under which more specific varieties of or approaches to the physical world—such as experimental philosophy, mechanical philosophy, and chemical philosophy—could be categorized.

Then the school in Broad Streel was named in 1683 one of 'natural history'. This was not quite what we think of as natural history today but certainly concerned the accumulation of empirical knowledge of the natural world as we find it, in a manner often associated with Francis Bacon, though he also had more specific and directed ideas of the method to be deployed in gaining understanding and material benefit from this knowledge. Perhaps the best examples of what 'natural history' meant in the specific context of the Old Ashmolean can be found in the two volumes by Robert Plot, the first keeper of the Ashmolean Museum and Professor of Chemistry, who wrote, or perhaps we might say assembled, *The Natural History of Oxfordshire* and *The Natural History of Staffordshire*.[93]

'Natural history' was a relatively neutral, uncommitted, anodyne term, but by the time a formal expression was needed in 1749 for the activity to take place in the school, 'experimental philosophy' was chosen for the future lectures; it did no more than describe the already-established practice in the school. This was very different from the old natural philosophy, was indeed purposely antipathetic to it. Knowledge was founded not on learned authority, and not even principally on the accumulation of empirical knowledge, but on the active pursuit of experimental intervention, often facilitated by instrumentation, and its assimilation was assisted by the repetition of such experiments and demonstrations.

A committed experimental philosopher was to become Sedleian Professor, after he had been championing this cause through many years of teaching. He had been abreast of contemporary practice while the university's official natural philosophy seemed not to be active in explaining the natural world in terms of mechanical causality or in basing any accounts on experimental investigations. Hornsby might have felt he could regard the appointment as the correction of a long-standing anomaly. What he seems to have done, however, was to extend the methods already established in the School of Natural History into topics appropriate to Sedleian lectures not covered by the existing programme, Thus, the Sedleian lectures and lecture demonstrations joined those of astronomy and of experimental philosophy 'at the Museum'.

A particularly telling example is Hornsby's extensive course of lectures on 'the different kinds of air'. In a printed notice of 30 May 1785 (Figure 4.6), Hornsby identified himself as the 'Professor of Natural Philosophy' and he pointedly characterized the eight lectures, to commence on 2 June, as follows:

> A Course of PHILOSOPHICAL LECTURES on *The different Kinds of Air, Natural and Factitious*, in which the principal Discoveries of Dr. Priestley and others will be introduced and proved by actual Experiment.[94]

He points out also that 'these Lectures are entirely distinct from the Course of EXPERIMENTAL PHILOSOPHY'. By choosing 'airs' as his subject and by referencing Priestly in particular, he brings the Sedleian lectures into the realm of contemporary research in chemistry. A 'short syllabus' promises lectures on 'fixed', 'nitrous', 'inflammable', 'dephlogisticated', and 'acid and alkaline' airs[95]—'Their several Properties proved by Experiment'. Thanks to a large manuscript volume of lecture notes, we can say that Hornsby has mastered a wholly unfamiliar area of work, not only as a narrative but also as a difficult manipulative practice, one for which Priestley was especially admired.[96] Hornsby refers also to the work of Black, Cavendish, Scheele, Lavoisier, and many others. He seems thoroughly at home in his new discipline. It took confidence as well as practiced competence to generate these 'airs' and demonstrate their properties before such an audience as could be expected in the School of Natural History. It was helpful, no doubt, that he had retained the disposition of 'ever agreeable & surprising':

> In the course of the Experiments shewn in treating of the nature of these permanently Elastic Fluids we have seen a variety of appearances not not [*sic*] less entertaining than instructive.[97]

There are further surviving lectures on topics not generally associated with Hornsby, which he has entered into extensively and, as they do not fit with the established agendas or published syllabuses of astronomy and demonstrational experimental philosophy, must be records of courses delivered from the Sedleian Chair.[98] Although it was not a development that survived beyond his tenure, Hornsby had brought the teaching, as he joined it in spirit and in practice to his other activities in the School of Natural History, into the contemporary scientific world.

OBSERVATORY, *May 30. 1785.*

THE PROFESSOR of NATURAL PHILOSOPHY begs Leave to acquaint the Gentlemen of the University that he propofes to begin, at the *Mufeum* on *Thurfday* next the 2d of *June*, A Courfe of PHILOSOPHICAL LECTURES on *The different Kinds of Air, Natural and Faſtitious*, in which the principal Difcoveries of Dr. *Prieſtley* and others will be introduced and proved by aſtual Experiment.

SHORT SYLLABUS.

LECT. 1. General Hiſtory—Apparatus deſcribed—Method of procuring Fixed, Nitrous, Inflammable, Dephlogiſticated Air—General Properties of each enumerated.

LECT. 2 & 3. Affeſtions of common Atmofpheric Air—Air attraſts Water—is attraſted by Water—neceſſary for Combuſtion—neceſſary for Animal Life—contaminated by Refpiration—neceſſary to Vegetation—to the Calcination of Metals, &c.

LECT. 4. Of Fixed Air.
    5. Of Nitrous Air.
    6. Of Inflammable Air—Theory of Balloons.
    7. Of Dephlogiſticated Air.
    8. Of Acid and Alkaline Airs.

    Their feveral Properties proved by Experiment.

As thefe Leſtures are entirely diſtinſt from the Courfe of EXPERIMENTAL PHILOSOPHY, it may be neceſſary to add, that each Perſon who propofes to attend this Courfe of Leſtures on *AIR* is to pay one Guinea at the Time of Subſcription.

Figure 4.6 Printed notice of Hornsby's Sedleian lectures on 'The different Kinds of Air'. Image: RT Gunther, Early Science in Oxford, Volume II, Oxford, 1937.

CHAPTER 5

George Leigh Cooke

CHRISTOPHER D. HOLLINGS

As we saw in Chapter 1, the teaching of natural philosophy within the university took a distinctly mathematical turn over the course of the eighteenth century, and by around 1800, the subject had become much more Newtonian than Aristotelean. Thus, when Thomas Hornsby died in 1810, it was natural that this change should be acknowledged in the duties set out for his Sedleian successor: to lecture on mathematical natural philosophy, as founded in Newton's *Principia*. From here on, the chair would gradually develop into one of the pillars of specialized mathematics teaching in Oxford, with particular responsibility for applied mathematics.

The remainder of the nineteenth century saw just two Sedleian Professors, who each held the post for over forty years. The first of these, George Leigh Cooke (1779–1853), was a university figure from a very traditional mould: a churchman first and foremost, who would eventually abandon his Oxford duties for those of his parish. Nevertheless, Cooke carried out his Sedleian teaching much more dutifully than most of his eighteenth-century predecessors, at least at the beginning of his tenure. At the same time, he was a defender of Oxford's traditional emphasis on classics, and placed a limit on the mathematics that an undergraduate ought to learn. We might therefore view him as a transitional figure in the development of the Sedleian Chair and of mathematical study in Oxford. In the following chapter, we will draw many a contrast between Cooke and his immediate Sedleian successor.

Where they mention him at all, previous surveys of science in nineteenth-century Oxford have not been kind to Cooke, usually characterizing him as 'inactive' or a sinecurist. Indeed, Cooke has fallen foul of the two major approaches that have been taken to the subject: the institutional and the mathematical. In the first case, he has been overlooked because he played only a minor role in the history of the university, whilst in the second, he has been dismissed because he was not a research mathematician and published no major works. In reality, the label 'inactive' only applies to Cooke if we insist on taking either Oxford or mathematics as the centre of our focus. If instead we cast our gaze a little wider, we find an energetic, and by no means indolent, clergyman at work, a representative of generations of similar Oxford academic churchmen.

Christopher D. Hollings, *George Leigh Cooke*. In: *Oxford's Sedleian Professors of Natural Philosophy*. Edited by: Christopher Hollings and Mark McCartney, Oxford University Press. © Christopher D. Hollings (2023). DOI: 10.1093/oso/9780192843210.003.0005

Figure 5.1 The geologist William Buckland lecturing to Oxford's nineteenth-century scientific association, the Ashmolean Society, in 1823. The figure at the right-hand end of the middle row is believed to be George Leigh Cooke—the only known picture of him.

Early life and career

George Leigh Cooke was born on 1 July 1779 and baptized on 12 September that year at Great Bookham in Surrey, where his father, the Rev. Samuel Cooke (1741–1820), was vicar.[1] Through his mother Cassandra Leigh (1744–1826), George Leigh Cooke was connected to the prominent Leigh family of Warwickshire and surrounding counties, which brought with it the links to the landed gentry and the aristocracy that would benefit him later in life.[2] There were a great many clerical connections within the wider family, and also ties to Oxford: Cooke's grandfather, the Rev. Theophilus Leigh (1693–1785), was a long-serving Master of Balliol College, where Cooke's father had also been a fellow from 1762 until his marriage six years later.[3] Although Samuel and Cassandra Cooke had eleven children together, only three survived into adulthood. George Leigh Cooke was the middle child, with elder brother Theophilus Leigh Cooke (1778–1846) and younger sister Mary Cooke (1781–post-1845). Theophilus pursued a similar career to that of his younger brother, as a fellow of Magdalen College,[4] and perpetual curate of the nearby parish of Beckley, to which the Cooke family had a strong connection, but Mary is more enigmatic, appearing only fleetingly in surviving records.[5]

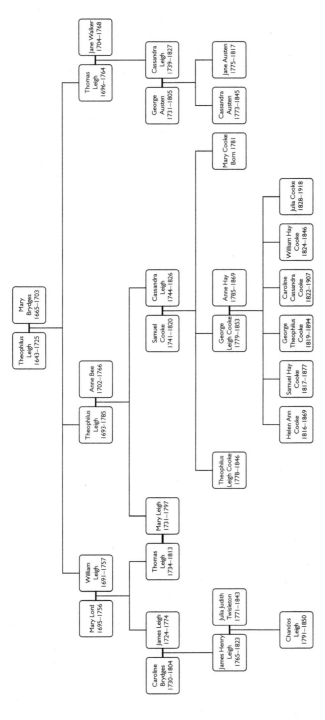

Figure 5.2 A much-simplified family tree, showing the interconnections between the Leigh, Cooke, and Austen families, including in particular those individuals mentioned in the main text and endnotes.

Presumably because of the connection afforded by his father and grandfather, George Leigh Cooke matriculated at Balliol College in 1797, at the age of 17, although he soon migrated to Corpus Christi College in order to take up a scholarship there.[6] A standard academic progress ensued: Cooke obtained his BA in 1800, followed by an MA in 1804; he was ordained first as a deacon and then as a priest in 1804 and 1805, respectively, elected to a fellowship of Corpus Christi in 1805, and obtained his BD somewhat later in 1812.[7]

We catch only brief glimpses of Cooke during these early years. On the academic side, a record survives of the books and stationery that he purchased from the Oxford booksellers Fletcher and Hanwell between 1799 and 1803.[8] Here we find an eclectic mixture of reading matter, consistent with the liberal education then offered in Oxford, and perhaps also pointing towards the broad intellectual grounding favoured by someone destined for a life in the Church. Cooke's earliest recorded purchase, of 9 April 1799, consisted of poetry and mathematics: the works of Virgil and Pindar (for 7s 6d and 4s, respectively), all six volumes of Dryden's *Miscellanies* (12s), and Wood's *Algebra* (5s). This last book, more fully *The Elements of Algebra, designed for the use of students in the University* by James Wood of St John's College, Cambridge, had first appeared in 1795 (with a second edition in 1798) as a textbook for the use of students at Cambridge University. It would go on to be one of the most successful algebra textbooks in nineteenth-century Britain, running to nineteen editions.[9] In 1799, it was the most advanced, up-to-date, and comprehensive textbook available in English on the solution, manipulation, and application of higher equations—a subject, moreover, that would not have been taught routinely as part of the basic mathematical education then offered by Oxford colleges, though it had at times been covered by the Savilian Professors of Geometry in their more advanced university-wide lectures.[10]

Cooke returned to Fletcher and Hanwell twice more in April 1799, carrying away a further collection of classical and poetical authors, including the two-volume *Complete Edition of the Poets of Great Britain*, Caesar's *Commentaries on the Gallic War*, Cicero's *Epistles*, *The Poems of John Cunningham*, and Oliver Goldsmith's *The Vicar of Wakefield*. Over the next three years, his book purchases were dominated by similar texts, though we do find a very small number of mathematical works mixed in among: Isaac Barrow's *Geometrical Lectures*, purchased for 7s in May 1799, Abraham Robertson's *Conic Sections* for a princely £1 1s two years later, and Charles Hutton's *Logarithms* for 16s in February 1802. Although this is the extent of Cooke's recorded outlay on mathematical books, we note that they are roughly consistent with the views that he would later express on what is an appropriate level of mathematics for Oxford undergraduates to learn. Moreover, his own biases of later life were clearly already in formation by 1803, when he was presumably preparing for a college fellowship, for his literary purchases then took a sharp turn towards books of sermons and theological texts, in combination with a series of philosophical works by members of the Scottish school of common sense.[11]

Other glimpses of the young Cooke are provided by his cousin, the novelist Jane Austen (1775–1817), to whom he was connected via the Leighs.[12] Besides being related by marriage, the Rev. Samuel Cooke was also Jane Austen's godfather, and her letters make several references to the Cookes, including visits to Great Bookham, and encounters with them elsewhere. It is in Jane's account to her elder sister Cassandra Austen (1773–1845) of a social occasion in Bath in April 1805 that we catch sight of George Leigh Cooke, then in his mid-20s (Jane was 29). She reported that—when he wasn't flirting with a Miss Bendish—'Cousin George was very kind & talked sense to me'.[14] She went on:

Figure 5.3 Jane Austen, distant cousin of George Leigh Cooke.

There was a monstrous deal of stupid quizzing, & common-place nonsense talked, but scarcely any Wit;—all that border'd on it, or on Sense came from my Cousin George, whom altogether I like very well.

Nor was this an isolated incident: a later letter, written from London in April 1811, makes passing reference to Jane having just spent the 'greatest part of the even^g very pleasantly' with George and Mary Cooke.[15] Indeed, earlier that month, Jane and Mary had gone sightseeing together in London on what Jane described as 'quite a Cooke day'.[16] George and Theophilus had also been present later in the day, though Jane was rather less enthusiastic about the elder brother, referring to 'his usual, nothing-meaning, harmless, heartless Civility'.[17]

Taking the natural first rung on the Church ladder, Cooke began his career as a fellow and tutor of Corpus Christi, and we are fortunate that he had a small number of students who saw fit to mention him briefly in their memoirs. For example, the judge Sir John Taylor Coleridge (1790–1876), who was at Corpus Christi between 1809 and 1812, later remembered Cooke as 'a most useful Tutor',

> inferior to no man in industry, or zeal, good common sense, patience, and excellent temper, to all which he joined a genuine sense of humour, and delightful simplicity of manner.[18]

It seems, however, that for reasons now lost, Cooke dissuaded Coleridge from further study in mathematics.[19] Cooke was also tutor to the historian and educator Thomas Arnold (1795–1842) and the churchman and poet John Keble (1792–1866).[20] His college teaching duties appear to have been confined to the classics, in connection with which he was appointed to the comparatively senior role of Lecturer in Greek.[21]

Cooke's major claim to fame within the college, however, may have come back to haunt him as a tutor. In 1797, as an undergraduate, he had founded the Corpus Christi Junior Common Room (JCR) as a club-room where undergraduates might more cheaply enjoy wine after dinners in the college hall. It quickly became the custom within the JCR to toast

Figure 5.4 Cooke's *alma mater*, Corpus Christi College, Oxford.

the health of the founder, and each year a 'poet laureate' was appointed to write a commemorative ode for the anniversary celebrations that took place every 20th November.[22] The JCR's 'Book of Songs and Odes' survives in the Corpus Christi archives, and records the odes written from 1812 onwards.[23] Cooke's name is frequently invoked, often—even in the earliest days—under his nickname 'The Codger'. A common theme is the elevation of Cooke's status as founder of the JCR above that of Richard Foxe (c.1448–1528), who merely founded the college. A sample stanza from the ode composed in 1843 by that year's poet laureate, J. C. Hughes, is representative of the way in which Cooke was typically cited:

> Codger Cooke when he founded our snug Common Room
> Ordained that its members should daily consume
> Such rich draughts of the grape as should banish all woe
> And drown sorrow & care in the goblet's bright flow.[24]

And an extract from the ode composed by Thomas Arnold thirty years earlier gives us what is perhaps our only hint about Cooke's appearance—or rather about one of his prominent features:

> And first we'll begin with the Knight of the Chin,
>   Our Room's most munificent Founder:
> To the Codger our debt let us never forget,
>   May his Health in a Bumper go round, Sir!
> Though now, as our Tutor, he perhaps may not suit, Sir,
>   Though he cram us with old Aristotle,
> The deeds we must praise of his juvenile Days,
>   For he gained us the Rights of the Bottle.[25]

It is possible that the JCR became a source of some embarrassment to Cooke once he had become a more senior member of the college, for it gained a reputation for raucous behaviour, and aside from praising Cooke, the annual odes often poked fun at the other dons. In 1836, the college authorities decided that they had had enough, and barred the JCR from holding its anniversary dinner. Cooke was no longer teaching at Corpus Christi by this time, nor living in Oxford, but the JCR committee learned that he would be visiting the city, and passed a resolution on 15 November to 'wait on the illustrious founder [. . .] George Leigh Cook [sic] B.D. of immortal memory [. . .] to petition him to use his influence [. . .] in behalf of the members of the J.C.R.'[26] However, there is no surviving record of any intervention by Cooke. Over the following years, friction with the college authorities continued, but the JCR managed to endure, in due course welcoming Cooke's son, William Hay Cooke (1824–46), as a member.[27] Nevertheless, the JCR's days were numbered and it was abolished in 1852, eventually to be re-established in the twentieth century as a less boisterous association. But the combination of conviviality and organizational ability that had presumably led Cooke to found the JCR in the first place remained with him for the rest of his life. He went on to establish a Literary Dining Club within the university, and when he died, an obituary notice described him as 'a most agreeable and facetious companion'.[28]

As Sedleian Professor

Cooke was elected Sedleian Reader on 27 July 1810, following the death of Thomas Hornsby on 11 April that year, but the precise circumstances surrounding his appointment are rather opaque.[29] No record survives of why he was chosen, nor do we know the names of any other candidates, though it is tempting to think that a tenuous link to Hornsby via Corpus Christi may have been a factor.[30] This being said, Cooke's appointment was by no means outlandish: although he was relatively young, he was already in the system and had a number of years of teaching experience behind him. Indeed, he was considerably better qualified for the post than many of his predecessors in the eighteenth century. Recall from Chapter 1 that the Sedleian Chair was now being turned in a more mathematical direction, and we know from his book purchases that Cooke had at least some mathematical interests. His recent experience as a university examiner (1808–10) takes on a particular relevance here when we consider who the chair's electors were at this time. Chief among them was John Parsons (1761–1819), Vice Chancellor and Master of Balliol, who had been the major driving force behind the introduction of the new examination statute in 1800, which had brought rigour to university assessments for the very first time (see Chapter 1).[31] Cooke's service as an examiner suggests an implicit support for the internal changes within the university that Parsons had sought to bring about, which may have made him appear a sensible professorial candidate in Parsons's eyes. There is little evidence to suggest that Edmund Isham (c.1744–1817), elector and Warden of All Souls, had any particular sway over the Sedleian election,[32] but the third elector, Martin Joseph Routh (1755–1854), was president of Cooke's brother's college (Magdalen), and this personal connection may have been useful.[33]

Like much of the rest of his life, however, Cooke's teaching as Sedleian Professor is rather difficult to trace, and indeed may not have been particularly extensive. According to the bond that he signed in taking up the post, Cooke's duties were to begin on 1 January 1811, whereafter he was obliged each year to deliver two thirty-lecture courses 'in natural Philosophy as grounded on Mathematical Principles, and particularly in the Principia Mathematica Naturalis Philosophiae of Sir Isaac Newton'.[34] The courses were to be announced in the Oxford newspapers at least ten days before they were due to begin, and would go ahead only if at least six undergraduates signalled their intention to attend. It was not until Saturday 22 June 1811 that Cooke's first such advertisement appeared, announcing 'a Course of Lectures on the Principles of Mechanics' for the following October.[35] It is not clear whether this course was ever delivered, however: there is no mention of it in the October newspapers. Over the following decades, announcements of Cooke's lecture courses, often on the principles of mechanics, but usually more specifically on Newton's *Principia*, appeared sporadically in the Oxford newspapers. Little evidence survives to say which of these courses was ever delivered, nor is there any evidence that Cooke appointed an assistant to give his lectures for him. Some of the lecture courses certainly went ahead, but Cooke probably suffered from the same lack of interest that Baden Powell had bemoaned (see Chapter 1).[36] In a letter of January 1820 to his patron James Henry Leigh (see the section 'Church career'), Cooke made a rare passing reference to his lectures that suggests that he took them seriously, at least in the early decades, and did not casually rehash them from previous iterations; in declining a dinner invitation, he noted:

> as I have rather a difficult Course of Lectures in Newton to give this Term, which commences in a few days, I find my time much occupied in the Preparation [...][37]

Figure 5.5 Martin Joseph Routh, long-lived President of Magdalen College, who had a hand in the election of both nineteenth-century Sedleian Professors, forty years apart.

Those historians who have labelled Cooke a sinecurist have thus been at least a little unfair,[38] but, as we saw in Chapter 1, he was sufficiently inactive by the end of his tenure for the geologist Nevil Story Maskelyne (1823–1911) to describe the Sedleian Professorship as 'practically obsolete' in his evidence to the government commissioners at the beginning of the 1850s.[39]

Indirect evidence of Cooke's engagement with Oxford mathematics teaching lies in the fact that he was pressed into service as a mathematical examiner following the creation of that role

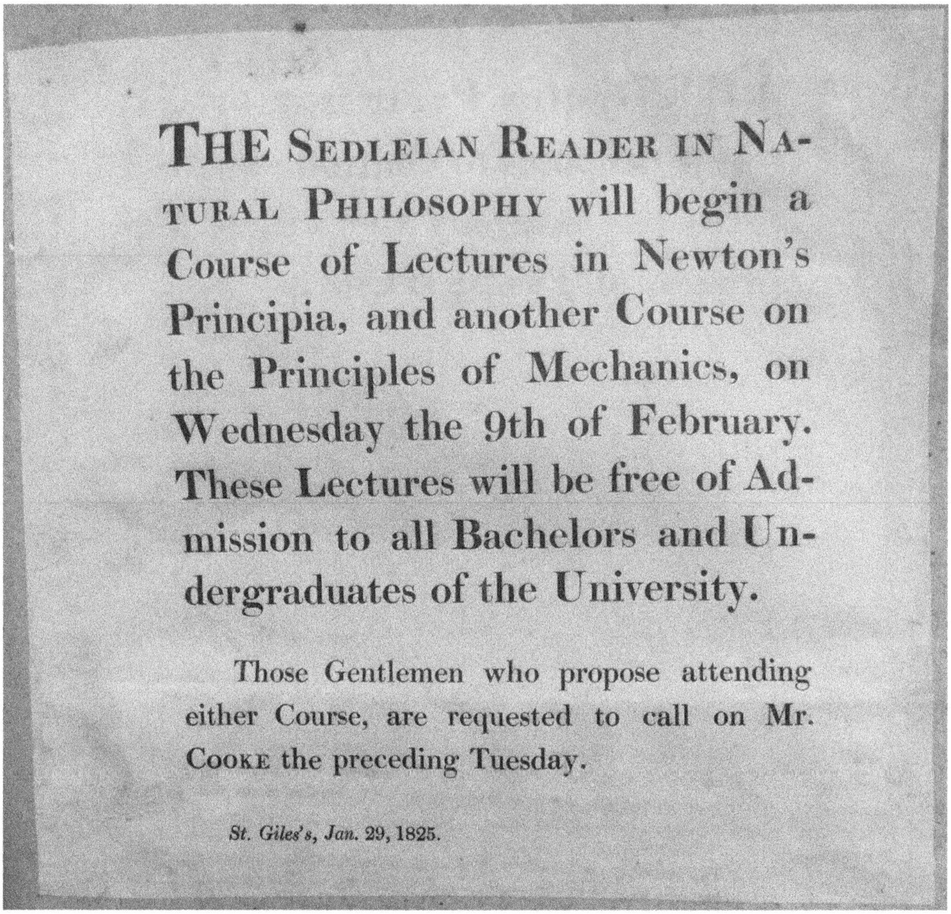

THE Sedleian Reader in Natural Philosophy will begin a Course of Lectures in Newton's Principia, and another Course on the Principles of Mechanics, on Wednesday the 9th of February. These Lectures will be free of Admission to all Bachelors and Undergraduates of the University.

Those Gentlemen who propose attending either Course, are requested to call on Mr. Cooke the preceding Tuesday.

St. Giles's, Jan. 29, 1825.

Figure 5.6 An example of a bill advertising Cooke's lectures in 1825. It is unknown whether the lectures actually took place.

in 1825.[40] As we have seen, over a decade earlier, he had acted for several years (1808–10 and 1815–17) as a more general examiner, responsible for both of the honour schools that then existed within the university: Mathematics and Physics, and *Literae Humaniores*. From 1825, the two honour schools were examined separately, and Cooke was brought in as a mathematical examiner during the first two years of the new system. He was pleased to report, in a letter to the Home Secretary Robert Peel (1788–1850) of June 1826, that the new arrangements had already given a boost to mathematical study in Oxford:

> Some stimulus, it appears to me, has already been given to Mathematical Studies by the appointment of Masters to conduct that Branch of the Examination solely. A more lively Interest is felt in the Subject by the Examiners, & the Examination is extended to a greater length. The consequence has been, that at the Examination just closed, the standard of merit was higher, & the general attainments of the Candidates were visibly improved, when compared with the state of things last June.[41]

However, he felt that there was still room for improvement at an institution where, in contrast to the situation in Cambridge, classical studies still took precedence, and progress in mathematics was often sacrificed to ensure 'the acquisition of high Classical Glory'.[42] To be clear, at no point did Cooke seek to dislodge classics from its premier position, but rather to ensure that mathematics was taken more seriously, even as the secondary subject. He therefore wrote to Peel, whom he had examined in 1808, in the hope that the latter, who had been the first Oxford undergraduate to attain a double first in classics and mathematics, might assist in establishing a prize to stimulate mathematical study:

> I venture to suggest to your consideration the establishment of a small annual Prize in Oxford for the encouragement of Mathematical Studies. The insight I have lately had into the state & progress of that Branch of Learning amongst us, strongly persuades me that some such measure would be highly beneficial.[43]

Peel responded in positive terms, noting that he had just established a similar prize at Harrow and promising to consult with his former Oxford tutor, Charles Lloyd (1784–1829).[44] However, the latter raised a number of practical objections to the suggestion, and so nothing came of it.[45] Nevertheless, Cooke was a subscriber to the fund for the mathematical scholarships that were eventually established in Oxford at the beginning of the 1830s.[46]

Whether or not he was actively lecturing in the final years of his life, Cooke was nevertheless still thinking about the teaching of mathematical natural philosophy. Apparently inspired by the 1849 approval of a new Oxford honour school in Natural Sciences, the following year Cooke published an introductory textbook concerning the early sections of the first book of Newton's *Principia*: specifically those dealing with the motion of bodies under centripetal forces, which were precisely those sections that were examined in Oxford.[47] Cooke's 'Compilation' (as he described it) follows the structure of the relevant sections of the *Principia* very closely, but with some expanded explanations; these appear largely in the same geometrical terms as employed by Newton, although Cooke did permit himself a much freer use of algebraic symbolism than Newton had allowed.

What is most interesting in assessing Cooke's approach to his subject is the preface that he affixed to the book, and which he also deemed important enough to publish as a stand-alone pamphlet, in the hope that it may

> contain matter of interest and utility to the general reader, and may assist in awakening the desire for further acquaintance with the branches of Natural Philosophy.[48]

Within the preface, Cooke located mathematics and natural philosophy as parts of a broader education, and extolled their virtues in encouraging 'those habits of close attention, abstraction, and patient investigation, which an University education is expected to generate'.[49] Nevertheless, he was clear about what the student's priorities ought to be: '[w]e must [. . .] not wish those studies in ancient Classical Literature to be in any way superseded, which have hitherto constituted the main feature in an Oxford education'.[50] Moreover, theological studies should not to be neglected either:

> Though sincerely rejoicing that the Book of Nature will no longer be a closed volume to many of our Students, we must remember the paramount importance of a correct acquaintance with the Book of Grace, and of a disposition to acquiesce in its revelations.[51]

Over the course of the preface, Cooke's views on the appropriate level of mathematical education for an Oxford undergraduate also gradually emerge. The heavily geometrical approach of

THE

THREE FIRST SECTIONS

AND

PART OF THE SEVENTH SECTION

OF

NEWTON'S PRINCIPIA,

WITH A PREFACE

RECOMMENDING

A GEOMETRICAL COURSE OF MATHEMATICAL READING,

AND AN INTRODUCTION

ON THE ATOMIC CONSTITUTION OF MATTER, AND THE
LAWS OF MOTION.

BY

GEORGE LEIGH COOKE, B.D.

SEDLEIAN READER IN NATURAL PHILOSOPHY,
AND FORMERLY FELLOW AND TUTOR OF CORPUS CHRISTI COLLEGE,
IN THE UNIVERSITY OF OXFORD.

OXFORD,

JOHN HENRY PARKER;
AND 377, STRAND, LONDON.
1850.

Figure 5.7 Cooke's edition of the early sections of Newton's *Principia*.

the subsequent text, with only occasional appeals to infinitesimal quantities, was not merely how Cooke had chosen to present this book, following Newton's original style, but was in fact a reflection of the extent of the mathematics that he thought an Oxford undergraduate ought to learn, at least in connection with natural philosophy. Since

> few Students will possess energy and talent sufficient to combine high attainments in [classical] studies, with such a progress in the higher Mathematical Analytics, as to be competent to carry on researches in Physical knowledge by the aid of the Differential Calculus[52]

it was mathematics—calculus in particular—that ought to be sacrificed to the other 'valuable and indispensable' subject:

> We must therefore be content in the great majority of cases to prosecute [mathematical] researches by aid of geometrical reasonings, and accordingly in such cases to limit our pure Mathematics to the four first, and sixth books of Euclid. Algebra to the end of Quadratic Equations, the general principles of Trigonometry, the general properties of the Conic Sections: and to crown these very moderate attainments with a knowledge of the three first Sections of Newton's Principia.[53]

With a restriction to these topics, there would be no 'unfair encroachment' by mathematics 'on the hours and energies required for Classical studies'.[54] It is quite startling to read these words from the hand of an Oxford professor of mathematics (in nominal duties if not in title), particularly when we consider that by the 1850s the analytical mathematics of continental Europe had long been making inroads into Cambridge examinations, as well as other British mathematical contexts.[55] Even in Oxford, it was far from unknown: Cooke's colleague in university mathematical teaching, Baden Powell (1796–1860), the Savilian Professor of Geometry, had published his analytical treatise on calculus in 1830, and Bartholomew Price (1818–98), who would soon succeed Cooke as Sedleian Professor, had similarly taken a self-consciously continental approach to the subject in a textbook of 1848 (see Chapter 6).[56] Thus, Cooke's views, long-held but only expressed in print in 1850, appear a little old-fashioned. Fortunately, even as early as the 1820s, other voices had prevailed in Oxford, as we see from the following anecdote relating the experience of the philosopher Francis William Newman (1805–97) under Cooke's conservative examination:

> He gained a Double First in 1826, being the first man who ever offered in the Schools the Higher Mathematics analytically treated. Cooke [...] pronounced that they could not [...] pass beyond the Geometry of Newton; but [Robert] Walker, Experimental Philosophy Professor, who probably of the three examiners alone knew the subject, persuaded his colleagues to let him examine Newman in the work he offered; and the candidate's answers were so brilliant, that the examiners, not content with awarding his First, presented him with finely bound copies of La Place and La Grange.[57]

In his later years, Cooke's active connections with Oxford University seem to have come largely in a political or religious context,[58] most visibly in connection with the 1847 general election, which took place in the aftermath of the repeal of the Corn Laws the year before.[59] Cooke's greater concerns, however, were for the threats that he perceived to the established Church, not least from the growing Oxford Movement; as he wrote to the university registrar Philip Bliss (1787–1857): 'I regard the Church as equally endangered on both flanks, by Romanizers on one, by Puritans on the other'.[60] And it is clear that he saw in William Gladstone (1809–98), who was standing for re-election as one of the MPs for the university,[61] someone who would defend the religious status quo: Cooke was a member of a committee formed in support of Gladstone's campaign, and was active in encouraging others to vote that way.[62] Cooke was certainly consistent in letting religious conviction guide his political actions: eighteen years earlier in 1829, he had been one of the few Oxford professors (six out of twenty) to vote against Robert Peel as MP for the university—their prior correspondence notwithstanding—in the by-election that followed Peel's sudden switch to support of Catholic emancipation, a policy that the notoriously anti-Catholic constituency had originally elected him to oppose.[63] Although Cooke was on the winning side (the Conservative Sir Robert Inglis was returned in Peel's place), he stood out from a professoriate that was generally less rabidly anti-Catholic.[64]

Church career

Although, as we have seen, it is unfair to characterize Cooke as a sinecurist, it is nevertheless true that the greater part of his energies over his lifetime were spent on clerical matters, rather than on Oxford teaching. In 1816, for example, he had served as public preacher within the university, and gained a reputation as 'an impressive preacher of earnest awakening sermons'.[65] But in order to find Cooke at his most active, we must turn away both from mathematics and from Oxford and examine his career within the Church.

In 1815, Cooke had already taken the common next step for an academic clergyman by giving up his fellowship at Corpus Christi in order to marry: in October that year, at St George's Church in Bloomsbury, he married Anne Hay (1785–1869), eldest daughter of William Hay of Russell Square in London.[66] Within a year, their first child, Helen (1816–69), was born, and a further five children followed over the next fifteen years, all of whom survived into adulthood.[67]

Where Cooke deviated from the usual pattern was in the fact that his marriage and the resignation of his fellowship did not coincide with his giving up his Oxford duties altogether and removing himself to a comfortable parish. At the time of his marriage, Cooke had already been the absentee rector of the parish of Wyck Rissington in Gloucestershire since 1811, and in 1813 he had taken on the same role for the parish of Broadwell with Adlestrop, also in Gloucestershire, upon the death of the previous rector, his uncle the Rev. Thomas Leigh (1734–1813).[68] Thus, it seems that the income from two parishes, together with the Sedleian stipend, was sufficient to enable Cooke to marry and start a family; from 1818, he supplemented his income with the £40 stipend of the Keeper of the University Archives, a position which he held until 1826.[69]

A major change took place in Cooke's life in 1820, however, when he again benefitted from a death in the extended family, this time of his cousin the Rev. James Austen (1765–1819), brother of Jane. Austen had been the absentee vicar of the parish of Cubbington near Leamington Spa, whose patron was Austen's and Cooke's mutual cousin James Henry Leigh (1765–1823).[70] Upon Austen's death, Cooke's mother successfully lobbied Leigh to grant the parish to her son, just as he had previously presented Cooke to Broadwell with Adlestrop, of which Leigh was also patron;[71] the perpetual curacy of the neighbouring Hunningham was subsequently also added to the gift.[72] To begin with, Cooke intended to follow James Austen's example by appointing a curate to carry out the majority of his Cubbington duties, but financial

Figure 5.8 St Mary's Church, Cubbington, where Cooke spent the greater part of his career.

Figure 5.9 Parish Church of the Blessed Virgin Mary, Beckley, near Oxford. Though not visible in the long grass, Cooke's grave is on the right-hand side of the frame, to the west of the church.

considerations quickly swayed him towards residency, as he indicated in a letter to Leigh in January 1820:

> The high stipend for Curates [. . .] is so powerful a Persuasive for residence, that Mrs Cooke & myself have been seriously meditating on transferring our home into Warwickshire, & only visiting Oxford for a few Weeks twice in the year, to enable me to give my Lectures.[73]

Over the next couple of years, the Cookes appear to have been frequent visitors to Cubbington (their next child, Caroline Cassandra, was born there), and then probably left Oxford permanently in 1823.[74] Cooke's care of the Cubbington parish and associated duties were to become his main concern.

When it comes to the surviving records, Cooke is much more visible in Cubbington and the surrounding areas than he was in Oxford. Indeed, after about 1820, he begins to look a lot like his eighteenth-century predecessors as Sedleian Professor in that his name crops up in a wide range of contexts, but only rarely in connection with Oxford teaching. In Cubbington, we find Cooke doing precisely the things that we would expect a nineteenth-century clergyman to be doing. For example, from the start of his tenure, he was very much concerned with the state of the local school: his happy discovery that Cubbington already had 'a roomy substantial Schoolhouse' was one of the main points he made in a letter to James Henry Leigh in May 1820, in which he also expressed the hope that schooling in the village may 'soon be put on a respectable footing'.[75] A history of Cubbington notes that Cooke 'was distinctly a pioneer in the cause of education in rural parishes'.[76] Indeed, concern for the school spanned Cooke's three decades in the village: in 1846, he closely oversaw the construction of new school buildings

Figure 5.10 Cooke's heavily weathered and overgrown tombstone in Beckley Cemetery.

on land donated for the purpose by Chandos Leigh (1791–1850), son of James Henry, and contributed several hundred pounds of his own money to the project.[77]

More generally, throughout the 1820s to early 1850s, we find Cooke supporting a number of charitable efforts and serving on many a local committee. For example, he was a subscriber to the Oxford Society for the Relief of Distressed Travellers and Others, and to the National Society for Promoting the Education of the Poor in the Principles of the Established Church.[78] Cooke was at times a patron of the Warwickshire Leamington District Temperance Society, and vice president of the Warwick and Warwickshire Horticultural Society.[79] In 1851, he addressed a joint meeting of local architectural and archaeological societies in Coventry on the subject of the museum holdings of the Warwickshire Natural History and Archaeological Society, of which he was a past president.[80] Elsewhere, Cooke's local eminence was such that his mere attendance of meetings was newsworthy: we know that he was present, for instance, at gatherings of the Society for the Propagation of the Gospel in Foreign Parts, and of the Governors and Friends of Birmingham Royal School of Medicine and Surgery.[81] The collective scope of the organizations with which he was involved was indeed quite broad. Cooke also appears, alongside his brother, in an early list of members of the British Association for the Advancement of Science, and would later become a life member.[82]

During the later years of his life, Cooke seems to have divided his time largely between Cubbington and the village of Beckley, just north of Oxford. The Cooke family had inherited the rectory and advowson (that is, the right to appoint clergy) of Beckley from the Bee family, via George Leigh Cooke's grandmother, Anne Bee, and had become significant landowners within the parish.[83] Both Samuel and Cassandra Cooke would eventually be buried at Beckley, and from 1802 until his death in 1846, Theophilus Leigh Cooke was vicar there; he was succeeded in that post by his nephew George Theophilus Cooke (1819–94), who had previously served as his uncle's curate.[84] Thus, George Leigh Cooke had good reason often to visit Beckley, which may indeed have been a convenient place to stay on those occasions (if any) when he delivered his Oxford lectures. In the end, this connection meant that Beckley was where Cooke was buried following his death in Cubbington on 29 March 1853.[85] His will, drawn up in 1847, distributed his substantial income from rents and investments between his widow and surviving children in such a way that they were apparently able to live out their lives in some comfort.[86] And the will also gives us one final glimpse of how Cooke viewed himself at this late stage of his life: whereas in earlier decades, he had signed himself Sedleian Reader and Vicar of Cubbington, it was now only the latter that appeared. Apart from passing references to shares in the canal and in the St Clement's Turnpike Trust, Oxford does not feature in the will; given Cooke's evident wealth, we might have expected to see at least a small bequest to Corpus Christi, for instance, but none such appears. Cooke ended his life firmly a country clergyman, rather than an Oxford don.

CHAPTER 6

Bartholomew Price

CHRISTOPHER D. HOLLINGS

At the beginning of the nineteenth century, the Sedleian Chair had passed smoothly into the hands of mathematicians. Its first occupant during the century, George Leigh Cooke, may not have been the most active of professors, but the teaching that he carried out, and his academic interests, such as they were, served to secure the chair as one in applied mathematics. When Cooke died in 1853, the university was in the throes of reforms imposed upon it by the government. As we saw in Chapter 1, the improvement of the professoriate and the provision of scientific teaching in the university were central to those reforms. Cooke's successor as Sedleian Professor, Bartholomew Price (1818–98), was of the pro-reform party, and had previously pushed for the changes in policy from which he now benefitted. Like Cooke before him, Price would go on to hold the chair for over four decades. During this time, he built a career as a prominent academic administrator, politician, and accountant, revived the fortunes of the university press, and eventually rose to the Mastership of his alma mater Pembroke College. Alongside these various other roles, he would carry out his Sedleian teaching with the greatest of diligence, and produce a series of highly regarded textbooks. Over several decades, Price worked gradually to raise the standard of mathematical study in the university. A few short papers aside, he was not a mathematical researcher, but as an administrator, teacher, and publisher he had an enormous impact on mathematics in Oxford, and in Britain more generally.

Early life and career

Bartholomew ('Bat') Price was born on 14 May 1818 at Coln St Dennis in Gloucestershire, the son of William and Mary Price. He was baptized on 25 June that year by his own father, who had been rector of the parish since 1810.[1] Little information survives concerning Price's mother, who died in 1828 at the age of 37.[2] The life of his father, William Price (1784–1860), on the other hand, is rather better documented.[3] Hailing from Wantage (then in Berkshire), he matriculated at Pembroke College, Oxford in 1799, and was a fellow there by 1806, though

Christopher D. Hollings, *Bartholomew Price*. In: *Oxford's Sedleian Professors of Natural Philosophy*. Edited by: Christopher Hollings and Mark McCartney, Oxford University Press. © Christopher D. Hollings (2023). DOI: 10.1093/oso/9780192843210.003.0006

Figure 6.1 Bartholomew Price.

he subsequently gave this up in favour of marriage and the rectorships of Coln St Dennis (which he held for 50 years) and Farnborough, near Wantage (from 1815). Bartholomew was the sixth of William and Mary's seven children, and the second of their two sons; his elder brother, William Henry (1813–89), also pursued a college-and-church career, as we shall see.[4]

Figure 6.2 Church of St James the Great, Coln St Dennis, Gloucestershire. A young Bartholomew Price is said to have been involved in the repair of the church clock.

We have just one small glimpse of Bartholomew Price's youth in Coln St Dennis: it is said that in 1839, aged 21, he worked with the local blacksmith to rebuild the parish church's ancient clock.[5]

By the time of the clock story, Price had been an undergraduate in Oxford for two years. Having been educated privately and at Westwood Grammar School in Northleach, he followed his father and elder brother in matriculating at Pembroke College, Oxford (in March 1837),[6] entering on a scholarship that had been established in 1683 for pupils from his school.[7] Price would spend the remaining sixty-one years of his life at Pembroke, in one capacity or another. He graduated BA in 1840, and even at this early stage he marked himself out as a member of a new generation of Oxford figures who were determined to take mathematics more seriously, even if—as George Leigh Cooke had feared—it meant sacrificing the traditionally dominant classics: Price sat for double honours, obtaining a first class in mathematics, but only a third in classics.[8] In 1842, he was granted one of the university's mathematical scholarships. These awards, made on the basis of an examination and carrying a £50 stipend, had been founded in 1831 for the promotion of mathematical study among candidates who had recently completed their BA.[9] During the middle decades of the nineteenth century, the scholarships provided a welcome boost to mathematics in Oxford, and Price would eventually serve as a scholarship examiner.[10] In the short-term, the scholarship enabled Price to remain in Oxford, where he obtained his MA in 1843. The following year, he took holy orders and was elected a fellow of Pembroke, where he also became a tutor and mathematical lecturer in 1845, marking the start of his long teaching career in Oxford. He was appointed to the Sedleian Chair quite early in that career.

Figure 6.3 Pembroke College, Oxford, where Price spent most of his life, in one capacity or another.

As Sedleian Professor

As we recorded in Chapter 1, Price had offered testimony on the state of Oxford mathematics teaching to the 1851 Oxford University Commission.[11] In making his specific recommendations for the Sedleian Professorship, Price had in fact defined the duties that he would soon take on.

George Leigh Cooke died on 29 March 1853, and the university moved remarkably quickly to appoint his Sedleian successor: as early as 14 April, Price received a letter from Richard Lynch Cotton (1794–1880), the Provost of Worcester College and Vice Chancellor at the time, announcing Price's election that day as Sedleian Professor.[12] In offering his congratulations, Cotton referred to the 'eminent nature of the competition'. Thanks to a newspaper report of Price's election, we know who that competition was: 'Mr. Dale, of Balliol, also a distinguished mathematician'.[13] This is almost certainly a reference to John Andrews Dale (1816–88), who had matriculated at Balliol in March 1834, and had gone on to hold a mathematical scholarship in 1839, just a few years before Price.[14] Thereafter, however, Dale does not appear to have been able to secure any formal permanent post or fellowship within the university, although he certainly assisted in the practical teaching of physics at Balliol, and served as a university examiner on several occasions from the late 1850s.[15] His name also appears in connection with Oxford's nineteenth-century scientific association, the Ashmolean Society, with which Price was also involved (see the section 'Original mathematics').[16]

In assessing why Price was elected to the chair over Dale, we might, as for previous elections, look to the electors, although this time no strong motives stand out on their part. Price's implicit support of the reforms that were being forced upon the university does not seem to have done him any harm, even though Cotton, elector and Vice Chancellor, had declined to answer the commission's questions.[17] The long-lived Martin Joseph Routh (1755–1854), aged 97 at the time of Price's election and still President of Magdalen, was similarly opposed to the government commission and suspicious of reform.[18] The third elector, Lewis Sneyd (1788–1858), Warden of All Souls, probably exerted little influence on, and perhaps took scant interest in, the selection of the Sedleian Professor: an obituary notice of him would later remark that '[t]he Rev. gentleman never took any active part in the affairs of the University'.[19] Thus, in considering the election, we must look, much more simply, at Price's credentials. He was already an active teacher of mathematics in a college setting, and had even taken the further step of writing up his college calculus lectures into his *Treatise on the Differential Calculus* (1848), a substantial new version of which had appeared in print just months before the Sedleian election.[20] Dale may well have been a 'distinguished mathematician', but he lacked Price's very visible experience. We might even speculate that Price had deliberately positioned himself for the chair, in the knowledge that the elderly and inactive Cooke would not be around forever.

The journalists who reported on the appointment of a new Sedleian Professor in Oxford all reached for the seventeenth-century Laudian Code to inform them about the chair's conditions, disregarding its more recent Newtonian history.[21] With reform in the air, however, it can have come as no surprise to Price to read the following in his letter from Cotton, which also hints that the latter was aware of Price's views on the chair:

> I have no doubt that you will be anxious yourself to render this Professorship as efficient as possible, but I may state that when the Committee now sitting on Univ^y Reform were considering the Professorships they appeared to anticipate some change in the discharge of the duties of the Sedleian.[22]

Discussion of the professor's duties appears, for example, in the university council minutes of March 1854, where we find the intriguing suggestion that the Sedleian Professor should 'teach the higher branches of applied as well as pure mathematics, mathematically'.[23]

As we saw in Chapter 1, some of the recommendations of the university commissioners were eventually followed in the second half of the 1850s, when the university revised the statutes governing the Sedleian Chair to decree that the professor should lecture in 'Physical Mathematics'.[24] The confirmation of these duties, however, can have had little practical effect on Price's teaching, as this was the type of mathematics that he was already covering. Almost immediately upon being elected to the Sedleian Chair, Price had advertised a course of lectures on analytical mechanics, to begin that October.[25] Price evidently had an audience for his lectures, which continued into 1854, and were followed, from May onwards, by lectures on dynamics.[26] In October that year, Price turned his attention to geometrical optics, and then in the following January to the motion of a rigid body.[27] Unlike Cooke, who had adhered quite rigidly to the instruction to teach Newton's *Principia*, Price allowed himself, and was soon officially permitted by the statutes, a greater flexibility in the courses that he lectured. In addition to the topics already mentioned, we find him offering teaching over the following decades in a range of other branches of applied mathematics, including hydrostatics and hydrodynamics, the construction of optical instruments, elasticity, and, indeed, Newton's *Principia*.[28] Price's lectures were first advertised in the newspapers in the traditional fashion, and then later on in the *Oxford University Gazette*, following its launch in 1870 as the official organ of the university. The record of advertised lectures, together with the brief annual reports of teaching that the professors were subsequently obliged to prepare for the *Gazette*, suggest that Price was

CORPUS CHRISTI PROFESSOR OF JURISPRUDENCE: Sir F. Pollock, Bart., M.A.

The Professor will lecture at All Souls College on the Early History of Contract, on Tuesdays **Juris-** and Fridays at Noon, beginning on Friday, October 21.   **prudence.**

---

3. MEDICINE.

REGIUS PROFESSOR OF MEDICINE: Sir Henry W. Acland, Bart., K.C.B., D.M.

The Professor will be glad to see Junior Members of the University on the subject of the Medical **Medicine.** Examinations on Tuesday, October 25, at Noon, in the Medical Department, or on appointment by letter on other days.

LICHFIELD LECTURER IN CLINICAL MEDICINE: W. Tyrrell Brooks, M.A.

The Lecturer will lecture at the Radcliffe Infirmary on Clinical Medicine, on Mondays at 5.30 P.M., **Clinical** beginning on Monday, October 24.   **Medicine.**

LICHFIELD LECTURER IN CLINICAL SURGERY: A. Winkfield (F.R.C.S.).

The Lecturer will lecture at the Radcliffe Infirmary on Bandaging and Minor Surgery, on Fridays at **Clinical** 5 P.M., beginning on Friday, October 21.   **Surgery.**

WAYNFLETE PROFESSOR OF PHYSIOLOGY: J. S. Burdon-Sanderson, M.A.
[*See* **NATURAL SCIENCE.**]

LECTURER IN HUMAN ANATOMY: A. Thomson, M.A.

A Course of Lectures on the Central and Peripheral Nervous System will be delivered at the Department **Human** of Human Anatomy on Mondays, Wednesdays, and Fridays, at Noon, beginning on Wednesday, October 19. **Anatomy.**
Demonstrations on the more technical parts of the subject will be given daily at 9.15 A.M.
Instruction on Osteology for junior members of the Class will also be provided.
The Morning Class will commence on Tuesday, October 18.
The Dissecting Room will be open daily between 9 A.M. and 4.30 P.M.
Members of the University who wish to attend any of the above Courses are requested to call on the Lecturer at his room in the Department on Monday, October 17, between the hours of 11 A.M. and 1 P.M.

---

4. NATURAL SCIENCE.

SAVILIAN PROFESSOR OF ASTRONOMY: Rev. C. Pritchard, D.D.

The Professor will lecture at the University Observatory on (1) the Theory and Computation of Eclipses **Astronomy.** and their Phenomena, on Tuesdays and Thursdays at Noon, beginning on Thursday, October 20; (2) Practical Work, on Tuesdays and Thursdays at 8 P.M., beginning on Thursday, October 20.
During the Long Vacation the Professor has arranged greatly improved facilities for the study of Practical and Descriptive Astronomy.

SAVILIAN PROFESSOR OF GEOMETRY: J. J. Sylvester, M.A., Hon. D.C.L.

The Professor will lecture at the University Museum on Invariants and Covariants of Systems of Conics, **Geometry.** on Mondays and Fridays at 11 A.M., beginning on Friday, October 21.

SEDLEIAN PROFESSOR OF NATURAL PHILOSOPHY: Rev. Bartholomew Price, D.D.

The Professor will lecture in the Mathematical Lecture Room, University Museum, on the Theory of Fluids **Natural** and Fluid Matter, on Tuesdays, Thursdays, and Saturdays, at Noon, beginning on Thursday, October 20. **Philo-sophy.**

PROFESSOR OF MINERALOGY: M. H. N. Story-Maskelyne, M.A.

The Professor will lecture at the Museum on Crystal-Symmetry and Crystallographic Equivalence, on **Mineral-** Tuesdays and Fridays at Noon, beginning on Friday, October 21.   **ogy.**

2

Figure 6.4 Advertisements of lecture courses from the *Oxford University Gazette* for 14 October 1892, including one of Price's last, on the theory of fluids.

quite diligent in carrying out his professorial duties.[29] He maintained his teaching right up to the point of his retirement from the Sedleian Chair at the age of 80.

At Pembroke

By the time that Price took up the Sedleian Chair, he had been teaching mathematics at Pembroke, as both lecturer and tutor, for eight years. We catch a glimpse of this early part of his career through the reminiscences of some of his students that were gathered in the 1890s by the Rev. Douglas Macleane (1856–1925), sometime fellow and chaplain of Pembroke, when he was compiling a history of the college.[30] The various memories collected by Macleane present Price as having been by far one of the best and more conscientious teachers at Pembroke, though this may have been influenced by an automatic respect for the man who was by this time the venerable Master of the college. Philip Hedgeland (1826–1911), who had entered Pembroke in 1846 and must therefore have been among Price's earliest students, recalled that

> [t]he only lectures in my time worth a straw, were Price's. He liked his work, & really helped those who were willing to be helped.[31]

In a similar vein, Hedgeland's contemporary, Edward Redman Orger (1826–1917), noted that

> Mr Price was conspicuous for his energy, and his constant efforts to induce men to read for Mathematical honours [...][32]

This last recollection is echoed in the memoirs of John Mitchinson (1833–1918), who would eventually succeed Price as Master of the college.[33] Determined to obtain a first in classics without the distraction of additional mathematical studies, Mitchinson recalled that he had had to stand his ground when Price had been adamant that he should also read for honours in mathematics. Sixty years on, Mitchinson claimed that a frustrated Price had told him: 'mark my words, Sir, to the end of your days you will be a half educated man'.[34]

As we will see later, Price became a very prominent and recognizable figure within the wider university, and this status seems to have been appreciated by his students. Orger noted, for example, that '[w]e used to be proud of his reputation in the University at large'.[35] Moreover, at least one of Price's former students, Edward Bartrum (1833–1905), recognized the efforts that he had made to raise the standard of mathematical study in Oxford:

> he was almost the only really able representative of the mathematical side of our studies for many years [...] Professor Price from the very first devoted himself to this special study with endless energy & his pupils for many years have been among the most distinguished men in Oxford.[36]

Price's eminence as a mathematician was recognized even by those students who did not seek mathematical honours; these students were nevertheless obliged to learn some Euclid and elementary algebra, and one of them remarked that to be taught these basic subjects by Price 'was a case of the "razor to cut the quartern loaf"'.[37]

Price's high status within the university was matched by his standing within the college, where on several occasions throughout his decades of service, he deputized for the Master in the role of Vicegerent.[38] Some grumbling within the college in one of these instances also highlights how Price became a representative of a new type of Oxford don: one who was married.

The question of whether to lift the centuries-old prohibition against college fellows marrying had been raised by the Commissioners in the 1850s, as a means of retaining able college teachers who might otherwise be lost to marriage.[39] Although the full relaxation of the restriction across the university would not come until 1882, certain smaller-scale measures were suggested and implemented in the 1850s.[40] One of these, apparently proposed by Francis Jeune (1806–68), Master of Pembroke, was that college fellows be allowed to marry if they also held a professorship.[41] In early 1857, Pembroke was among the first colleges to adopt this revised principle, alongside other reforming measures.[42] Fellows who held professorships or lectureships within the university were to be permitted to marry, subject to a two-thirds vote of the other fellows. Price evidently won such a vote, for in August 1857 he married Amy Eliza Cole (1835–1909) at Littleham, near Exmouth in Devon; the ceremony was performed by Jeune.[43] Following his marriage, Price moved out of his rooms in Pembroke and into a house on St Giles', where he and his wife raised a large family.[44] One of the objections that had been raised against the marriage of fellows was that they would be much less visible and active within the college community, and this is certainly the complaint that was levelled against Price in connection with his Vicegerency of 1873. In November that year, Edward Moore (1835–1916), Principal of St Edmund Hall, and a former Pembroke student, wrote a gossipy letter to John Mitchinson, who had just departed Oxford:

> I forgot almost to tell that the Master again nominated Price as Vice Gerent, which some of the Fellows are not best pleased at, as Price is not only now entirely non resident, but never thinks of performing any of the duties of the office, & in the Master's absence of course it causes great inconvenience that his representation sh^d be out of the way, & as often as not in London.[45]

Non-residency was not the only factor in Price's absence from Pembroke: by 1873, he was heavily involved with the university press, and it was these duties that were often taking him to London.

Mastership

Price's Vicegerencies of Pembroke must have prepared him for his eventual Mastership. But his appointment as head of the college so late in life was not his first attempt at the position.[46] In 1864, Jeune had resigned the Mastership to become Dean of Lincoln, and there emerged two natural candidates to succeed him: Price and the classics tutor Evan Evans (1813–91).[47] An account of the subsequent election appears in the memoirs of John Mitchinson, who expressed clear opinions of Price and Evans.[48] Although Mitchinson considered Evans 'a thorough gentleman', he compared him unfavourably with Price on academic grounds: 'Evans was an idle man, devoid of scholarly or literary tastes, designed more for social life'.[49] Price was Mitchinson's preferred candidate, owing to his eminence as a mathematician and his standing as Sedleian Professor. Mitchinson was not the only interested party to focus on Price's mathematical credentials: the latter's former student, Edward Bartrum, later recalled that upon hearing of the vacancy in the Mastership in 1864, he had written to Mitchinson

> urging the election of Professor Price as Master, that Oxford might thereby have one College at all events in which Mathematics should be allowed to hold their own if not to reign supreme.[50]

On a more practical level, Price's supporters also believed that the Mastership would bring him firmly back into college life.[51] However, it was not Price's time, owing to certain complications, which are not mentioned in Mitchinson's somewhat sanitized account of what happened. According to Mitchinson, the two rival camps were so evenly balanced and entrenched that the contest could only be decided by the casting vote that Evans wielded as Senior Fellow. Not wanting to reduce Evans 'to the odious necessity of voting himself into the Mastership', Price's supporters therefore persuaded him to withdraw.[52] What Mitchinson tactfully omitted to mention is that Price had suffered an embarrassment at the hands of his elder brother. William Henry Price had served as a fellow of Pembroke from 1840 to 1860, and during his later years there he had acted as the college's bursar. Although the details are murky, it seems that somehow by 1864 William Henry Price owed Pembroke the substantial sum of £1,257 10s, for in February that year Bartholomew Price agreed to repay this amount to the college on his brother's behalf.[53] It is not unreasonable to suppose that this affair damaged the younger brother's chances of election as Master.

There are also hints—though they are a little oblique and come from a single source—that questions of religion also entered into the Mastership contest, either from genuine conviction or as a pretext upon which Price's opponents could attack him. At the end of February 1864, Price wrote a long letter to Mitchinson concerning the election.[54] The many crossings-out and indistinct words that the letter contains suggest that it was hastily written, and the tone is certainly an impassioned one.[55] Price wrote of the 'extreme pain' that he had felt at the 'unfounded reports and calumnies' as to his sympathy with the so-called 'rationalizing School'.[56] This is probably a reference to the ideas of liberal theology then circulating in Oxford, which sought to reinterpret Christian teachings in light of modern knowledge of the world. As such, the emphasis was placed on reason and experience over scriptural revelation, a controversial position in mid-nineteenth century Oxford.[57] Although it now seems perfectly reasonable that a teacher of the newest mathematical and physical ideas of the time should adopt such a rationalist stance, Price was keen to deny that he had been 'compromised by confederacy with that School in other matters'.[58] No 'confederates' were named, but Price was perhaps tainted by association with the late Savilian Professor of Geometry and prominent liberal theologian, Baden Powell (1796–1860), with whom he had collaborated closely in university mathematical teaching.[59] In his letter to Mitchinson, Price sought to take a middle path, arguing for the teaching of the newest knowledge, whilst distancing himself from controversial points of view and offering his critics the texts of two sermons as evidence of the 'safeness' of his position.[60]

With Price's forced withdrawal from the contest, Evans was swiftly elected Master of Pembroke, but there are no indications of ill feeling between the two. In writing to Mitchinson, Price had been clear that his opposition to Evans as Master had not been personal: he simply did not think that Evans was up to the job.[61] In the long term, these concerns proved to be justified: although apparently an amiable figure who was popular with the undergraduates, Evans failed to provide the strong leadership that the college needed, resulting in a disgruntled and fractious fellowship.[62] Price drifted further away from college affairs, resulting in the grumbling later recorded by Moore, and yet there remained a strong 'Price faction' in Pembroke: according to Moore, the nomination of Price as Vicegerent in 1873 had been 'to avoid a row'.[63] Evans struggled to maintain order in college meetings, and was often only able to exert his authority by use of the double vote that, as Master, he wielded in any ballot.[64] But even this came under attack: the Shakespearean scholar Alfred Thomas Barton (1840–1912), referred to by one commentator as 'the cantankerous Barton',[65] used the opportunity of the revision of the college statutes at the end of the 1870s to attempt to abolish the Master's double vote. The proposal was defeated—thanks, ironically, to the very device it was intended to eliminate—and Barton wrote bitterly of this to Mitchinson, painting Price in a particularly negative light. According to Barton, Price himself had once spoken out against the double vote:

> I reminded him of his opinion, he admitted that on general grounds he approved the change, but contrived all the same to vote against it. No doubt he was thinking of his own chances of using the double vote as Master.[66]

Barton was thinking of Price in connection with the Mastership because it was rumoured at the time that Evans was about to be promoted to some high ecclesiastical office, leaving Pembroke behind. Writing of the young faction of four fellows who had attempted to abolish the Master's double vote, Barton asserted that

> we unanimously agreed that Price as Master would be a veritable tyrant; and I doubt if any of the four would elect him could a more generous-minded man be found.[67]

Whilst Price was universally praised as a competent man of business and a shrewd politician, this is not the only hint that he had an authoritarian, and perhaps ungenerous, streak in his character. As the years passed, however, he seems to have mellowed: writing of kindnesses done by the elderly Price in the mid-1890s, a friend of the family made a contrast by observing that '[h]e was not "Bat Price" of 1840/1860'.[68]

Barton's manoeuvrings at the end of the 1870s turned out to be premature: Evans remained at Pembroke until his death in November 1891.[69] At this time, however, the college factions were still very much in place, so much so that when discussions turned to the question of Evans's successor, some fellows felt that it was time to give the college a fresh start by electing an external candidate.[70] The voting began at the end of December 1891, and ran over several days to four inconclusive rounds. At each stage, the votes were permuted among internal candidates including Price and Mitchinson, and external figures such as Thomas Hodge Grose (1845–1906), classicist and fellow of Queen's.[71] But in the absence of the necessary absolute majority, the only way to break the impasse, in accordance with the college statutes, was to call in the Visitor, the Chancellor of the university, who was at that time the Marquess of Salisbury, then also Prime Minister.[72] Over New Year 1891, while the college awaited the marquess's decision, letters were exchanged on the matter between Mitchinson and the historian and churchman Mandell Creighton (1843–1901). The latter had no formal connection to Pembroke, but confessed 'a deep interest' in the politics of the situation, 'because of its value in illustrating papal conclaves'.[73] Creighton found it curious that elements of the fellowship should be intent on 'going outside the College': '[n]ow Grose is in no sense a distinguished man, not comparable to Price or yourself'.[74] He believed that Salisbury would be conscious of the regard in which Price was held more widely, and so would conform to 'the general sense of the University'; within this broader setting, Price would be 'unexceptionable', and moreover was 'not likely to hold office long'.[75] Salisbury's decision, appointing Price Master of Pembroke, was communicated to the college on 14 January 1892.[76] Within the university at large, the appointment seems to have been welcomed,[77] and in the reminiscences that he communicated to Macleane, Bartrum asserted that Salisbury had 'done justice to one (as it seems to me) of the most, meritorious men Oxford has had in the century'.[78] An obituary of Price would later put a positive spin on the circumstances surrounding the election by asserting that the appointment had 'the character of a public as well as a collegiate recognition'.[79] Price would govern Pembroke as Master

Figure 6.5 Price as Master of Pembroke College.

until his death in 1898, sometimes with the assistance of a Vicegerent; in what was perhaps a diplomatic move by a practiced man of university and college politics, Price appointed A. T. Barton to that role.[80]

Within the wider university

Price did not only play the political game at Pembroke, but also more widely: by the end of his life, he had sat on almost every board or committee within the university. Foremost among these was the governing Hebdomadal Council, an elected body created in 1854. Price was elected to the council in the autumn of 1855 in what was regarded as a victory for pro-reform elements within the university.[81] Over the following decades, he was re-elected time and again, ultimately remaining on the council for forty-three years. As one obituary of Price observed:

> Elections came and went; figures rose up and disappeared. There was never any question on either side or in any caucus of omitting or opposing the Master. He was the one safe and indispensable man.[82]

Price was valued for his perspective: by the end of his life, he was one of the few remaining links at the high levels of the university to the pre-Commission days of the 1830s and 1840s.

Price filled many other positions within the university. For example, he served on several occasions as an examiner (see the section 'The promotion of mathematics'), and was also a Curator of the University Chest. Indeed, it is in financial matters that he appears to have excelled, regarding a proper grasp on finances as being the key to university reform. One obituary commented that '[b]usiness men were often astonished to find a Professor and an Oxford Don with so good a head for business',[83] while Mitchinson suggested that 'Price ought to have been a business man. He would have built up a huge concern and died a millionaire'.[84] Price served for several years as bursar of Pembroke, during which time he reorganized the college's accounts, and he sought to carry the same changes across to the university. In 1867, he reported to a House of Commons Select Committee on the confused state of university accounting.[85] Later that same year, he was asked by a committee of Hebdomadal Council to prepare a detailed report on university finances, including an outline of sources of income, as well as suggestions for how that income might best be collected, recorded, and distributed.[86] Price advocated the adoption of a more rigorous system of accounting, to be overseen by a financial board and a carefully appointed secretary.[87] In considering the university's accounts, Price had necessarily noted the income derived from the university press, and nowhere were his financial skills more in evidence than in his long-standing connection with this branch of the university.[88]

By the nineteenth century, the university press had receded into little more than a publisher of Bibles,[89] but its activities were revived and its remit re-broadened in the mid-century under the Vice Chancellorship of Francis Jeune (1858–62).[90] Then, as now, the press was governed by a panel of academics known as the Delegates, and it was probably through Jeune's influence that Price was appointed to the Board of Delegates in 1861. In his new role, Price was a key figure in pushing through changes to the way in which the press operated, eventually turning it into an important source of revenue for the university. In 1868, as a member of the press's Finance Committee, he penned a memorandum calling for the engagement of a 'Secretary of Accounts' as an executive officer for the press. Price was duly appointed to the new post,

Figure 6.6 The Oxford University Press printing offices, completed in 1830.

stepping down as a Delegate, in March 1868 at an annual salary of £500.[91] Over the following two decades, his powers as Secretary (not to mention his salary) steadily increased to make him the central controlling figure of the press.[92] Since this was a full-time job, Price appears to have stepped back from many of his normal college duties, though he maintained his Sedleian teaching as before.

As Secretary, Price served as a mediator between the academic and commercial sides of the press.[93] He also became the main point of contact between the press and its authors, and increasingly demanded a more professional attitude from them: his patience with authors who were slow in producing manuscripts was limited.[94] Price brought a financial hard-headedness to his duties with the press, rejecting the suggestion that it ought to subsidize academic works that might not otherwise be printed.[95] Instead, he was 'somewhat disinclined to publish learned but un-remunerative books'.[96] Under Price, the press expanded into the rapidly growing and profitable market for schoolbooks.[97] Price was particularly keen to expand the provision of scientific texts, for both schools and universities, in light of the continuing growth in the teaching of those subjects. As early as May 1863, he had presented a proposal to the Delegates for what became the Clarendon Press Series of scientific educational titles.[98] The series, which went on to include such classics as W. Thomson and P. G. Tait's *Treatise on Natural Philosophy* (1867) and James Clerk Maxwell's *Treatise on Electricity and Magnetism* (1873), 'helped to raise the standing of the Press as a scientific publisher'.[99] Sadly, however, Price's plans for an expanded and coordinated list of mathematical titles never came to fruition.[100]

In June 1885, Price formally retired as Secretary to the Delegates, but his duties with the press did not end there: by a special vote of the university, he was appointed Perpetual Delegate and became the paid chairman (£500 p.a.) of the press's Finance Committee, with much of the responsibility for commissions remaining in his hands.[101] The board appointed Philip Lyttelton Gell (1852–1926) as its new Secretary, but this was not to be a happy arrangement. A relative Oxford outsider, Gell soon fell out both with the Delegates and with Price; he appears to have felt undermined by the power that the latter still exercised within the press.[102] Gell became a rather isolated figure, and his health finally broke down in 1896; after a series of leaves of absence, the Delegates dismissed him as Secretary. Gell's downfall appears to have been engineered by Price, the hostile Delegates, and the Vice Chancellor J. R. Magrath (1839–1930), so that the latter's favoured candidate, an Oxford insider, Charles Cannan (1858–1919), could be appointed to the Secretaryship.[103] During Gell's absences, the duties of Secretary, if not the title, had fallen once again on Price, though he relied heavily on the assistance of Cannan and the mathematician H. T. Gerrans (1858–1921) as Pro-Secretaries. The continued robust health of the octogenarian Price was remarked upon in Oxford circles,[104] and it is probably because of the vigour that he enjoyed almost to the end of his life that family tradition, rather unfairly, blamed Gell and the work created by his illnesses for Price's death in December 1898.[105]

As author

Among the expertise that Price brought to his roles at the university press was that of an experienced author: around the time that he first became a Delegate, he was completing the fourth and final volume of his *Treatise on Infinitesimal Calculus*. These books, which emerged from Price's Pembroke and Sedleian teaching as a near-comprehensive overview of applied mathematics, were intended in the first instance for Oxford students, but also contain material that went far beyond the undergraduate curriculum.

Price's earliest textbook was *A Treatise on the Differential Calculus, and its Application to Geometry: Founded Chiefly on the Method of Infinitesimals*, first published in 1848. The book derived from Price's calculus lectures in Pembroke, a fact that he acknowledged in his Preface by way of explaining the 'colloquial style' that he had adopted.[106] It is arranged in two parts: the first ('analytical investigations') provides a comprehensive introduction to the principles of differential calculus, whilst the second considers the applications of these to geometry; in outlining the structure of the book, Price noted that

> [w]ere the writer's system carried out to the full, it would be necessary to add a third part, containing mechanical applications[,][107]

and this is precisely what appears in Price's later books.

Despite the book's origins, Price does not appear to have intended it solely for Oxford students, but sought a wider readership.[108] On some level though, he kept the immediate needs of university students in view, but stressed the value of going beyond purely examinable material:

> it is thought that whatever tends to present to the student the principles in a sensible form, and thereby enables him the better to grasp the matter, is not foreign to the purpose.[109]

Price's treatise was certainly more advanced, and considerably more detailed, in its treatment of differential calculus than the prior text written for Oxford students by Baden Powell at the end of the 1820s.[110]

As his subtitle indicates, Price took the notion of infinitesimal quantities as the basis for his discussion. These appeared in the traditional form of differentials, denoted using the continental notation for calculus, rather than the Newtonian dot notation which still persisted in Britain, particularly in mechanical applications.[111] In what was then already a slightly old-fashioned position, Price took differentials to be the more fundamental concept than derivatives, though this focus was realigned in later editions.[112] The concept of a limit also had a role to play in Price's text, but with much less fanfare than infinitesimals, and certainly not in the central position that had been granted to it by Augustus De Morgan (1806–71) in his treatise on calculus a decade earlier.[113]

Price was not at all parochial in his academic life, and enjoyed extensive contacts with people elsewhere in the British Isles, and in continental Europe. At this early stage, we also see that he was reading foreign mathematical literature, for in his Preface he acknowledged a debt to many foreign authors, but singled out only Augustin-Louis Cauchy (1789–1857), now acknowledged as the source of much of modern mathematical analysis, for explicit mention.[114] On the other hand, he disavowed any particular obligation to writers in English, with the exceptions of De Morgan and the mathematician William Spottiswoode (1825–83), a friend of Price's who provided material on the curvature of surfaces.[115]

A few years later, Price set to work preparing a new edition of the book, this time to be published by Oxford University Press.[116] The new version appeared in print in 1852, but with a prefatory note from Price to say that he had made so many changes that it should be considered a new book, rather than a new edition. The overall structure remained broadly the same, but there was much rearrangement and expansion of material within this: the 279 pages of the original version had almost doubled to 540 in the new one. The greatest expansion occurred within the 'applications' part of the book, with the whole now taking the reader 'close upon the boundaries of our knowledge'.[117] In particular, Price introduced material on the geometrical interpretation of infinitesimals, with a view to 'guarding against' the merely 'superficial knowledge' that the student would gain from purely symbolic manipulations, 'which is useful neither for its results nor as an intellectual exercise'.[118] This was probably a barb aimed at the ongoing debate in British mathematical circles about the place of symbolic algebra within mathematical education, and the validity of its conclusions.[119] Pedagogical issues remained central to the construction of Price's new book, certainly in connection with the colloquial style, which he found it

> expedient to retain, under the conviction that it invests a book with a personal and living character more akin to the explanations of a speaking teacher, and thereby infuses life into what might be otherwise dry text.[120]

One small but significant change to the book was the loss of its short final chapter on the basic principles of integral calculus.[121] This was not because Price aimed to cut this material altogether, but rather because he planned to deal with it at greater length elsewhere: this new edition of his *Treatise on the Differential Calculus* carries a second title page, identifying it as volume I (differential calculus) of *A Treatise on Infinitesimal Calculus; containing Differential and Integral calculus, Calculus of Variations, Applications to Algebra and Geometry, and Analytical Mechanics*. The next instalment in this ambitious programme appeared in 1854 as volume II on integral calculus and the calculus of variations.[122] Again, Price's purpose was to present an up-to-date account of the subject that differed substantially from prior treatments in English. As he observed in his Preface, previous textbooks on integral calculus had tended

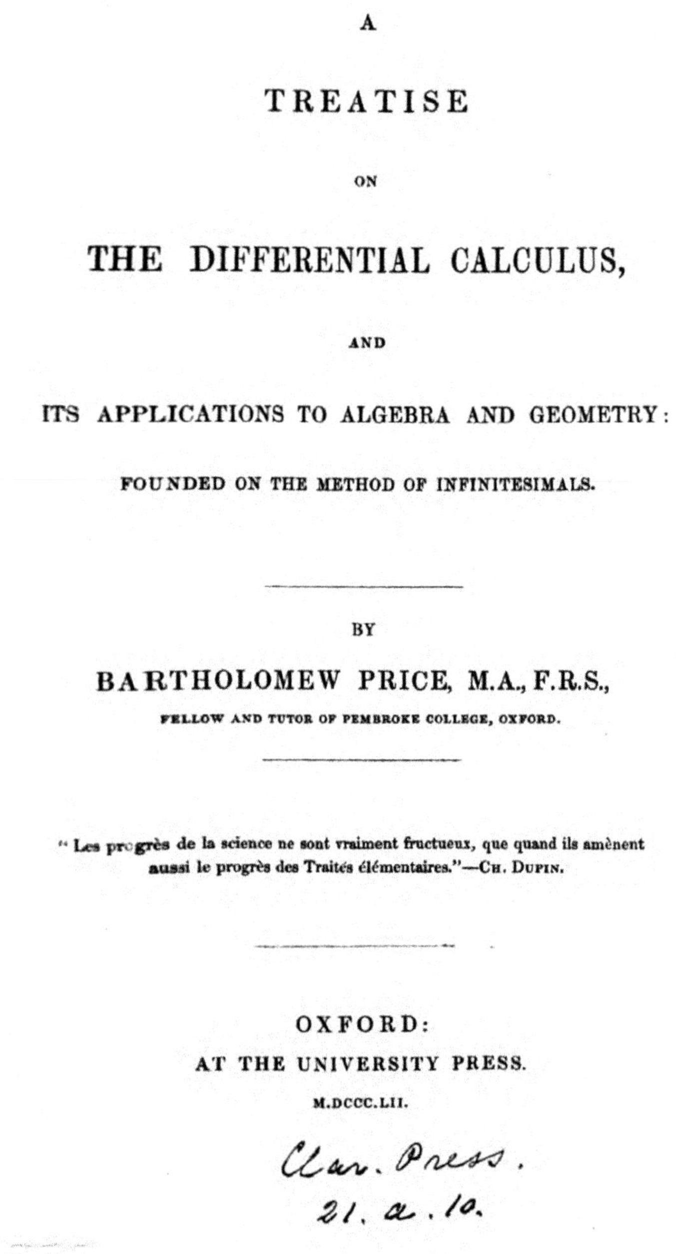

A

TREATISE

ON

THE DIFFERENTIAL CALCULUS,

AND

ITS APPLICATIONS TO ALGEBRA AND GEOMETRY:

FOUNDED ON THE METHOD OF INFINITESIMALS.

BY

BARTHOLOMEW PRICE, M.A., F.R.S.,

FELLOW AND TUTOR OF PEMBROKE COLLEGE, OXFORD.

"Les progrès de la science ne sont vraiment fructueux, que quand ils amènent
aussi le progrès des Traités élémentaires."—CH. DUPIN.

OXFORD:
AT THE UNIVERSITY PRESS.
M.DCCC.LII.

Clar. Press.
21. a. 10.

Figure 6.7 Title page of Price's *Treatise on the Differential Calculus* (1852).

to develop the subject as an inversion of differential calculus, rather than building it from its own first principles, and had also 'obliged [the student] to burden his memory with certain rules which he mechanically applies'.[123] In works by other authors, Price asserted, it was only when geometrical examples appeared at a late stage that the student was given any intuition for what the integral calculus is about, and even then, matters could be obscured by certain algebraic sleights of hand. For Price, integral calculus was too important a topic for this situation to be permitted to stand. Once again, he turned to foreign authors, who 'have been alive to the defects, and have succeeded in remedying them'.[124] He acknowledged De Morgan as having been the first writer in English to have provided a sound basis for the integral calculus, but questioned his introduction of integration so early in his treatise (on page 97 of 640). In Price's view, differential calculus ought to be thoroughly mastered first—hence the arrangement of his own volumes. The goal of this second volume was to develop integral calculus upon the basis of infinitesimals.[125] There was almost certainly a foreign influence on Price's writing here: although he gave no source for his definition of a definite integral, his notation was strikingly similar to that employed by Cauchy in his famous 1823 treatment of the subject (and is quite different from that used by De Morgan).[126] On the basis of this definition, he built up a toolkit of techniques for performing integration, and applied these to geometrical problems; an introduction to the calculus of variations followed, as did a discussion of the solution of various types of differential equations.

Price's first two volumes were reviewed in the literary magazine *The Athenæum* by the fellow of Magdalen and headmaster of Magdalen College School James Elwin Millard (1823–94).[127] Millard began his review with a sentiment that we will see again: that there are those who insist that 'Oxford mathematics do not generally rank very high', but Millard's response to such people is that they should 'take up Mr. Price's two volumes, and they will soon be convinced of their error'. There follows much praise of Price's books, which are compared favourably with works produced for Cambridge students. The Cambridge works are criticized for being tailored to the greater culture of cramming that existed there, with the implication that the reader might learn calculus *properly* from Price's books, rather than simply use them to pass exams. Millard credited the books' 'ease and familiarity' of style to their origin in the lecture room.

The next volume in Price's series appeared in 1856, and arguably marked a new phase in his writing. Whereas volumes I and II were based, at least at their core, on Price's college teaching, and comprised a detailed introduction to calculus, volume III, on statics and the dynamics of a material particle, was probably much closer in its content to his Sedleian lectures.[128] This volume concerns the general principles of mechanics, presented in the language of infinitesimals, which Price found 'peculiarly appropriate' to the subject.[129] Once again, Price was keen to point out that his method was 'counter to that of most English authorities on the subject', and that 'it is rather in accordance with that of foreign, and chiefly French, writers'.[130] The discussion is based firmly on analytical methods, rather than a more traditional fluxional or geometrical approach, something that was praised by the volume's reviewer for *The Athenæum*, Augustus De Morgan, who held that, thanks to books such as Price's, Oxford would soon overtake Cambridge as a mathematical university.[131]

Readers, such as De Morgan, who anticipated a fourth volume of Price's *Treatise* (on the dynamics of material systems) had to wait until 1861,[132] perhaps because Price temporarily diverted his efforts into producing a second edition of volume I, published in 1857. Second editions of the subsequent volumes followed during the 1860s. When a second edition of volume IV eventually appeared, rather later, in 1889, it was met with yet more positive commentary. Reviewing the book for *Nature*, the mathematician Alfred George Greenhill (1847–1927) again reached for the Cambridge comparison, and saw Price's book as one for learning a subject, not just for passing exams.[133] Greenhill's review also commented upon recent changes in the attitude towards mathematics in Oxford. Since 1883, James Joseph Sylvester (1814–97) had been Savilian Professor of Geometry, in which role he had sought to nurture a culture of mathematical research like the one that he had established during his years at Johns Hopkins University.[134] Although a research ethos would not become fully established in Oxford until into the twentieth century,[135] such early strides were being made in the second half of the nineteenth century, and Price had a small role to play in these.

The promotion of mathematics

We have seen that in advocating the improvement of mathematical study in Oxford, Price practiced what he preached: his teaching as Sedleian Professor was considerably more advanced and was performed more conscientiously than that of his predecessor, and was carried out within a more organized system that accommodated advanced mathematics more comfortably than it had in the past. For Price, mathematics was an end in itself, to be taken as far as possible. Even by teaching calculus, Price was going far beyond the material covered by Cooke.

Another respect in which Price's teaching activities differed from those of Cooke was in his taking on private pupils as a mathematical coach. The presence of mathematical coaching in nineteenth-century Cambridge has been well documented; it was a widespread practice at that institution because of the extremely competitive nature of the Tripos examinations.[136] There was no comparable culture of competition in Oxford, but coaching still took place, perhaps in some cases prompted by changes in the examination arrangements (see below), and to make up for the inadequacies of the mathematical instruction at individual colleges—or at least to give it an expert boost. Private coaching was thus a means of raising the general standard of mathematics among Oxford undergraduates, as well as being a source of extra income. Former Pembroke students later recalled, often with some bemusement, how large numbers of 'outcollege' men 'used to stream in, at all hours of the day & night' to seek Price's help with mathematics.[137] By the early 1860s, Price's private coaching was apparently taking place on such a scale that it was deemed necessary to include account of it in proposed changes to tuition and financing arrangements at Pembroke.[138] Private coaching also arguably made Price the central mathematical figure of the university by extending his influence beyond his specialized Sedleian teaching and beyond Pembroke into the fundamental mathematical education of students of other colleges.[139] One of Price's most famous private pupils was Charles Lutwidge Dodgson (a.k.a. Lewis Carroll, 1832–98), who studied across the road from Pembroke at Christ Church.[140] Beginning in the mid-1850s, Dodgson's letters and diaries contain several references to Price, to coaching, and to their varied mathematical conversations.[141] During the Long Vacation of 1854, Dodgson joined a 'mathematical reading party' that Price organized in Whitby.[142] Other participants would later cite the storytelling that took place there as being the incubation of the *Alice* books.[143] Upon completing his studies and becoming a mathematical lecturer himself, Dodgson often called upon Price's advice and experience.[144] The two became firm friends and colleagues, bonding over their love of recreational mathematics, and perhaps also because of shared speech impediments.[145] Dodgson was a frequent visitor to the Price household, sometimes capturing the family in photographs.[146] Price encouraged Dodgson's original mathematical investigations, helping him to present his work to the Ashmolean Society, and also communicating a paper on Dodgson's behalf to the Royal Society.[147]

Figure 6.8 Charles Lutwidge Dodgson, a.k.a. Lewis Carroll, private pupil and then colleague of Price.

Figure 6.9 Bartholomew Price, as photographed by Charles Dodgson, c.1855.

In Dodgson's writings, Price is immortalized in one of the Hatter's verses in *Alice's Adventures in Wonderland*:

> Twinkle, twinkle, little bat!
> How I wonder what you're at!
> Up above the world you fly,
> Like a teatray in the sky.

The rhyme is said to capture something of Price's 'energetic involvement in University life, flying from committee to meeting, and from lecture to tutorial', with a flapping black academic gown no doubt adding to the effect.[148]

Starting in the late 1840s, Price served on a number of occasions as a mathematical examiner for the university, and his experience in this role seems to have fed into his desire to raise the standard of mathematical study in Oxford.[149] Debate about the format of Oxford examinations, and about the standing of mathematics exams within the university system, had been rumbling on since the beginning of the century, with such figures as Baden Powell having had much to say on the subject.[150] Until 1850, the mathematics examinations were voluntary assessments that candidates could take for the extra accolade—though, unsurprisingly, few did.[151] Thereafter, however, it became a requirement for all candidates to sit exams in one other honour school (the options being mathematics, or one of the two newly created schools: natural science, and law and modern history) alongside the compulsory *literae humaniores* (classics) exams.

Price took a break from his examining duties in 1855, but this was the year in which he became embroiled in a brief public exchange with one of his successors over the encouragement of mathematics candidates. The starting point was a letter that Price published in *The Times* in December that year, and which was also printed as a stand-alone pamphlet.[152] In the letter, Price addressed a candidate who had been disappointed by the results of the most recent mathematical examinations. This anonymous candidate was probably not alone in having been disheartened: of the seven candidates who had sat for mathematical honours, five were placed in the third class and one in the fourth; only one candidate attained the second class, and there were no firsts.[153] An eighth candidate had withdrawn from the honours examinations even before the exams had begun, in anticipation of such an outcome. Price was sympathetic to his addressee, expressing his 'surprise and sorrow' upon seeing the list:

> sorrow, I say, for yourself and the other candidates who are so much disappointed by it, and on account of the discouragement which it gives to mathematical studies at Oxford.

It was this last sentiment that prompted Price to make so public a criticism of the examination as follows in his letter. From personal knowledge of the candidates, he knew that the class list was not a fair reflection of their abilities, and believed instead that the examination papers had been made unnecessarily difficult. Price feared that the class list would undo the efforts that he and others had already been making to encourage mathematical study in Oxford, and that the subject could be 'well nigh extinguished' by the discouragement caused by such a poor class list. Price was keen not to impugn the *character* of the examiners, all of them 'gentlemen of principle and the highest intellectual attainment', and all of them his former pupils. In his view, their error was one of *judgement*: they had set the level of the examination too high, and had not tailored it sufficiently to the teaching that had come before it. They had failed to keep in view the typical attainments of 'a student of ordinary calibre' as a benchmark for setting the level of the exam. Upon investigating the examination papers, Price had found them 'on the whole to be much above the capacity of ordinary students', consisting largely of unseen problems and very little bookwork.

The candidate's response to Price's letter is unrecorded, but we do have a clear reaction from one of the examiners. The three mathematical examiners for 1855 were William Spottiswoode (mentioned in the section 'As author'), Francis Ashpitel (1828–97), and Francis Harrison (1829–1912). Spottiswoode had already built a reputation as a mathematician on the basis of his treatise on determinants and a series of pamphlets entitled *Meditationes analyticae*;[154] he would later serve as president of the London Mathematical Society, the British Association for the Advancement of Science, and the Royal Society.[155] Both Ashiptel and Harrison, on the other hand, were academic clerics of a more traditional type. As of 1855, Harrison was a fellow and mathematical lecturer at Oriel College, and appears to have been quite active in the administration of the wider university, though he would end his life as rector of North Wraxall in Wiltshire.[156] Ashpitel taught mathematics at Brasenose College, and would similarly leave Oxford for parish work, in his case at Flitwick in Bedfordshire.[157] It was Ashpitel who issued a strongly worded response to Price's letter; written directly to Price four days after the appearance of his letter in *The Times*, Ashpitel's reply was eventually printed in the same forum in January 1856.[158]

Ashpitel appears to have taken Price's letter as an attack on himself, finding it 'ungenerous, and full of misrepresentations'. Perhaps not unreasonably, he was upset that Price had raised concerns about the examinations so publicly, rather than first writing to the examiners directly. Ashpitel also took exception to what he saw as insinuations of his 'unfitness' as an examiner, particularly from the man who had recommended him as an examiner in the first place. As far as Ashpitel was concerned, he had applied the standards that he had learnt from studying with Price, having obtained a first class in mathematics six years earlier. Writing in quite general terms, Ashpitel disputed several of Price's claims, focusing in particular on the supposed lack of bookwork in the exam papers, and the assertion that high achievement early in the degree course is an accurate marker of success in finals. Unfortunately, Ashpitel seems to have missed the point of Price's original letter: nowhere did he address the issue of the encouragement of mathematics candidates.

The importance that Price attached to an accurate discussion of the issue at hand is demonstrated not only by the fact that he almost immediately responded to Ashpitel's letter with a long and extremely detailed critique of the exam papers and of the letter itself, but that he also kept the matter within the public sphere by publishing a pamphlet containing (with permission) Ashpitel's letter and his own response thereto, dated 1 January 1856.[159] In the response, Price was keen to justify and reassert the reasons for his original letter:

> It was with much pain that I felt myself obliged to comment on Examination Papers set by you and by two other Examiners, all of whom had been my private Pupils, and for whom I had great regard; and especially because one is a most highly valued friend. There are, however, public duties which must sometimes supersede the feelings of private friendship.[160]

First of all, Price cited his duties as Sedleian Professor and his need to encourage the study of the subjects of his professorial teaching. Secondly, he noted his interest in the promotion of mathematics in Oxford, and the discouragement that the recent examination results had given: he was already aware 'that some students immediately proposed to discontinue their Mathematical studies'.[161] Price reiterated that the publication of his original letter had not been done lightly, but in a spirit of '[j]ustice to the disappointed Candidates'.[162] He emphasized that the examiners had done nothing that was contrary to the strict letter of the examination statutes, but suggested that their inexperience *as* examiners (they had all been appointed in 1855), not to mention their collective lack of teaching experience, had left them without any 'practical knowledge of the capacities and attainments of ordinary students'.[163] Only Ashpitel had any substantial examination experience, mainly as Price's co-examiner for mathematical scholarships, rather than as a finals examiner. In that capacity, Price had found Ashpitel a 'diligent and careful' colleague, who framed questions that were 'original, although hard', and he was keen to stress that he had not accused Ashpitel of being an unfit examiner, a slur that the latter had taken from Price's original letter.[164] Price praised Ashpitel's mathematical abilities, but found in these the heart of the problem with the examination papers:

> You must not make your reading the measure of the extent of that which Candidates for Honours usually reach; any more than it is fair for you to make your ability the standard of theirs.[165]

Ashpitel was evidently a mathematical high-flyer, whose own reading as an undergraduate had 'greatly exceeded the limits which Candidates for Honours in the Public Examination usually attain to', and who had perhaps therefore as examiner contributed to the skewing of the expected standard.[166]

Price's reply to Ashpitel goes into considerable detail in responding to the comments and counter-accusations made by the latter. In most cases, however, Price was able to demolish Ashpitel's points by demonstrating that he had misquoted or misrepresented Price's words. Price also returned to the examination papers, standing by his original assessment of them, and now also finding errors, which he described in detail. He ended by defending the examination system, agreeing with Ashpitel that the examiners' decisions must be final, but that the value of the class lists 'will be impaired, if the students themselves do not generally recognise them as fair criteria of their merits'.[167]

Quite when Price's pamphlet appeared in print in relation to the other instalments in this saga is not clear, though it was probably after Ashpitel's first letter of December 1855 was printed in *The Times* in late January 1856.[168] The latter event prompted a further short public response from Price, dated the same day as the publication of Ashpitel's letter, and printed two days later.[169] In a single paragraph, Price denied the accusation that he had singled out Ashpitel from the other examiners for particular criticism (though, upon reading the letters, it is easy to see how Ashpitel might have felt this way), and yet again took the opportunity to stress the magnitude of the issue:

> The question is one now of great public importance; for when examinations, both competitive, and as tests of qualification are being introduced into all branches of public service, it is necessary that the result of them should merit public confidence.[170]

This back-and-forth between Price and Ashpitel rumbled on a little further, now exclusively in pamphlet form. In a pamphlet dated February 1856, Ashpitel responded to Price's point-by-point analysis of the examination papers with an attempt at a point-by-point rebuttal of his own.[171] Unfortunately, Ashpitel's pamphlet is more of a personal attack on Price than a sober contribution to the discussion. Rather than seeking to address Price's specific criticisms of the examination papers, Ashpitel played semantic games, trying to catch Price out in self-contradiction. And, evidently focused on countering what he saw as a direct attack on his character, Ashpitel still did not tackle the central point of the discouragement of mathematical studies in Oxford. When Price responded with his own further pamphlet in March 1856, it was clear that the correspondence had run its useful course, and that communications were breaking down.[172] This time, Price's tone was more waspish, with disapproving comments about the 'asperity' of Ashpitel's own writings.[173] Once again, he made detailed comments on the contents of the exam papers, but here criticized the 'carelessness' of the examiners in setting erroneous questions much more directly.[174] The inaccuracy of the claims in Ashpitel's letters also came under fire.[175] Price's final comments, in which he brought this correspondence firmly to a close, provide us with an indication that a parallel but much less public version of this discussion had been going on within the university:

> The time for another Examination is drawing near; and it is desirable that the confidence of all [. . .] should be restored. The Authorities have thought right to nominate into the Board of Mathematical Examiners, in the place of one whose term of office has expired, a Gentleman of mature years, great experience, and sound judgment, and the University may place confidence in that judgment.[176]

The 'Gentleman of mature years' was the Reader in Experimental Philosophy, Robert Walker (1801–65), whom we encountered in Chapter 5 as one of Cooke's co-examiners in the 1820s. Having previously been an examiner on several occasions, and a mathematics tutor at Wadham College between 1828 and 1853, Walker was a prominent figure in Oxford mathematics teaching, and a natural person to replace Harrison as examiner.[177] Ashpitel served out his two years as a mathematical examiner, but never filled the role again.[178] The same is true of Spottiswoode, though this was probably because much of his subsequent career was spent away from Oxford. This affair does not appear to have damaged the relationship between Price and Spottiswoode, his 'most highly valued friend': when Price married in August 1857, Spottiswoode was one of the witnesses, and their intimate correspondence continued until the end of Spottiswoode's life.[179]

Original mathematics

Although Price was a mathematics educator first and foremost, he did also write a very small number of original mathematical papers. The earliest of these were delivered to the Ashmolean Society, Oxford's nineteenth-century scientific association, founded in 1828.[180] Elected a member in February 1844, Price was on the society's committee by December, and, true to form, would go on to become its treasurer.[181] In December 1846, Price delivered his first paper, 'On the principle of virtual velocities', which explored a notion from Lagrange's *Mécanique analytique* (1788–9).[182] Price formalized certain ideas and demonstrated how to apply them to analyse mechanical problems. The paper appeared in print the following year in Spottiswoode's *Meditationes analyticae*,[183] and seems to have given rise to two further lectures on this topic by Spottiswoode and Powell.[184] Another paper by Price, delivered in December 1847, on 'Geometrical Principles, and the means of interpreting symbols of quantity' appears to foreshadow the comments on symbolic reasoning and the geometrical interpretation of infinitesimals that he would soon include in the 1852 version of his *Differential Calculus*.[185] Somewhat later, in the 1860s, Price became interested in the theory of probabilities, and presented a paper on that subject to the society in October 1864.[186] A particularly interesting paper, which Price presented on his thirty-first birthday in 1849, concerned 'some points in the theory of Mathematical Science'.[187] The paper dealt with the organization of mathematical ideas and the extent to which they are informed by, and match up with, physical reality, leading to the notion of two processes for the development of ideas: the abstract and the empirical. Price considered the interplay of pure and applied mathematical ideas. His argument was that the mechanical philosophy that had been developed over the preceding centuries by Newton and his successors provides us with an example of an abstract theory that can be developed

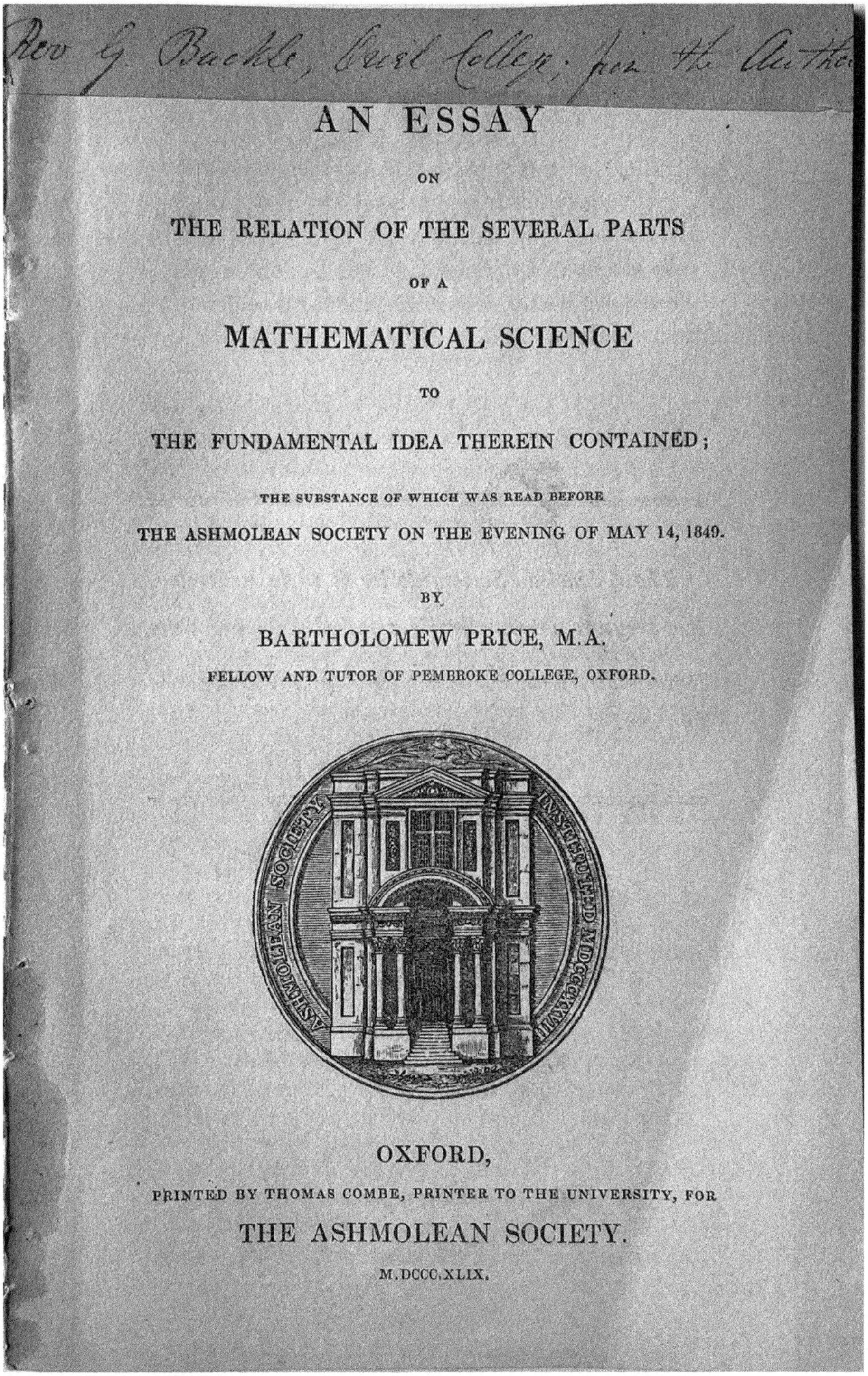

Rev G. Buckle, Oriel College; from the Author

AN ESSAY

ON

THE RELATION OF THE SEVERAL PARTS

OF A

MATHEMATICAL SCIENCE

TO

THE FUNDAMENTAL IDEA THEREIN CONTAINED;

THE SUBSTANCE OF WHICH WAS READ BEFORE

THE ASHMOLEAN SOCIETY ON THE EVENING OF MAY 14, 1849.

BY

BARTHOLOMEW PRICE, M.A.

FELLOW AND TUTOR OF PEMBROKE COLLEGE, OXFORD.

OXFORD,

PRINTED BY THOMAS COMBE, PRINTER TO THE UNIVERSITY, FOR

THE ASHMOLEAN SOCIETY.

M.DCCC.XLIX.

Figure 6.10 Price's lecture on the relationship between mathematical ideas and physical reality, delivered to the Ashmolean Society in May 1849.

independently of the physical world but that is nonetheless applicable to it. The Ashmolean Society's *Proceedings* record that 'Professor Powell expressed agreement with Mr. Price's view of the two distinct processes',[188] and also enables us on this one occasion to place Price and his Sedleian predecessor in the same room: the *Proceedings* contain the tantalizingly vague note that 'some further remarks were made by Professor Cook [*sic*]'.[189]

Price was also active in the British Association for the Advancement of Science, both administratively and mathematically. His name was first included as a subscriber to the association in its annual report for 1847, when it met in Oxford, and he appears thereafter in a range of roles, including as a member of its council and various other committees, and as a Vice President.[190] Over the years, he delivered four technical communications: one in 1847, 'On a new proof of the principle of virtual velocities' (probably a repeat of his Ashmolean Society paper),[191] one in 1861, 'The influence of the rotation of the Earth on the apparent path of a heavy particle' (with ballistic applications),[192] and two in 1865, 'On the extension of Taylor's Theorem by the method of derivatives' and 'On some applications of the theory of probabilities'.[193] When the association met in Oxford in 1860, Price served as President of Section A (Mathematics and Physics) and delivered a presidential address, the essence of which was a defence of the use of mathematics in the natural sciences.[194] In the address, Price pointed to the variety of subjects contained within Section A, noting that some had been 'perfected' whilst others had not. The substance of the ensuing text suggests that to Price, 'perfected' meant 'mathematized'. He stressed the importance of mathematical subjects even if they seem dry to some, and spoke out against recent criticism of what some people viewed as the unnecessary precision of mathematical argument. But he ended on a positive note: that not very long ago, the state of mathematical knowledge in Britain had been deplored by some, but now, thanks in part to the regular meetings of the association, the standard had been raised.[195] Price named a few of the great British mathematical names of the age, but the fly in the ointment was probably the fact that he could not yet—quite—include any Oxford figures in his list.

Beyond Oxford

Oxford dons are not generally renowned for their acknowledgement of the world beyond the city, but we have seen ample evidence that this sentiment does not apply to Price. Early in his career, he had no qualms about stepping into government circles by giving evidence to the University Commission of the 1850s, and he moved in increasingly higher places as his career progressed.[196] Prominent administrative positions within the university, the eventual Mastership of Pembroke, and the connections afforded by his Secretaryship of the university press meant that Price was in regular contact with the great and the good of Victorian Britain.[197] As early as 1853, he was in correspondence with William Gladstone (1809–98) about his evidence to the University Commissioners,[198] and would later be invited by Gladstone, because of his 'known experience & ability in College business', to join the Cleveland Commission of the 1870s, which was tasked with investigating the property and income of the Universities of Oxford and Cambridge.[199] At other times, the two were in contact over Gladstone's works for the press, and other publishing questions.[200] And Gladstone was not the only political figure to draw Price into matters of policy: a few years later, for example, he gave evidence to a Royal

Commission on Scientific Instruction (the Devonshire Commission), once again arguing for the raising of the standard of scientific education in Oxford.[201]

Price would also have been visible beyond Oxford in a more specifically scientific capacity. We have already noted his involvement with the British Association for the Advancement of Science, but this was far outstripped by his activities at the Royal Society. Price was elected to fellowship of the Society in 1852 with the citation:

> The Author of a Treatise on the Differential Calculus & other Papers printed by the Ashmolean Society of Oxford. Eminent as a Mathematician.[202]

Among the signatories of Price's proposal for fellowship were his Oxford colleagues Robert Walker and Baden Powell.

We have seen that Price was rarely a passive member of any body with which he was associated, and this is no less true of his involvement with the Royal Society. For a total of eight years over his several decades as a fellow, Price sat on the society's council, twice serving as Vice President, and was also, unsurprisingly, a member of various committees.[203] He relayed several papers to the Royal Society, and also exerted a quiet influence on the British mathematical landscape by acting as a referee for papers that were submitted to the society for publication.[204] It was through the Royal Society that Price was appointed Visitor to the Royal Observatory, Greenwich in 1865.[205] Though never active as an astronomer, he had been a fellow of the Royal Astronomical Society since 1856, and his connection to Greenwich, together with a parallel Visitorship of the Oxford University Observatory (whose accounts he audited from 1874 until his death), gave Price a reputation as a promoter and defender of the subject.[206]

In looking at Price's connections beyond Oxford, we see that, like many of his contemporaries, he sat at the heart of an extensive network of correspondence. Much of it was related to the university press,[207] but we also find Price engaged in mathematical discussions of a kind that would have been alien to his Sedleian predecessor. From an early stage of his career, Price became a go-to figure for general mathematical enquiries, with one correspondent of the 1850s addressing him erroneously as 'Professor of Higher Mathematics', giving us a glimpse of how Price was viewed outside the university.[208] Eminent mathematicians sought Price's advice, particularly on those topics that were covered by his textbooks, and he similarly sent them his own mathematical problems.[209] It is clear that Price was thoroughly embedded within the British mathematical, and more general academic, community at large. In addition to the memberships already mentioned, Price joined the London Mathematical Society in 1866 within a year of its foundation, and would also go on to be a member of the Physical Society of London, an honorary member of the Cambridge Philosophical Society, and a founder member and early Vice President of the Oxford Mathematical Society.[210] In his correspondence, he is well represented in the archives of such major figures as G. G. Stokes (1819–1903), John Herschel (1792–1871), and William Thomson (later Lord Kelvin, 1824–1907).[211] A few letters have also survived from William Whewell (1794–1866) and Augustus De Morgan.[212] With his many correspondents, Price engaged in networking, discussed mathematical problems as well as university and national politics, and often furnished them with the latest versions of his textbooks.[213] We also find a nice example of scientific philanthropy: in the mid-1860s, Price was involved in efforts to secure a pension to support George Boole's widow and children.[214] Nor were Price's wider contacts confined to Britain—his reach extended beyond the usually very local concerns of his Sedleian predecessor. In compiling his textbooks, Price consulted continental European sources, but he wasn't simply a passive reader of foreign books: he was also in correspondence with mathematicians on the continent.[215] Traces survive of a connection with the French mathematician Charles Hermite (1822–1901), and also with the Italian geometer Luigi Cremona (1830–1903).[216] In 1891, Price may perhaps have met Henri Poincaré (1854–1912), who was visiting Sylvester in Oxford.[217] Price's major international contacts all came at the end of his life, after Sylvester had arrived to take up a post in Oxford in 1883, and they might therefore have been brought about via the more vibrant mathematical research community that Sylvester sought to promote.

Final years

Although Price stood down as Secretary to the university press in 1885, his workload hardly reduced. Like many an academic (particularly in an age before regularized pensions and retirement ages), Price simply kept working into his late 60s and beyond. In the 1890s, he was still involved in the organization of science teaching in Oxford,[218] and he was still delivering his Sedleian lectures, even after becoming Master of Pembroke and taking on the additional duties, such as a Canonry of Gloucester Cathedral, that came with it.[219] But, upon reaching his 80th birthday in May 1898, he decided that the time had come to resign the professorship that he had held for over half his life. At the instigation of Thomas Fowler (1832–1904), President of Corpus Christi College and a long-standing friend and associate of Price's,[220] a plan to honour Price upon his retirement soon emerged. At the end of May, the following note was circulated around Oxford:

Figure 6.11 Astronomical clock dedicated to Price in Gloucester Cathedral.

Figure 6.12 Price's tombstone in Holywell Cemetery, Oxford.

Several friends of the MASTER OF PEMBROKE are desirous of showing their sense of the great services which he has rendered to the University during the last fifty years, and think that the present is an appropriate occasion for so doing, as he has recently completed his eightieth year and is about to resign the Professorship which he has so long held. It is proposed, therefore, to invite him to a Complimentary Dinner in the Hall of Queen's College, at which the VICE-CHANCELLOR has consented to preside, on Friday, June 24th, at 7 o'clock for 7.30 precisely. We shall be glad to know whether it will be agreeable to you to be present on the occasion.[221]

The price of a ticket to the dinner was one guinea, and Queen's was chosen as the venue since Price had been an honorary fellow there since 1868, thanks to the Sedleian connection (see Chapter 1).[222] The college's Provost, J. R. Magrath, was the Vice Chancellor mentioned in the note above.[223] The twelve signatories of the note included three college heads, four professors, and Price's old nemesis (and Pembroke Vicegerent) A. T. Barton.[224] Around 100 people were subsequently present at the dinner, where Fowler led the congratulatory toasts. In reply, Price 'expressed the great obligations he owed to his pupils', praised the high public spirit that he had always found in Oxford committees, and asserted that the future of the university was in good hands.[225]

In his toast, Fowler expressed the hope that even after retiring as Sedleian Professor, Price 'might still lead an active, useful, and vigorous life, and continue to give his great services to the University'.[226] This is precisely what Price did, but sadly only for a few months more. During the remainder of 1898, his health declined, and he died at the Master's House at Pembroke on 28 December.[227] His funeral took place on 3 January 1899 in the college chapel, and

notwithstanding its occurrence in the middle of the vacation, the gathering was one of the largest that [had] been seen in recent years in the University.[228]

Like the dinner in Queen's just a few months earlier, the funeral was attended by senior figures from across the university and elsewhere. The funeral party processed from Pembroke across Christ Church Meadow and round to Holywell Cemetery, where Price was buried in a grave already occupied by his daughter Rose. Two weeks later, in a letter to Stokes to thank him for attending the funeral, Price's widow, Amy Eliza Price, wrote of her late husband with perhaps the highest praise that could be given to a Victorian public figure: '[w]e have to be thankful for his life of active usefulness'.[229]

Price's career followed a very different path from that of his Sedleian predecessor(s). For Price, college and university work were not simply stepping stones on the path to a life within the Church, but rather they were a career in themselves, and one that Price took very seriously. Thanks to the university reforms of the mid-nineteenth century, Price was able to pursue an academic career in something like its modern sense, but still with the traditional ecclesiastical associations and duties.[230] In using his position and influence to promote mathematics as a respectable area of study, and to improve its teaching in Oxford, Price paved the way for his research-active successors.[231] With his network of contacts, both in Britain and abroad, Price broke the Sedleian Chair away from its parochial Oxford origins, turning it into a post whose subsequent occupants would exert an international influence on science policy and scholarship.

CHAPTER 7

Augustus Love

JUNE BARROW-GREEN

When Augustus Love, a Cambridge wrangler, was elected to the Sedleian Chair in 1898, all his predecessors bar one—Thomas Millington who held the chair from 1675 to 1704 (see Chapter 3)—had been educated in Oxford. To date, no details of Love's election have come to light, so it is unknown who, if anyone else, was in the running or even whether there was an election, or indeed if it was contentious to look beyond Oxford at the time. Be that as it may, Love set a trend for non-Oxford educated Sedleian Professors which continued until 2019 when Jon Keating, a graduate of New College, was appointed.

Love was born in 1863 in Somerset, the son of a surgeon and the second boy in a family of three boys and two girls, and spent his boyhood in Wolverhampton. His elder brother, Ernest, who obtained second-class honours in the Natural Sciences Tripos at Cambridge, became a lecturer in Natural Philosophy at the University of Melbourne. The elder of the sisters followed Ernest out to Melbourne where the two of them shared a house. Love's own domestic life followed a similar pattern, with the younger of the sisters keeping house for him in Oxford.

Love and his brothers were educated at Wolverhampton Grammar School where the headmaster, Thomas Beach, was known for sending 'a succession of good scholars to both Universities'.[1] Both Beach and the mathematical master, Henry Williams, were graduates of the Cambridge Mathematical Tripos. The former, a Junior Optime in 1856, had previously been mathematical master at Lancaster Grammar School, while the latter had been 28th Wrangler in 1864.[2] With such teachers, and with his brother Ernest already in situ, one might have expected Love to have a firm idea of his future at Cambridge when he successfully sat the examination for Minor Scholarships, but, when he went up to St John's College in 1882, he was undecided whether to read classics or mathematics. Having chosen mathematics, his progress was rapid, and it was said that 'no one with any personal acquaintance could fail to recognize his extraordinary cleverness'.[3]

In common with others aspiring to a high place in the Mathematical Tripos, Love employed a coach to prepare him for the rigours of the examination.[4] His choice, R. R. Webb, was the senior wrangler of 1872, and the most successful coach of his day.[5] Webb was also a college lecturer with a reputation as an excellent teacher, notably in elasticity and dynamics, both

June Barrow-green, *Augustus Love*. In: *Oxford's Sedleian Professors of Natural Philosophy*. Edited by: Christopher Hollings and Mark McCartney, Oxford University Press. © Oxford University Press (2023). DOI: 10.1093/oso/9780192843210.003.0007 subjects in which Love would shine.[6] As expected, Love excelled in the 1885 examination, coming second in the order of merit (Figure 7.1), having been pipped to the top spot by Arthur Berry.[7] The following year, he was one of six students to be placed in Division 1 of Part III of the Tripos, and a year later he outdid Berry to win the First Smith's Prize.[8]

Having been elected a Fellow of his College in 1886, Love became a college lecturer in 1888. In 1898 he was appointed to one of the five newly founded University Lectureships in

MR. E. H. LOVE
(St. John's College)
Second Wrangler

Figure 7.1 Augustus Love, Second Wrangler

Mathematics, giving courses on elasticity, wave motion, and optics. Meanwhile, he was also engaged in coaching, his most famous student being G. H. Hardy, who, in his *Mathematician's Apology*, wrote appreciatively of his tutor:[9]

> My eyes were first opened by Professor Love, who taught me for a few terms and gave me my first serious conception of analysis. But the great debt that I owe him—he was, after all, primarily an applied mathematician—was his advice to read Jordan's famous *Cours d'analyse*.

Several of Love's other students gained high places in the Tripos, the most notable of whom was Charles Godfrey who was placed fourth in 1895, and who made a successful career as a textbook author and reformer of mathematics education.[10]

Oxford and teaching

Bartholomew Price resigned from the Sedleian Chair in the summer of 1898, having been in post for 45 years (see Chapter 6). Love was appointed not long afterwards, and arrived in Oxford in 1899. It seems Love's decision to move to Oxford had not been an easy one, but with little immediate prospect of being able to further his career at Cambridge, he took the plunge.[11]

On taking up the professorship, Love was made a member of the Senior Common Room at The Queen's College where, on his first visit, a memorable exchange took place, or at least so the story goes:[12] a stranger to the assembled company, Love introduced himself: 'I'm Love.' 'Ah' said one of the fellows '*Erως* or *Agaρη*?'[13] History does not relate what he replied. This was not the only occasion on which Love's name gave rise to amusement. In December 1898, a Melbourne newspaper, possibly due to the presence of Love's brother at the local university, published the following item:[14]

> Mr AEH Love of Cambridge has written a very learned book on a very learned subject, it being a mathematical treatise on elasticity. A few days ago an undergrad dropped in on one of our leading booksellers and airily enquired whether they had a copy of 'Love's Elasticity'. The shopman eyed him doubtfully for a few moments, and probably imagining him to be a pupil of the Ormond professor of music, asked doubtfully, 'Er—do you happen to remember the author's name?'

In common with his predecessor Price, Love lectured in Oxford continuously for over forty years,[15] beginning and ending with courses on gravitational attraction and potential theory. In the intervening years he covered a wide range of topics in applied mathematics, from the well-established analytical dynamics, electricity and magnetism, waves and sound, to the more modern (and more advanced) relativity theory and tensor analysis.[16] Love was recognized as a great lecturer with a talent for seeing things from the student point of view.[17] The distinguished statistician David Kendall, who graduated from Oxford in 1939, considered Love to be 'the best mathematician lecturer in Oxford'.[18] As one obituarist put it, Love had 'a happy way of clearing up a difficult point with an illuminating phrase'.[19]

Love was the author of two textbooks, the first of which, *Theoretical Mechanics. An Introductory Treatise on the Principles of Dynamics*, appeared while he was at Cambridge.[20] Gilbert Walker, who reviewed it for the *Mathematical Gazette*, considered it to be 'Among the most interesting textbooks that have appeared in recent years [. . .] conspicuous for the thoroughness of its treatment of fundamental principles as well as for the excellence of its style'.[21] As well as examples, it contained numerous problems—Walker counted over 660—which would have made the book attractive for those involved with the Mathematical Tripos, student and teacher alike. A second much revised edition appeared in 1906. In mathematical content it was not dissimilar from the first, but it was easier to navigate, being much better laid out, with fewer digressions and fewer problems, and as such was probably much more appealing to Love's Oxford students than its predecessor. A third edition followed in 1921 but it was little changed, apart from a further reduction in the number of problems, although, at 30 shillings, it was deemed rather expensive.[22]

For his second textbook, which appeared in 1909, Love turned his hand to the calculus, a subject he had been teaching to Oxford chemistry and engineering students for several years. In *Elements of the Differential and Integral Calculus*, his objective was to make the subject as widely accessible as possible, for as he said, 'The principles of the Differential and Integral Calculus ought to be counted as part of the intellectual heritage of every educated man and woman in the twentieth century no less than Copernicus or Darwin'.[23] It was certainly a laudable aim, if somewhat unrealistic. Nevertheless, the book hit the mark with one reviewer who thought 'Nothing could be simpler than the exposition, and the arrangement is all that can be desired. In a word, it is admirably done'.[24]

Research

Love's research publications span almost his entire career, with the first of his fifty-seven papers appearing in 1887 and the last in 1939.[25] They were mostly theoretical, but some had obvious applications such as his paper on the collapse of boiler flues.[26] From early on he was drawn towards the subjects of hydrodynamics and elasticity, choosing hydrodynamics for his first publication which appeared in 1887,[27] and elasticity for his Smith's prize essay, a version of which was later published.[28] The former, which was on English research in vortex motion, had been commissioned by Felix Klein for *Mathematische Annalen*, and showed a broad knowledge of work in Germany as well as in England.[29] Klein also hinted to Love that he would like a comparable article on elasticity. But that was not to be. Not only had Love already been invited by Webb to collaborate on a treatise on the subject, but, as Love told Klein, much of the most important recent research had been done by continental, rather than English, mathematicians.[30] Later, Klein did succeed in commissioning two articles from Love, but these were on hydrodynamics and were for his famous *Encyklopädie*. They appeared in 1901.[31]

Some five years after the approach from Klein, the first volume of the treatise on elasticity appeared with Love as the sole author, Webb having handed him the project. Now considered Love's major work, *A Treatise on the Mathematical Theory of Elasticity* was originally published in two volumes in 1892 and 1893 (Figure 7.2). A second edition, substantially altered from the first—Love described it in the preface as 'a new book'—was published as a single volume in 1906, with a third edition in 1920, and a fourth in 1927 (reprinted in 2013). Each new edition contained results obtained since the previous one, all carefully referenced so that it provided an accurate picture of the current state of research, a quality for which the book was widely praised.[32] Containing much original work by Love, it was designed to appeal to physicists, engineers, and mathematicians alike.[33] Although appropriate for students,[34] unusually for Cambridge-produced texts of the time, it didn't contain examples but instead results were

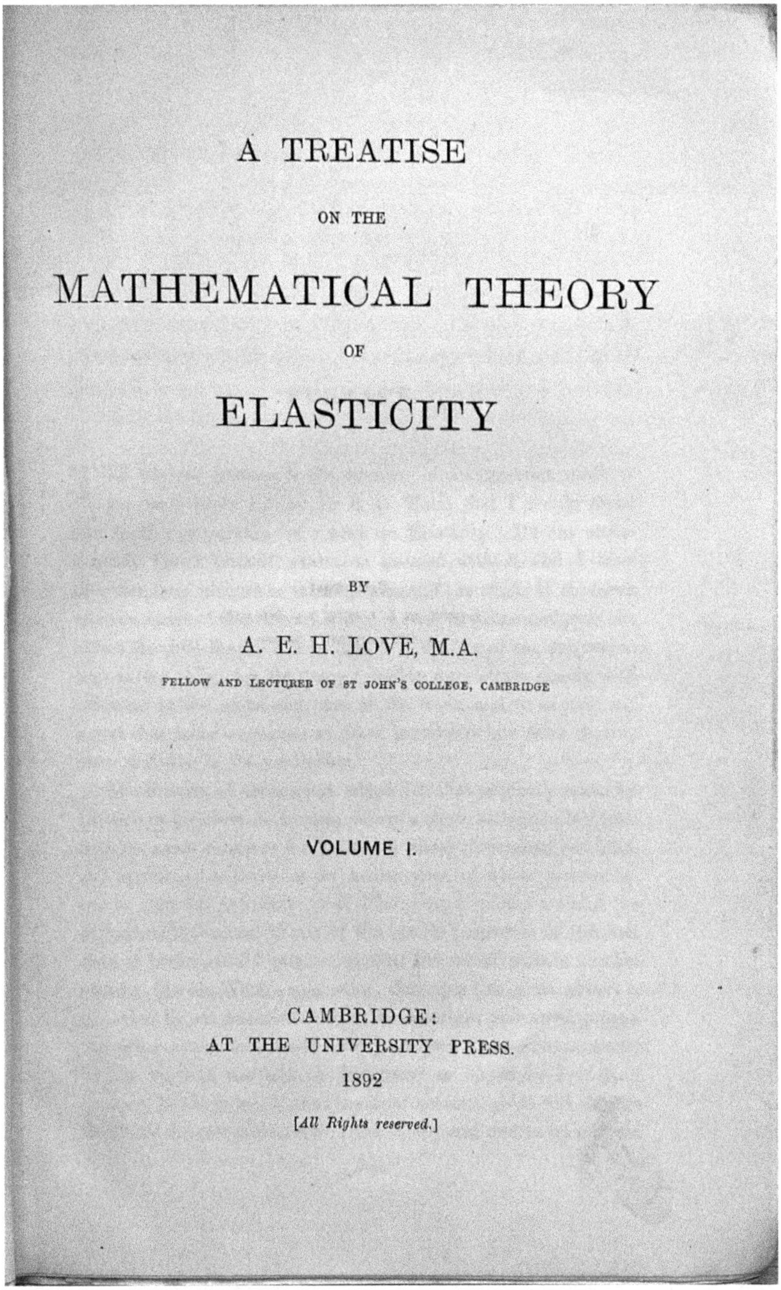

A TREATISE

ON THE

MATHEMATICAL THEORY

OF

ELASTICITY

BY

A. E. H. LOVE, M.A.

FELLOW AND LECTURER OF ST JOHN'S COLLEGE, CAMBRIDGE

VOLUME I.

CAMBRIDGE:
AT THE UNIVERSITY PRESS.
1892

[All Rights reserved.]

Figure 7.2 Augustus Love, *Elasticity*, volume I (1892).

stated without proof and students encouraged to work them out for themselves—Love didn't want students to waste time 'problem-grinding' as he called it.[35] The book was an immediate success—George Greenhill, characteristically employing a military turn of phrase,[36] described it as 'an elegant and modern artillery of analysis',[37] while another reviewer commended its 'rigour free from pedantry'[38]—and it soon became established as a classic in the field of mathematical physics, alongside Horace Lamb's book on hydrodynamics.[39] As an American reviewer remarked of the fourth edition:[40]

> Those of us who are interested in mechanics have cause to be very thankful that Love's *Elasticity* and Lamb's *Hydrodynamics* are kept in print and kept up to date. There is not an equivalent of either in any other language.

Certainly in Germany there was no need of equivalents, since, on the instigation of Klein, German translations of both books had long since appeared.[41] Discussions between Klein and Love about the translation of *Elasticity* had begun in 1900, with Ernst Zermelo named as the translator.[42] But when the translation was eventually published in 1907 (Figure 7.3), it was a translation of the second edition, and the translator was one of Klein's assistants and a former doctoral student, Aloys Timpe.[43] Love helped Timpe with the translation, and in many places the 'richer' (*reicheren*) English terminology was kept in preference to the 'clumsy' (*schwerfällig*) German terms.[44] To allow for students to use the English and German versions simultaneously, the German text was kept as close as possible to the English.[45]

Love's reputation as Britain's leading exponent on elasticity was cemented further in 1909 when the Royal Society awarded him one of its most distinguished prizes, a Royal Medal, for his work in the field. Elasticity was the subject that above all defined him as mathematician. Even after death he was still referred to as 'the elastic theory man'.[46] In 1911, he combined his broad knowledge of the subject and his skill as a communicator in an extensive article for the eleventh edition of the *Encyclopaedia Britannica*.[47]

Love's work in elasticity covered a wide range from the equilibrium of beams and plates of different shapes to the theory of vibrations in many difficult cases, as well as applications to problems connected with the shape of the Earth. Questions concerning the latter became increasingly central to his research and in 1907 they were the focus of his Presidential Address to Section A of the British Association for the Advancement of Science (BAAS) in Leicester (Figure 7.4).[48] Two years later, when the BAAS met in Winnipeg, Canada, his opening of the 'Discussion on *Earth Tides*' brought to a wider public his recent work which contained his introduction of what are now known as Love numbers, dimensionless parameters which characterize the elastic response of the Earth to the tides, and which are today key numbers in tidal theory.[49]

When the 1910 Adams Prize of the University of Cambridge called for 'Some investigation connected with the physical constitution or motion of the earth',[50] so well did the topic fit in with Love's research, one could almost imagine it had been chosen with him in mind. His winning essay, *Some Problems of Geodynamics*, appeared in book form the following year.[51] Basing his investigation on the hypothesis of isostasy,[52] he employed elaborate harmonic analysis to tackle the various problems, leading the reviewer in the *Bulletin of the American Mathematical Society* to aptly describe it as 'a mass of intricate analysis interspersed with very readable comment'.[53] Perhaps surprisingly, the book was also reviewed in *The Athenaeum*, a literary journal, with the anonymous reviewer, who concentrated his review on the 'readable comment', fittingly concluding that it would be 'a revelation to any reader who may be inclined to think that there is no more room for mathematical research of the first order', i.e., pointing out to the uninitiated observer that what might seem to be abstruse mathematics for its own sake can lead to important practical results.[54] As well as the fine-tuning of his earlier research, the book also contained what is considered Love's most important discovery and now named for

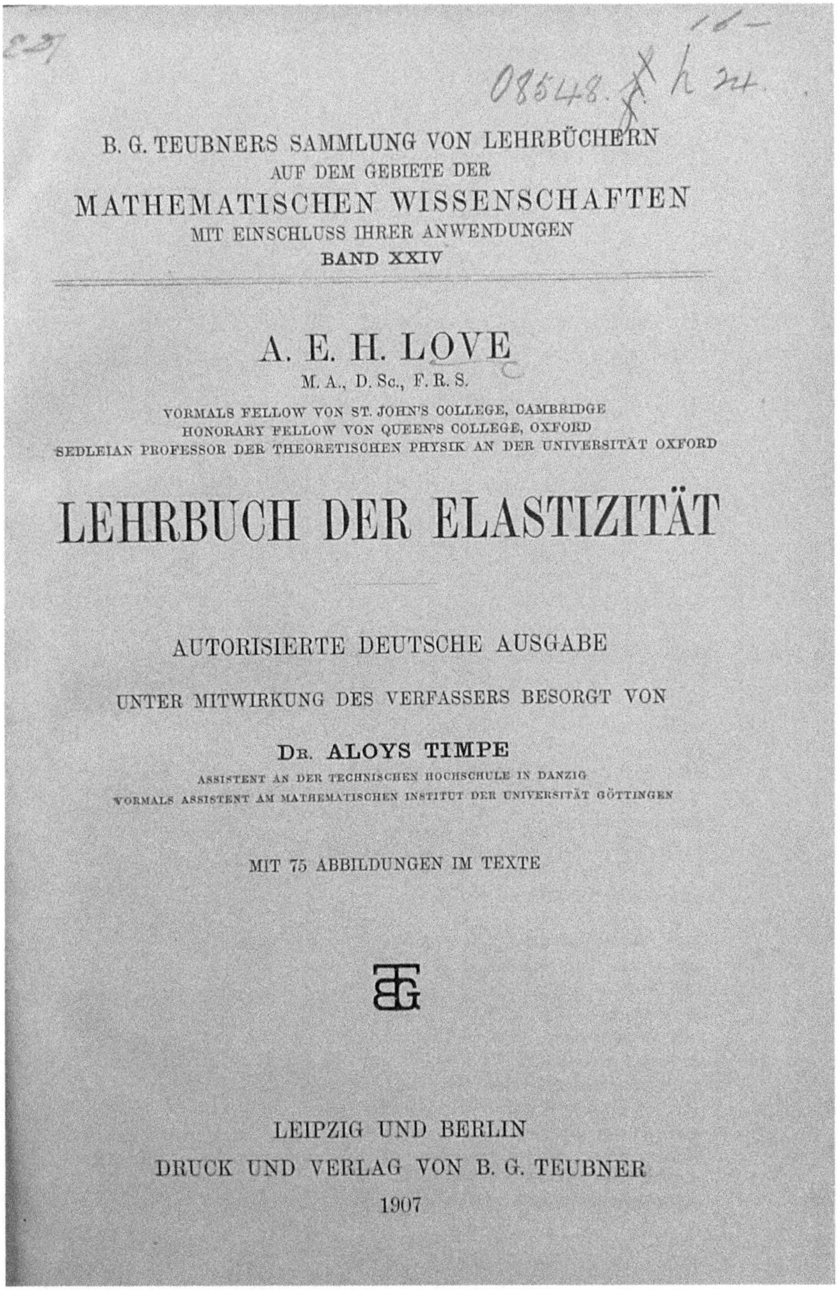
B. G. TEUBNERS SAMMLUNG VON LEHRBÜCHERN
AUF DEM GEBIETE DER
MATHEMATISCHEN WISSENSCHAFTEN
MIT EINSCHLUSS IHRER ANWENDUNGEN
BAND XXIV

A. E. H. LOVE
M. A., D. Sc., F. R. S.
VORMALS FELLOW VON ST. JOHN'S COLLEGE, CAMBRIDGE
HONORARY FELLOW VON QUEEN'S COLLEGE, OXFORD
SEDLEIAN PROFESSOR DER THEORETISCHEN PHYSIK AN DER UNIVERSITÄT OXFORD

LEHRBUCH DER ELASTIZITÄT

AUTORISIERTE DEUTSCHE AUSGABE
UNTER MITWIRKUNG DES VERFASSERS BESORGT VON
DR. ALOYS TIMPE
ASSISTENT AN DER TECHNISCHEN HOCHSCHULE IN DANZIG
VORMALS ASSISTENT AM MATHEMATISCHEN INSTITUT DER UNIVERSITÄT GÖTTINGEN

MIT 75 ABBILDUNGEN IM TEXTE

LEIPZIG UND BERLIN
DRUCK UND VERLAG VON B. G. TEUBNER
1907

Figure 7.3 Augustus Love, *Lehrbuch der Elastizität* (1907).

him: 'Love waves', which are fundamental in seismology.[55] Prior to Love's discovery, the theory of waves in an elastic medium without a boundary had been worked out by Siméon Denis Poisson (1830) and George Gabriel Stokes (1849). They had shown that given a disturbance, the wave motion will separate itself out into two types: faster compressional longitudinal waves

Figure 7.4 Augustus Love.

and slower distortional transverse waves. When a boundary is present, the situation becomes more complicated because each type of wave can then give rise to repeated reflections of both types of waves, and this makes it difficult to analyse what happens in the neighbourhood of a boundary. In 1885 Lord Rayleigh mathematically predicted the existence of a type of surface wave in which each particle moves in an ellipse against the direction of propagation (in contrast to surface waves in water in which each particle makes a circular motion in the direction of propagation), and conjectured that they might play an important part in earthquakes (Figure 7.5).[56] However, when surface seismic waves were first detected in the studies of earthquake records in the early 1900s, it was found that they had features which were not consistent with the characteristic features of Rayleigh waves. Love postulated that these inconsistencies were due to another type of surface wave which propagates as a result of the difference in density between the Earth's crust and its underlying mantle. These waves vibrate horizontally and transversely to the direction of motion (the amplitude decreasing with depth), and account for the horizontal shifting of the Earth during an earthquake. Love waves travel faster than Rayleigh waves and cause the greatest structural damage outside the epicentre of an earthquake (Figure 7.6), and are fundamental in the understanding of plate tectonics.

In one of his last papers on elasticity, Love successfully attacked the problem of finding a solution to the biharmonic equation subject to boundary conditions over the perimeter of a rectangle, thereby solving the two-dimensional form of a problem described by the French mathematician Gabriel Lamé as 'le plus difficile peut-être de la théorie de l'élasticité'.[57] As Love described in a lecture to the London Mathematical Society, it was a result that could be used in practical applications such as in determining the thickness of glass for a shop window able to withstand a certain pressure of wind.[58]

During the First World War, Love remained in Oxford but diverted time towards research in ballistics, developing a small arcs method for calculating high-angle trajectories, work which was needed for the creation of range tables for anti-aircraft gunnery.[59] It seems very likely that he was drawn into this work by the Cambridge geometer George Richmond, who was Love's exact contemporary at Cambridge and was one place below him in the Mathematical Tripos, and who during the war was one of the leaders of the Anti-Aircraft Experimental Section of the Ministry of Munitions.[60] After the war, Love continued his work on ballistics, collaborating with his Oxford colleague Frederick Pidduck on Lagrange's internal ballistic problem, a problem concerning the interaction between the propellant gases and the projectile inside a gun barrel which had been first published by Poisson from the manuscripts of Lagrange.[61]

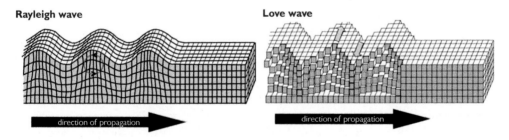

Figure 7.5 Rayleigh wave and Love wave.

Figure 7.6 Twisted railway lines in the aftermath of an earthquake.

During the war, Love also had the opportunity to interact with scientists at the Royal Aircraft Factory at Farnborough who were concerned with the development of the fledgling science of aeronautics. On one occasion, the physicist Frederick Lindemann, later Lord Cherwell, a courageous pilot and renowned as a calculator, was proposing to test a fender-like device for pushing aside barrage balloon cables by flying an aircraft into just such a cable, having worked out that the maximum force the cable would exert on the device, which was fixed to the front of the aircraft, was 120lbs. The physicist Reginald Jones, to whom Lindemann had told the story, recounted what happened next:[62]

> Just as he [Lindemann] was about to make the test a telegram arrived to say that he was on no account to do it because his calculation had been shown to the distinguished applied mathematician, Professor A.E.H. Love, and Love considered that he had ignored some second-order terms that could be very important. He must therefore delay the test until Love had completed a rigorous calculation. After some weeks, Love came up with his answer—121 pounds.

Although in the end Lindemann's estimate for the force differed little from Love's, the incident highlights the esteem in which Love was held across the scientific spectrum.

Love also made contributions to classical electrodynamics, as well as authoring several sensitively written obituaries of colleagues. He was well known for his versatility as a mathematician, albeit one who 'rejoiced in algebra rather than geometry',[63] and his explanatory power and proficiency in pure mathematics can be seen in his articles in the *Encyclopaedia Britannica*.[64] In his final paper, published in the year before he died, he tackled a case of Boussinesq's problem.[65] The name of the problem derives from the work of the French mathematician and physicist Joseph Boussinesq, who in 1885 was the first to consider the problem of the indentation of the plane surface of a semi-infinite elastic solid by pressure applied normally to the plane boundary. The particular cases successfully addressed by Boussinesq are of practical importance in the construction of roads and pavements, and in connection with Boussinesq's solutions, Love himself made reference to a recent report of the Road Research Board. In the case treated by Love, the elastic solid is temporarily deformed by the pressure against it of a perfectly rigid right circular cone with its axis normal to the indented plane and its vertex penetrating the region originally occupied by the solid. It is of importance in various areas of applied mechanics, including in soil mechanics where the cone is the conical head of a cylindrical pillar, and the solid is the soil on which the pillar rests. Love's success in solving the problem rested on the fact that he had the skill to guess a combination of potentials which satisfies the boundary conditions.[66] The success of his investigation, which involved clever analysis, shows that even towards the end of his life, his analytical powers remained undimmed.

The careful attention to detail which characterizes Love's work, combined with the modesty Love displayed in connection with his own work, is captured by the applied mathematician Richard Southwell in his review of a paper on elasticity Love submitted to the Royal Society in March 1929. Southwell recommended that the paper be published in full, adding the remarks [67]

> I have [no material modifications] to suggest, except that references should be given to the author's 'Theory of Elasticity': he seems to have hesitated to do this anywhere.
>
> The paper seems to me to be a valuable discussion on an important class of problems. I have had no time to verify the details of the analysis, but it is not in the habit of Prof. Love to make mistakes.

Love in the community

Love had been in Oxford for three years when, in September 1902, he was invited to Kristiania (now Oslo) to be the university's representative at the four-day meeting to mark the centenary of the birth of the mathematician Niels Henrik Abel. It was a glittering affair, designed as a national Norwegian showcase, and attended by the King of Sweden and Norway. Fridtjof Nansen, the Arctic explorer, was president of the reception committee, and among the attendees were the writer Henrik Ibsen and the composer Edvard Grieg. Some 80 foreign mathematicians attended the celebrations. Among the other British delegates were A. R. Forsyth, Sadleirian Professor of Mathematics at Cambridge (who received an Honorary Doctorate, and who on behalf of the English delegation gave a speech celebrating Abel), George Greenhill, Professor of Mathematics at the Royal Military Academy at Woolwich, and E. W. Hobson, a lecturer at Cambridge. Love clearly relished the occasion, writing to his friend Joseph Larmor at Cambridge:[68]

> The Christiania [the name for Kristiania until 1897] people did us very well & I should have had a thoroughly enjoyable holiday if I had not had the misfortune to crush my left thumb in a cabin door on the steamer coming home.
>
> I learnt from the *Times* that Forsyth flushed with his Christiania triumph, which was genuine, rushed to Belfast to set the world right in the matter of teaching mathematics.[69]

Love enjoyed travel, and whether it was on this occasion, or more probably on a later one, he often recounted the story of driving across Norway together with Hobson, the two entertaining each other by singing.[70]

In August 1912 the International Congress of Mathematicians (ICM) came to Cambridge on the invitation of the Cambridge Philosophical Society. It was the fifth ICM to be held and the largest to that date, with 574 full members and a total of 708 attending. Love, together with Hobson, who in 1910 had succeeded Forsyth as Sadleirian Professor, were the joint Secretaries for the Congress. They were responsible for the local organization which was no small task—the President of the Congress, G. H. Darwin, estimated it at no less than 16 hours a day each.[71] Love and Hobson were also responsible for editing the two volumes of the Congress *Proceedings*. But Love's role did not stop there. He was one of the organizers of Section III of the Congress (Mechanics, Physical Mathematics, Astronomy), he gave a talk on tidal theory, he organized a day trip to Oxford,[72] and he actively participated in the discussion of Carl Runge's report on the teaching of mathematics to physicists.[73] Regarding the latter, he made his own views very clear: in an ideal world, every mathematician would be a physicist and vice versa. Science students should not be limited to the useful parts of mathematics, for that, he said, would destroy the logical unity of the subject and relegate it to a subservient position not in line with its importance.[74]

Love was a member of several scientific organizations and played an active role in scientific life in Britain, beginning with his election to the Council of the London Mathematical Society (LMS) in 1890, three years after he had been elected to the Society's membership. He served continuously on the LMS Council until 1920, and again from 1922 to 1925, and for fifteen of those years, from 1895 to 1910, he was an Honorary Secretary, and he twice served as Vice-President. One of his roles as Honorary Secretary was to find reviewers for articles submitted to the *LMS Proceedings*, an undertaking which could prove quite challenging, and Love would often consult with his friend Larmor (who was the Society's Treasurer for much of the time Love was in office), such as he did on an occasion in 1909:[75]

> Many thanks for Hassé's paper & your report. I knew it was too bad to trouble you but these things are difficult to get through or get rid of. Niven wants to be sure of the details before making up his mind, Macdonald keeps a paper for six weeks in spite of reminders, Lamb won't look at this stuff, and we have no confidence in Bateman or Cunningham.

Love was President of the LMS for the customary two years from 1912, his term ending during the early months of the war. For his Presidential address, he discussed the question of value and importance in mathematical research, or, as he put it, 'Wherein lies the difference between valuable research and laborious trifling?'[76] In answer, he proposed four qualities which characterize valuable research: novelty or creativity, relevancy, definiteness, and generality (not to be confused with vagueness). In the particular case of mathematical physics, he added a fifth: adherence to fact.

Gratifyingly for historians of mathematics, he made a plea for their subject:[77]

> I would plead for more attention to the history of mathematics. Towards elucidating this history Great Britain has not done very much; the study of it receives here but little encouragement. Yet it seems to me to be extremely desirable, if not actually indispensable, for entering into the heritage that has been bequeathed to us, and for seeking to enhance its value.

Love's interest in history of mathematics is also evident in his correspondence with Larmor. He was a subscriber to the collected works of Christiaan Huygens which had begun publication in

1888 with a volume of the correspondence and which, by 1901, had reached the ninth volume but it was still on correspondence. Publication of Huygens' actual works was yet to begin. This was a model which did not appeal to Love, as he made plain to Larmor when discussing a proposed edition of Newton's works:[78]

> I think it would be a great misfortune to begin an edition of Newton with the letters. People get sick of them, at any rate if they run to anything like the same length as Huygens's did. The Huygens' Committee are printing a very large number letters to Huygens as well as those which he wrote, and in any case where letters are published it seems necessary to have those both sides. Sampson seems to have the instinct of the historian and biographer and I think that he will make an excellent editor but I should have preferred a good translation of the more important works such as the *Principia* and the *Curvae tertii ordinis* to a collection of letters.

Love was elected to the Royal Society in 1894 at the young age of 31, with a distinguished cast of proposers including Arthur Cayley, George Howard Darwin, Andrew Russell Forsyth, George Greenhill, and James Joseph Thomson, all of whom knew him personally (Figure 7.7). Among the society's most assiduous Fellows, he was elected to the council in 1902, was a member of many committees, including the Royal Society Catalogue of Scientific Papers Committee and the Stokes Memorial Committee, exhibited at the society soirées, and was often called upon to represent the society, even at such events as the 1902 Coronation Naval Review.

As chair of the Royal Society's Mathematics Committee, Love was heavily involved in the reviewing process of mathematical papers submitted to the society, acting as a referee himself for over thirty years. It was a task he undertook conscientiously, often writing long reports, especially on papers he felt should be rejected. For example, in the case of a paper on torsion by Karl Pearson, submitted in the summer of 1900 to the *Philosophical Transactions*, to which he gave a decision of 'I think not', he included over two closely written pages of detailed remarks.[79] Having first observed there was nothing new in the methods employed in the paper, he made it clear that although he personally was not impressed by the results, he was prepared to concede to others more knowledgeable than himself that they may have value. But even so, he considered Pearson's 'prolixity', of which he gave several examples, made the paper unpublishable. In short, he said, should the paper be thought to have merit then it should be substantially condensed, which in effect meant it would have to be rewritten. Love's verdict won the day, and Pearson had to make do with a two-page abstract of his paper being published in the *Proceedings of the Royal Society* the following year.[80]

In the case of a paper by George Hartley Bryan and William Ellis Williams on the stability of flight, submitted in June 1903, just few months before the Wright Brothers made their historic flight, Love was positive but again critical of the paper's length—'too much "penny a line" padding'—noting that, for example, readers of the *Philosophical Transactions* do not need to be told 'that an equation of the fourth degree "is a called a biquadratic equation"'.[81] This time he recommended publication, in the *Philosophical Transactions* or, if 'properly condensed' in the *Proceedings*, adding the prescient remark that the conclusions 'may prove to be important in the development of flying machines'.[82] The paper, despite being published after the Wright Brothers' flight, was virtually ignored when it appeared, but a few years later Bryan's ideas achieved wide recognition with the publication of his 1911 book *Stability in Aviation*, in confirmation of Love's judgement.[83]

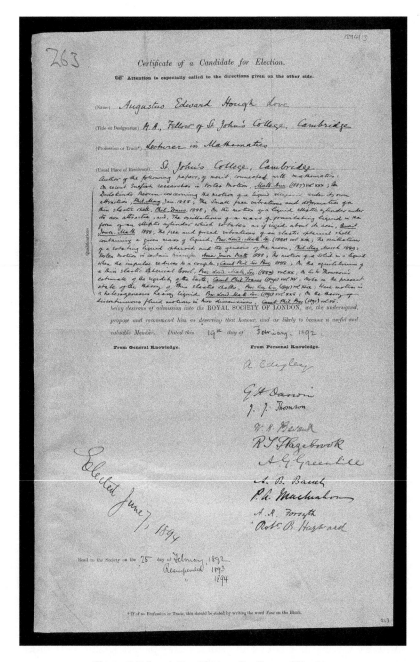

Figure 7.7 Love's Royal Society Certificate of Election.

In 1937 the Royal Society awarded Love its most prestigious prize in the field of mathematics, the Sylvester Medal, in recognition of his sustained contributions to elasticity and hydrodynamics.[84]

Another arena in which Love increasingly made his presence felt was the British Association for the Advancement of Science, having first given a paper at the annual meeting as early as 1888 when the meeting was held in Bath.[85] As well as being the President of Section A in 1907, he also served continuously on different BAAS committees between 1902 and 1931, many of the committees running for several years. These included committees on the Teaching of Mathematics, Engineering Research, Calculation of Mathematical Tables, Seismological Observations, Geophysical Discussions, and the Determination of Gravity at Sea Committee (which he also chaired). The remits of these Committees were all very different and the variety of their subject matter holds a mirror to the breadth of Love's mathematical and scientific knowledge, his willingness to serve demonstrating his dedication to the scientific community.

Unfortunately, no archive of Love's papers exists, but a glimmer of the humorous side of his nature can be seen through some of his correspondence with Larmor which is preserved in the St John's College Library archives.[86] In 1899, not long after his arrival in Oxford, he wrote to Larmor to inform him that[87]

> Stout is an awful warning; he is trying to do 'what is usual', otherwise he is unchanged and as you may expect his efforts in the direction of normality are not very successful. So far I have no reason to suppose that the malady from which he is suffering will presently attack me also, but doubtless there are 'dangers lurking unseen'. [. . .] Stout and Weldon are both coming over to celebrate their benefactors.

Quite what Love was getting at here with respect to the philosopher and psychologist George Stout, who had recently arrived in Oxford as a reader in mental philosophy, is somewhat enigmatic but there was clearly camaraderie between them. Both Stout and the evolutionary biologist Raphael Weldon, who had just been appointed to a chair in zoology at Oxford, had been Cambridge contemporaries of Love's and Larmor's, all four being fellows of St John's College.

Love enjoyed keeping up with the St John's gossip and a few years later, in 1907, he had heard some rather unexpected news about Charles Taylor, the Master of St John's:[88]

> I hope you are all pleased with the Master's engagement. Is 'better later than never' true? Or must a man as Foxwell said make the mistake of his life sometime or other? Did the Master omit to 'pack his portmanteau'?

Taylor had been Master of St John's since 1881, so was well known to both Love and Larmor. He married for the first time in October 1907 at the age of 68, and died suddenly the following July. The reference to Foxwell is a reference to another fellow of St John's, the economist Herbert Foxwell who held the chair of economics at University College London while maintaining a college lectureship at St John's.

Aside from music and travel, another of Love's extra-curricular interests was the game of croquet and he was a familiar figure in the Parks in Oxford where 'as regularly as the swallows brought the summer' he could be seen wielding a mallet 'with doughty energy and enjoyment'.[89] No doubt he could be readily identified by his remarkable moustache which was described by the astronomer Thomas Cowling, a student of Love's in the 1920s, as 'charmingly reminiscent of a frozen waterfall'.[90]

Conclusion

Although Love was initially hesitant at taking up the Sedleian Chair, his migration to Oxford was an unequivocal success as witnessed by both his mathematical productivity and service to the scientific community. After the move, he kept close ties with Cambridge, especially with his friends Larmor and Hobson, and it must have given him great pleasure when St John's College made him an Honorary Fellow in 1927, the same year he was elected to a fellowship at The Queen's College in Oxford.[91]

Love died in Oxford in 1940 after an operation. He was aged seventy-seven and had been active in his role as Sedleian Professor until very shortly before his death, having occupied the chair for twelve years beyond the normal retirement date. As his Oxford colleague E. A. Milne wrote,[92] 'Certainly, if compulsory retirement from formal University business is in general wise, the case of Love shows that it would be still wise to provide for exceptional relaxation of the rule'.

The sketch of Love (Figure 7.8) was drawn by one of his students, Kenneth Thornhill, during a lecture c.1938, when Love was in his mid-seventies. Thornhill, in recounting the episode, bears witness to Love's success as a teacher:[93]

> At first [Love] was somewhat affronted that anyone should do that sort of thing at his lectures; but when I explained that I was there for the second year so as not to miss anything, his eyes twinkled and his face glowed as it always appeared to do at any words of appreciation; and he immediately acknowledged, if only a little grudgingly, that the likeness was good enough to sign.

Love was remembered by his colleagues not only as a man of mathematical talent and versatility, and as an outstanding teacher, but also as a friend. Testament to his modesty, generosity, and kindliness is woven through the many tributes that were paid to him after his death.[94] He was, in Milne's words, indeed 'truly simple and so truly great'.[95]

Figure 7.8 Augustus Love drawn by C. K. Thornhill, c.1938, and signed by Love.

CHAPTER 8

Sydney Chapman

PETER CARGILL

Sydney Chapman was born on 29 January 1888 in Eccles, Lancashire (Figure 8.1). His early years are summarized in his obituaries,[1] and other aspects of his career can be found in the Further reading section for this chapter. He was admitted to the University of Manchester in 1904 to study engineering, graduating in 1907, but, finding mathematics more congenial, he was advised to go to Cambridge, gaining a scholarship from Trinity College in 1908, completing the mathematics degree in 1911 and being awarded a Doctor of Science in 1913.[2] He held appointments at the Royal Greenwich Observatory, Trinity College Cambridge, Professor of Mathematics at Manchester (1919), and Head of Mathematics at Imperial College (1924). He was appointed to the Sedleian Chair in 1946, retiring from Oxford in 1953: this appointment is covered in detail below. He subsequently held positions in the USA: at the University of Alaska in winter and at the High Altitude Observatory (HAO) in Boulder, Colorado in summer, where he died on 16 June 1970. He was elected a Fellow of the Royal Society in 1919 and awarded many prizes and honorary degrees. He is remembered by the Royal Astronomical Society (RAS) and American Geophysical Union (AGU) through their Chapman medal and Chapman conferences, respectively.

Chapman was regarded at the time of his death as one of the great figures of twentieth-century geophysics, an assessment that remains true.[3] His work on the kinetic theory of gases, the upper atmosphere, ionosphere, magnetosphere, and geomagnetic storms stand out. Another of his achievements was the initiation and implementation of International Geophysical Year (IGY) in 1957–8, which saw the birth of space exploration using satellites. All of these are discussed in this chapter. Finally, we will refer to the topic of magnetohydrodynamics (MHD) frequently. A short summary for the interested reader can be found here.[4]

Peter Cargill, *Sydney Chapman*. In: *Oxford's Sedleian Professors of Natural Philosophy*. Edited by: Christopher Hollings and Mark McCartney, Oxford University Press. © Oxford University Press (2023). DOI: 10.1093/oso/9780192843210.003.0008

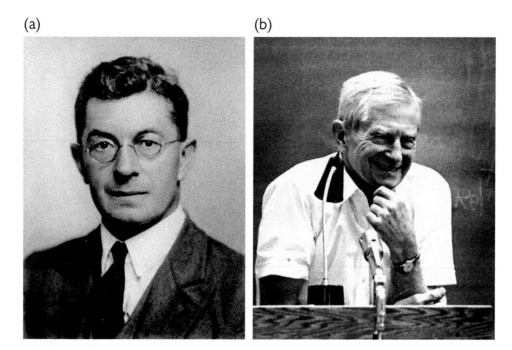

(a)   (b)

Figure 8.1 Sydney Chapman. (a) His formal photograph as President of the Royal Astronomical Society 1941–3. (b) At the University of Alaska, dated June 1967.

1911–46: Cambridge, Manchester, and Imperial

While in Cambridge, Chapman was offered a post by the Astronomer Royal ((Sir) Frank Dyson[5]) as a chief assistant at Royal Greenwich Observatory (RGO). Chapman then returned to Cambridge with a five-year Trinity fellowship, though also spending time at RGO. He received an exemption from conscription during the First World War due to his pacifist views. At RGO the Astronomer Royal had asked him to lead the construction of a new magnetic observatory and this was his introduction to a lifetime interest in geomagnetism.[6] There was also work on stellar observations, a series of papers on the kinetic theory of gases, and his initial model for geomagnetic storms as discussed later in this chapter.

Chapman's methodology was well established by 1920. He was a pioneer in the use of large data sets, as demonstrated by his early work on the solar and lunar aspects of geomagnetic activity.[7] But he was, at heart, a mathematician, reducing complex problems to ones that could be solved with the tools available at the time. Sometimes this led him up the wrong path, as in the models for the currents associated with geomagnetic activity. In these older papers, at times he wrote with great clarity, such as the 1931 Bakerian lecture on ozone, and his pioneering work on the magnetosphere, when he expressed the basic idea in a single page in *Nature*.[8] But other papers can be harder work to read, in part due to their length; one of the merits of his two volumes published with Bartels is that when topics were revisited, a greater clarity was evident.[9]

One should not underestimate the difficulties confronting Chapman and others. Computers as we know them were nonexistent, making both calculations and plotting difficult tasks.

Computational modelling in plasma physics and MHD did not arrive until the late 1960s, so everything involved some sort of approximation. In the late 1930s and 1940s the physics framework needed to understand geomagnetic storms did not exist. These were formidable handicaps and it is to Chapman's credit that he got quite a lot right. The data limitations were also severe, being entirely ground-based.[10]

S. I. Akasofu, who was a close collaborator with Chapman after 1960, wrote:

> Although Chapman had many deep insights into physical processes, he tended to become an applied mathematician when he encountered mathematical uniqueness issues. Mathematical rigor was his life and it was part of the reason which caused some friction with Alfvén.[11]

T. Gold also comments on Chapman's reactions to Alfvén's formalism for MHD:

> Many others, especially those with a puritanical mindset like Sydney Chapman rejected these (ideas) out of hand. I had many discussions with Chapman ... and he insisted that nothing short of a complete calculation could ever be used to solve a problem in MHD: as Chapman put it, 'You might be mislead terribly by such approximations.'[12]

Indeed we are still trying to do complete MHD calculations on very large computers with approximations that perhaps do not always bear scrutiny. Perhaps Chapman had a point!

After five years in Manchester, Chapman moved to London to become head of the Imperial College Mathematics Department. He was given the task of reviving the department from a low spell. Once this had been accomplished, people who had subsequent successful careers moved through the department, including George Temple (1901–92), who was Chapman's Sedleian successor at Oxford in 1953, (Sir) Bill McCrea (1904–99), a distinguished astronomer, and (Lord) Bill Penney (1909–91), leader of the UK atomic bomb project and subsequently rector of Imperial College. Chapman had also married in 1922 and established his family in Wimbledon, cycling to work on occasion: his bicycling exploits were legendary.[13]

In these years Chapman began his involvement with UK and international scientific organizations such as the International Union of Geodesy and Geophysics (IUGG). He led the UK delegation to the IUGG general assembly in Washington between 4 and 11 September 1939 and was able to maintain contact with his friend Julius Bartels until the end of the assembly.[14] Chapman was also involved with the RAS. He was nominated as RAS president by his predecessor, H. C. Plummer, in February 1941, and took up the role in April that year.[15] He was the first president in the twentieth century of whom it could be said unequivocally to represent geophysics. The war years were not the time to attempt major changes in the RAS (however much it may have needed them) but D. H. Sadler in Tayler's history of the RAS[16] draws attention to efforts in these years to formalize various procedures of council and the meetings of the fellows.[17] In December 1942 Chapman nominated his old friend E. A. Milne as his successor with Milne taking over in April 1943. As was customary, Chapman was then supposed to take on the role of a vice-president for a year, but he persistently missed council meetings, and resigned in July 1943. The council minutes show the cause of resignation as 'pressure of war work'. His earlier pacifism was diminished by the rise of Hitler but, beyond a mention in Cowling of incendiary bombs and 'military operational research', it is unclear what roles he took on.[18]

Theory of nonuniform gases

Before looking at Chapman's time at Oxford, we present in detail the major achievements of the first part of his career. The first concerns the kinetic theory of nonuniform gases, triggered by discussions with Sir Joseph Larmor, and which led to several lengthy papers.[19] Nonuniform here means that there are spatial gradients of, for example, the temperature and velocity. The basic theory of gases was developed in the second half of the nineteenth century, leading to the Boltzmann equation and the Maxwell–Boltzmann distribution function. The moments of the Boltzmann equation lead to the hydrodynamic (or MHD) equations. These hydrodynamic equations also involve terms that arise from an integration over the collision terms in the Boltzmann equation, leading to expressions for viscosity, thermal conduction, electrical resistivity, and thermal diffusion when more than one gas species is present, generally referred to as transport coefficients. The difficulty in solving the problem lies to a degree in what model is used for the molecules in the gas, and can lead to different scalings of the coefficients with temperature.

Chapman's work involved the study of a strongly collisional gas, as defined by the Knudsen number (K_n) being small.[20] He used solutions of the Boltzmann equation to obtain estimates for the transport coefficients, although (Sir) James Jeans, who was the first referee of Chapman's 1912 paper, felt that these results lacked general validity. This weakness was dealt with in a later paper to the eventual satisfaction of Jeans, who again acted as the referee. In parallel, David Enskog in Sweden undertook a different calculation that reached the same results: the expansion procedures are known as the Chapman–Enskog method.[21]

In the late 1920s, Chapman began to pull these ideas together with the view of writing a book and at this stage, T. G. Cowling appeared.[22] He was under the supervision of E. A. Milne at Oxford in 1928, but Milne approached Chapman for a problem that Cowling could work on. Chapman had earlier claimed there was a radial limitation to the solar magnetic field; however, on working through Chapman's paper, Cowling realized that his result was incorrect,[23] which impressed both Milne and Chapman and led to Cowling moving to Imperial College. Cowling's involvement with Chapman's proposed book began as follows:[24]

> Chapman then had a manuscript containing the skeleton of a book on gas theory which he wanted to see completed by a collaborator who had more time than he. . . . Chapman lent me his manuscript: when I expressed my pleasure at its novelty and power, he after due thought invited me to be the collaborator. I of course accepted. He largely left me a free hand, apart from drawing my attention to a number of developments that had occurred since his manuscript had been written. When I had largely completed my task in 1935, I had to wait over a year for him to go through the new manuscript and suggest improvements. Our joint book came out in 1939.

This is the famous Chapman and Cowling book *The Mathematical Theory of Non-uniform Gases*,[25] one of Chapman's great achievements. It is his best-cited piece of work: the updated 1970 edition currently averages around forty citations a year on the Astrophysics Data System, a testament to its longevity. The book uses Enskog's approach to calculate the transport coefficients rather than Chapman's. Chapman and Cowling in their Introduction freely acknowledge the greater clarity of Enskog's methods, a generous gesture from someone of Chapman's standing, in effect saving Enskog's work.[26] Cowling also benefitted from this collaboration, subsequently writing important papers concerning electrical conductivity in the presence of a magnetic field.

Ozone and Chapman layers

Chapman's work on the formation of ozone is well described in his 1931 Royal Society Bakerian lecture.[27] Making use of recent data concerning the presence of ozone and higher ionized layers (Chapman layers), he proposed a series of reactions that led to the formation of ozone, confined to a layer a few tens of kilometres above the Earth's surface (Figure 8.2).

Considering solar UV radiation with wavelengths around and longer than the then-observable lower limit of 290 nm, Chapman argued that the incident radiation led to a series of reactions involving the creation and destruction of ozone level (right-hand column of Figure 8.2) that led to an equilibrium estimate of its level. Here M is a catalyst such as a molecule. The first reaction uses photons with wavelengths < 240 nm to create oxygen radicals from an oxygen molecule, the second leads to the formation of ozone, the third uses longer wavelength photons (320 nm) to destroy ozone while the fourth also removes ozone and creates molecular oxygen.

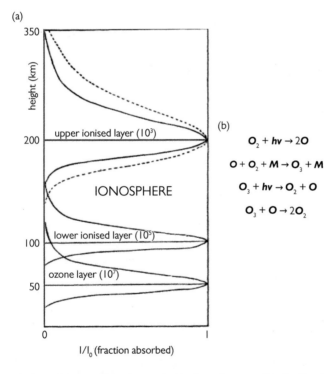

Figure 8.2 (a) The fraction of incident UV radiation absorbed as a function of height. The three different ionized (Chapman) layers are shown and correspond to different wavelength bands. The lowest is where ozone is formed. N_0 is the maximum rate of dissociation of molecules per cm³ per sec. in each layer. (b) The reactions involved. hv is a UV photon and M a catalyst.

The modelling has been updated in the past decades, as well as the ozone layer being lower in the atmosphere than Chapman proposed, yet the basic ideas remain. The upper two layers correspond very roughly to the present E and F layers of the ionosphere, and Chapman noted that they would need to be formed by different radiation from the ozone layer. The relevant wavelengths are in the extreme ultraviolet (EUV) and Chapman outlined some possible processes due to EUV radiation in his paper, though today's picture is more complex.

Chapman and geomagnetic storms

Geomagnetic activity in the form of the aurora has been known for millennia while in the nineteenth century the ability to measure perturbations to the Earth's magnetic field led to the idea that currents would flow in the very upper atmosphere (the ionosphere). Modern studies began on 1 September 1859 when, at 11:15 GMT, Richard Carrington and Richard Hodgson observed a large solar flare. A few minutes later, magnetometers at Kew Observatory recorded a modest change.[28] Eighteen hours after the flare, a large disturbance in the Earth's magnetic field was recorded. Now known as 'the Carrington Event', this geomagnetic storm was the first recorded event that demonstrated the link between solar and geomagnetic activity.[29] The initial magnetic field change on September 1 was due to the abrupt ionization of the very upper atmosphere, leading to localized electric currents. The arrival of the interplanetary disturbance on September 2 produced first a magnetic field enhancement and subsequently a deep depression of order 400 nT.[30] The average velocity between Sun and Earth of whatever caused the main storm phase was in excess of 2000 km/s. The direction of the currents associated with the field disturbances can be deduced from Ampere's law. The initial enhancement on September 2 is due to a current flowing in the west/east direction and the depression due a current in the opposite direction. Seen from above the poles, these currents are counter-clockwise and clockwise, respectively, and are indicated in a modern sketch (right panel of Figure 8.6c) as the magnetopause and ring currents.

A modern example of a major geomagnetic storm is shown in Figure 8.3, the cause being the 'Bastille Day' flare of 14 July 2000. Figure 8.3a shows the Sun's corona (the hot, 1 MK outer atmosphere) as seen in the extreme ultraviolet with the flare at disc centre. Figure 8.3b shows an artificial eclipse created by a coronagraph on the SOHO spacecraft. The field of view extends to 30 R_S (2.1×10^7 km, where R_S is the solar radius). Figure 8.3c shows a spatial average of the change in the surface magnetic field, as defined by the storm (D_{st}) magnetic index.[31] The eruption began with a major flare at 10:24, which in turn expelled part of the corona into interplanetary space. This is termed a coronal mass ejection (CME) and is the faint halo in the centre of Figure 8.3b: the 'snowstorm' is due to mildly relativistic particles accelerated in the eruption hitting the instrument detectors. The CME is the agency by which the solar disturbance is communicated to the Earth, as would have been the case in the Carrington event.

In the early twentieth century, the Norwegian scientist Kristian Birkeland (1867–1917) made major contributions to geomagnetism with work published in the Proceedings of Norwegian Polar Expeditions. He proposed that the aurora and associated surface magnetic field changes were due to electric currents in the upper atmosphere. To address the origin of such currents, he conducted an experiment where a beam of electrons was fired at a terrella,[32] and

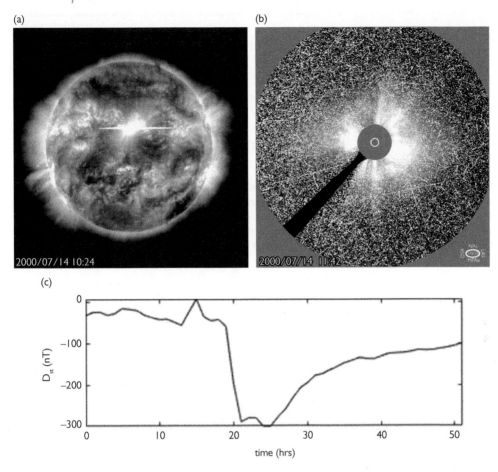

(a)

2000/07/14 10:24

(b)

2000/07/14 11:42

(c)

Figure 8.3 The Bastille day flare and associated geomagnetic storm. (a) and (b) show, respectively, the Sun in EUV with the flare in the centre and a coronagraph image showing the CME (faint halo). (c) shows the disturbance to the Earth's magnetic field, as measured by the D_{st} index, that begins 30 hours after the CME initiation. Data from July 15 and 16 is shown.

it was found that they concentrated in regions around the poles. He then extended his experimental result to suggest that a stream of solely electrons emanating from the Sun caused the aurora.[33] Such a stream cannot exist due to electrostatic repulsion, but Birkeland revised his model to have a neutral stream (a roughly equal number of electrons and protons). He suggested that a large-scale circuit would be set up, with electrons streaming into the auroral region along the terrestrial magnetic field, and closing through a cross-field current in the ionosphere (the Birkeland current, see Figure 8.4a). Thus, by 1910 a picture of geomagnetic activity being associated with solar activity and having some sort of large-scale current flow in near-Earth space had been proposed. Birkeland's early death led some to see his work as being 'forgotten': perhaps it is better to say that it did not receive the credit it merited.

We mentioned earlier Chapman's initial interest in geomagnetic activity with studies of solar and lunar diurnal variations attributed to atmospheric tidal-induced ionospheric currents. Involving the study of large data sets, this interested him for most of his life. However, in

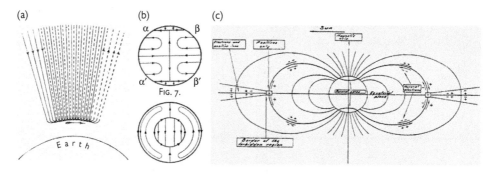

Figure 8.4 Three possibilities of storm-associated electric currents referred to in the text. (a) Birkeland's scenario: currents flow Earthward (solid lines) and outward (dashed lines) along the Earth's magnetic field. The currents close across the magnetic field in the ionosphere (lower arrow). (b) An example of Chapman's current scenario: the upper and lower panels are the current system as viewed from the Sun and the poles, respectively. (c) Alfvén's model including field-aligned currents indicated on the right.

1918 he proposed a theory of larger (geo)magnetic storms linked to major solar activity, which he argued were separate from the tidal-induced variations.[34] Like Birkeland, an important aspect was the idea of a stream (or beam or cloud) of solar particles being responsible for the storm. Also like Birkeland, Chapman proposed that the solar stream must involve charged particles of the same sign, arguing that otherwise recombination of positive and negative charges would lead to a neutral gas. Lindemann noted that it is the impossible for a stream of single charges to survive in the interplanetary medium due to electrostatic repulsion between the particles.[35] Instead, Lindemann proposed a stream with a roughly equal number of positive and negative charges (a plasma). This is consistent with a plasma being quasi-neutral with an almost (but not exactly) equal number of positive and negative charges. He also refuted Chapman's argument concerning the formation of a neutral gas. Lindemann suggested that the beam/cloud is accelerated by radiation pressure, attaining velocities of several hundred kilometres per second.

Through the 1920s Chapman continued to develop a picture of geomagnetic activity based on large data sets and characterized by various indices and in particular calculated the ionospheric currents using a simple shell model (see Figure 8.4b for an example). Unlike Birkeland, there is no connection to solar particles and the current patterns show smooth large-scale properties. The nature of storms as individual, distinctive events was neglected. By 1929 he was ready to return to the theoretical aspects of geomagnetic storms. This was in collaboration with a student at Imperial College, Vincent Ferraro.[36] The initial announcement of the work appeared in *Nature*,[37] followed by a series of papers which were in effect one long paper split into five parts with a length of seventy-six pages.[38] Chapman was the lead author on all the (sub)papers, though Ferraro was listed as the author of Section 9 in the second 1932 paper. The initial discussion benefits from the figure that appears in their *Nature* paper as reproduced here.

Having taken on board Lindemann's criticism, a transient quasi-neutral unmagnetized fully ionized stream approaches the Earth, travelling with a velocity of 1000 km/s through interplanetary space and with a plasma density at the Earth of 60 particles cm−3. (These values are a bit larger than contemporary measurements for storm-causing solar events, but not excessively

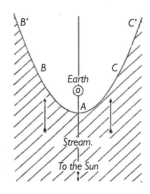

Figure 8.5 The Chapman–Ferraro model. The original orientation is preserved with the Sun at the bottom and the Earth in the middle, and its magnetic field carves out a cavity in the solar stream.

so.) Since the electrical conductivity is effectively infinite, the solar stream cannot penetrate the dipolar terrestrial magnetic field, and compresses it on the dayside. Thus, a boundary (B'BACC'), which is a layer of electric current, is set up between the solar plasma and terrestrial field a few R_E (where R_E = 6390 km is the radius of the Earth) on the dayside, and a cavity is carved out by the terrestrial field in the solar stream, the cavity now being called the magnetosphere (Figure 8.5). This initial compression of the terrestrial field accounts for the observed positive excursion of the surface field at the start of the storm. The estimate of an upstream distance of 5 R_E for the boundary is a bit smaller than today's estimates, but again the difference is not excessive.[39] This part of Chapman and Ferraro's model is of the highest importance and is the foundation of the concept of a magnetospheric cavity as well as opening an entirely new aspect of solar–terrestrial relations.

The first four longer papers focus on the dayside boundary layer and add little to what was said in the *Nature* letter. In places they are a good example of Chapman's mathematical approach in reducing a 3D problem with an (analytically) undeterminable boundary shape to a 2D slab with an infinitely long and wide boundary between the solar stream and terrestrial field. It is in the fifth paper that the question of the main phase of a storm arises. It is argued that an (ion) current flows between the surfaces BB' and CC' forming a 'current ring', though significant questions were left unanswered, especially how this relates to the 2D currents calculated by Chapman elsewhere. This part of the model is best forgotten.

Chapman returned to much of this work in his two-volume *Geomagnetism* co-authored with Julius Bartels.[40] The origin of these volumes lay in Chapman's winning submission for the 1928 Adams Prize essay sponsored by the Cambridge Department of Mathematics and St John's College. Thereafter, Chapman and Bartels agreed to collaborate on these volumes, which took almost a decade to see publication. Over 80 years later, the books read well and, in the view of the author, possess a clarity not seen in the original papers. They also address the work of Birkeland.[41] The preface is dated 1940, and is unusual in that the nations of the two authors had been at war with each other for several months.

Alfvén as a critic

The Swedish scientist and Nobel laureate Hannes Alfvén received his PhD in 1934 and worked on a variety of topics throughout his career.[42] He developed a theory of geomagnetic storms in 1939, submitting the paper to *Terrestrial Magnetism and Atmospheric Electricity* (a journal of the AGU, subsequently incorporated into *Journal of Geophysical Research* and hereafter referred to as *Terrestrial Magnetism*). The paper was rejected and Alfvén would have suspected that the referee was one of Chapman's circle.[43] The work subsequently appeared in the *Proc. Royal Swedish Academy*, a journal now defunct. The AGU published a synopsis of the important aspects in 1970.[44]

Alfvén's papers, like those of Chapman and Ferraro, introduces important concepts. These include particles both 'trapped' in the terrestrial magnetic field as well as drifting around the Earth as a ring current and the presence of a 3D current system connecting the particle 'belts' and the ionosphere/aurora along the lines of Birkeland (Figure 8.4a). All these aspects have subsequently been proven to be present although the way both sets of authors argued how the ring current formed does not stand up to inspection. The required tools and data did not exist in 1940. However, Alfvén's model did not include the magnetospheric cavity of Chapman and Ferraro and required solar particles with rather a high energy.

Chapman did not respond to Alfvén's paper, but a comment came from Cowling who wrote a short paper at Chapman's instigation:[45]

> Chapman sent me copies of two of Alfvén's papers. . . . Chapman had already, at a private meeting prior to 1939, pointed out to Alfvén certain aspects of the latter's theory which he thought were inconsistent with the observations. . . . His idea was that I should produce a critical review of the papers. . . . The review that finally appeared in Terrestrial Magnetism in 1943 [*sic*] was a third version: Chapman rejected the earlier two because, in my hurry to get on with my criticism, I had failed to give an adequate and fair account of what Alfvén had actually said.[46]

The paper presents a brief but fair summary of Alfvén's work but is severe on the field-aligned currents. It is not difficult to imagine Alfvén's reaction. His papers submitted to *Terrestrial Magnetism* has been rejected and yet here in the same journal was Chapman's colleague criticizing his model! The delay in the appearance of Alfvén's ideas in a widely read journal is to be regretted.

Although Alfvén attempted to discuss their respective models with Chapman after the war, the latter remained reticent, not enjoying controversial discussions.[47] Alfvén was then silent on storms until 1951 when he published a scathing note in *Nature*,[48] responding to a short article by David Martyn on the Chapman–Ferraro model.[49] In the first paragraph Alfvén states

> It is not my intention to review here the objections to this theory (Chapman–Ferraro)—objections which I believe to be fatal—nor is it worth while to discuss the curious superstructure which Dr. Martyn tries to erect on this weak ground.

Befitting a Scottish-born Australian, Martyn responded robustly while Chapman was more polite. Alfvén followed this up with papers in the 1950s and it was difficult to see any peaceful resolution.[50] This 1957 paper also criticizes Chapman (correctly) for neglecting the interplanetary magnetic field: 'It is pointed out that the fundamental assumption about a non-magnetized beam is in conflict with cosmic ray evidence' and here Alfvén uses the language of MHD, something Chapman never seemed to feel comfortable with. Incidentally Alfvén's 1951 objections related to the need to verify any theory with experiment, such as those conducted by Birkeland, who Alfvén regarded highly. Chapman, on the other hand, held the opposite viewpoint about Birkeland. There are accusations that Birkeland's work was ignored but in fact Chapman and collaborators give an account of it.[51] At that time they formed the opinion that the 2D currents in Chapman's work fitted the available magnetic field data better than Birkeland's 3D scenario, although whether Chapman should have retained this viewpoint for a further three decades is another matter. And so arguments raged between 'the Nordic school' and 'the Chapman school'. Lest the reader think that this is just another academic disagreement, this was an argument between two gigantic scientific figures of the twentieth century, with others of comparable status such as Cowling being involved. Personality differences did not make any resolution easier.

Eventually in 1966 space-based observations suggested that the currents proposed by Birkeland did exist,[52] and with the encouragement of A. J. Dessler, quite suddenly the paradigm changed. Southwood documents an incident at a symposium commemorating Birkeland's birth centenary in 1967 at which Chapman was a speaker and comments on how he, to the astonishment of the audience, belittled Birkeland's contributions.[53] In fact, Chapman gave a fair assessment of Birkeland, but rather ruined things by stating at the end of his written article that 'Birkeland ... made little direct contributions to our observational knowledge of the aurora', though crediting 'indirect influence'! If nothing else, this was bad manners. It does though seem that towards the end of his life Chapman began to realize he had missed some aspects of Birkeland's work, especially the three-dimensional currents.[54] Incidentally, it strikes the author as remarkable that for two decades after 1940, Chapman published no significant developments of his original model.

Sadly these disputes continued for long after Chapman's death in 1970. In 1985 Cowling wrote:[55]

> I now regret that my first encounter with Alfvén was as a bespoke critic. Nevertheless, as the last survivor of Chapman's original group, I *feel* it incumbent on me to defend their ideas against the assaults which Alfvén *still* continues to make on them.

This was forty-three years after Cowling's paper was published at a time when new observations had rendered much of the debate pointless. Cowling was regarded as a kindly though austere person, but the reason for this comment is clear. On pages 1 and 2 of his monograph *Cosmic Plasma*,[56] Alfvén launches into another criticism of the Chapman–Ferraro theory, with other comments on the theory later in the book. The book also criticizes the approach of many other scientists, though not by name. After his Nobel Prize, Alfvén advocated various concepts that were not at the time part of mainstream plasma physics, some being of interest, and also indulged in bouts of iconoclasm. Cowling reviewed the book in stark terms, though others praised Alfvén's approach.[57]

In the past decades, some historians have presented a balanced account of the disagreements.[58] Others such as Fukushima do their best to be impartial, but let slip their real view.[59] Unfortunately Borowitz, for reasons best known to himself, attributes the ill-feeling between the Chapman and Nordic schools to Chapman's xenophobia, an absurd accusation given Chapman extensive travel and range of collaborators.[60]

Chapman and Oxford

As the Second World War approached, Chapman's pacifist views had moderated due to his exposure to what was happening in Europe, and especially to young German scientists, and, as noted earlier, he participated in the war effort.[61] He returned to Imperial in 1945 but nearing age 60 did not feel up to building up the department again. Hyman Levy (1889–1975) took over, having been acting head of department during the war.

Meanwhile, on 5 June 1940, A. E. H. Love, the incumbent Sedleian Professor since 1898 and an authority on wave propagation and elasticity, died, aged 77.[62] The war delayed replacing him. There was an initial discussion in October 1944, but a subsequent delay was at the insistence of the university's governing Hebdomadal Council due to the need to review 'the future of physics in the university' (which sounds like a bureaucratic tactic to do nothing). H. M. Margoloiuth (Secretary of the Faculties) in January 1945 presented cogent reasons why the chair should be filled and the appointment board was concerned that other prestigious chairs would be advertising. The Sedleian Chair was advertised in March 1945 with an annual salary of £1200 and a retirement age of 65, the latter a new feature.

Input was sought from those responsible for election of the chair. In a letter dated 30 May 1945 to (Sir) Douglas Veale (1891–1973) at the university registry, Sir Geoffrey (G. I.) Taylor apologized for his absence from the election meeting, being in the US on Manhattan Project-related duties, and resigned from the board of electors. On May 31, E. A. Milne contributed a lengthy handwritten letter. He would be in Leningrad (St Petersburg) on the date of the election meeting and offered no specific suggestions. Instead, the purpose of the letter seems to be to say that he had been awarded the Bruce Medal of the Astronomy Society of the Pacific, for which he was duly congratulated by Veale. There followed a correspondence with the long-suffering Veale as to whether Milne would stay on the election board.

Letters of reference were taken out for Robert Stoneley (1894–1976, University of Leeds, solid Earth science) and Leslie Howarth (1911–2001, University of Bristol, boundary layer theory). Nothing further is heard of these applicants. Another applicant was Sydney Goldstein (1903–89, a fluid dynamicist and Howarth's PhD supervisor) who then withdrew on moving to Manchester. On July 7, letters were sent to (Sir) Nevill Mott (1905–95, University of Bristol and 1977 Nobel Prizewinner) and to (Sir) Harrie Massey (1908–83, then occupied with the Manhattan Project at the University of California-Berkeley and due to return to UCL to become a pioneer of UK space research). The letter to Mott 'invites him to accept the professorship' while that to Massey 'whether you would be willing for your name to be considered'. Mott curtly declined on July 10. On 20 August 1945, a formal offer was made to Massey, who eventually declined in September 1945.

The next board of electors meeting in October 1945 asked Milne whether Chapman was interested.[63] They also authorized approaches to Drs Penney, Peierls, and Heitler. These are Bill Penney (nominally at Imperial College but then the UK lead on the Manhattan Project), (Sir) Rudolf Peierls (1907–95, University of Birmingham, also involved in the Manhattan Project), and Walter Heitler (1904–81, then in Dublin). Chapman, Penney, and Heitler all responded positively, though later Penney and Peierls withdrew. So on 17 December 1945, the Sedleian Chair was offered to Chapman who accepted and arranged to start on 1 April 1946. The difficulty in filling the Sedleian Chair ought to have sent a warning to the university. A number of those approached expressed the view that they were happy where they were and, importantly, would have the resources they felt they needed, such as lectureships and facilities. Oxford was unwilling to offer any such carrots.

A different Chapman emerges after the war. His major monographs with Cowling and Bartels sum up many of his achievements between 1912 and 1940 and draw a line under the first part of his career. In the immediate post-war years and indeed up to roughly 1960, there are fewer long papers, more short notes, and an increasing emphasis on what today one would call an 'elder stateman' role. But by most standards he was still a productive scientist, though Cowling comments on the lack of exciting advances. There is a lack of strong collaborations such as he had enjoyed in the 1930s with Ferraro, Cowling, and Bartels. He investigated a variety of problems, including Blackett's hypothesis concerning the ratio of the angular momentum and magnetic moment of large rotating bodies being given by a combination of fundamental physical constants.[64] More important were his 1946 and 1950 papers on upper atmosphere nomenclature.[65] The term 'aeronomy', now used to describe upper atmosphere and ionospheric studies, originated in the first of these papers. Chapman in fact proposed that this term supersede 'meteorology', but that has not been adopted. The latter of these papers broke up these regions into 'spheres': thermosphere, mesosphere, ionosphere, though much of this was a summary of earlier proposals. Chapman's main Oxford achievement, the International Geophysical Year (IGY), is discussed later in that eponymous section.

While Cowling noted that Chapman 'enjoyed the gracious living in Oxford', especially welcome at a time of rationing, there were difficulties. At Imperial College, he had good secretarial support, not available at Oxford. In 1948 he approached the university through Veale about this. The bureaucracy rolled into action, precedent was scoured for advice, initially determining that only heads of department were entitled to secretarial support. Chapman was then asked to make his case to another committee, which he duly did, making the point that his time was being spent on minor matters that were a waste of his abilities.[66] This committee then recommended to another committee that support for Chapman be found within the general support of the Mathematics Department. Sadly the outcome is not recorded in the available files, but the prolonged nature speaks for itself.

Chapman soon found the need to travel to India and then in the winter of 1949–50, he wrote to Veale asking for approval to spend a year at Caltech. A salary of $10k and travel expenses of $2k were offered. The visit was initiated by Caltech and was part of an effort by Vannevar Bush and Lloyd Berkner to strengthen US collaboration with Western Europe: a scientific equivalent of the Marshall Plan.[67] This was one manifestation of the dominance that US science was to take on post-war. Much of continental Europe had been devastated: the UK—being obsessed with its delusional role as a great power, sterling as a reserve currency, and following the passage in 1946 by the US Congress of the McMahon Act, building its own atomic bomb—budgeted accordingly.[68] Having seen the use of a strong science base in the war, the US military and then the US government (through the National Science Foundation and then NASA) invested in science.

The Caltech visit went ahead and subsequently Chapman approached Veale about spending four weeks a year at the Institute of Geophysics in Fairbanks in the (then) territory of Alaska. The institute had recently undergone some upheaval due to a dispute between its director and the head of the university, and was looking for a prestigious name to help it rebuild. In 1952 Chapman wondered whether he was behaving properly with these absences and volunteered to resign from the Sedleian Chair if needed. In the event, he retired from Oxford in 1953 and the Sedleian Chair was filled by George Temple.

Chapman had no intention of going into a quiet retirement. A move to the USA was attractive for many reasons: no rationing (abolished in 1954 in the UK), travel without worrying about exchange controls (abolished in 1979), and the ability to actually observe the aurora. He subsequently spent winters in Alaska and summers at the High Altitude Observatory (HAO) in Boulder, Colorado where its founder, Walt Roberts, was beginning an expansion of activities.[69]

But there was more to it than that; after his death, the Alaska Geophysical Institute director Keith Mather wrote a very nice tribute including the words:

> He spent three months of each year at College (Alaska)—usually the winter months when the aurora was overhead, and he also preferred walking in the snow—and a goodly part of the remainder of the year at HAO/NCAR in Boulder, Colorado. But where Sydney Chapman might be at any given time involved an uncertainty principle of the Heisenberg kind, evidenced only by the postage stamps on the meticulously written letters that arrived from London, Cambridge, Moscow, New Delhi, Sydney, Tokyo, Ann Arbor, Minneapolis, Mexico or Ibadan.[70]

International Geophysical Year (IGY)

There had been the International Polar Years (IPY) in 1882 and 1932. The birth of the IGY has been attributed to a dinner hosted by James Van Allen (1914–2006) on 5 April 1950, attended by Chapman and others, which led to the proposal of another IPY that subsequently developed into a broader IGY.[71] Chapman credits the suggestion of the IPY/IGY at this dinner to Lloyd Berkner.[72] However, in a 1998 interview, Van Allen thought that Chapman already had the idea of another IPY and was waiting for Berkner to suggest it![73] It was to be held in 1957–8 (25 years after the last IPY and coinciding with the maximum of the 11-year solar activity cycle) with the aim of coordinating geophysics observations around the world. Korsmo provides a good and concise retrospective.[74] By 1953 roughly 12 people had formed a steering group, with Chapman and Berkner as president and vice president, respectively. Uncertainty about a Soviet contribution was removed by 1955, although they did not commit to rockets or satellites. The USA invested extensively in IGY and committed to launching an IGY satellite. Beyond the USA and the USSR there was enthusiasm, but probably little money. Chapman's role can best be summarized by Van Allen:

> And so, all the rest of us spoke up (at the 1950 dinner) with the feeling that it sounds like a great idea to us and why don't we get going on this. Well after this occasion, Sydney Chapman took the international leadership in going ahead with it . . . he sensed the timeliness and he had the international prestige of a geophysicist and scientist, very well earned clout to make it go. He spent a good part of his life for many years on the IGY. . . . For the next several years (after 1950), Chapman spent a good part of his time traveling internationally, going to various countries all over the world developing the idea of appointing committees for individual specialties, enlisting the support of influential scientists in all the major countries for this undertaking, and organizing the institutional structure for coordinating the work. He was, I think, essentially the unanimous choice for president of the IGY. . . . That was one of his major professional undertakings but he, of course, continued to do personal research throughout that period. . . . But certainly his most important contribution to the science of geophysics during that period was making a success of the IGY. It was a non-governmental enterprise. That was one of its distinguishing features. The adherent bodies were not government agencies in the countries, but independent scientific organizations of one kind or another.[75]

Two major achievements from the IGY changed the field of space science and solar–terrestrial relations. One was the need for coordinated observations, both ground and (near-)space based. This was partly based on observing campaigns using rockets and 'rockoons'. The rockoon was an ingenious idea due to Van Allen and others. In order to get further into space, a balloon was launched and, at a predetermined height, a rocket was launched from the balloon.[76] It was also recognized that data should be shared between scientists. This led to the establishment of World Data Centres (WDC), locations where ranges of data were archived and available to users. This pioneered the way to the present-day large-scale data archives.

The second major outcome was the launch of orbiting satellites. *Sputnik 1* was launched in October 1957. The US response has been well documented: the Vanguard launch failures, and then the successful launch of *Explorer 1* and *3* via a Juno/Jupiter rocket which led to the discovery of intense belts of charged particles (the Van Allen belts) by the small Geiger counter built by Van Allen.[77] Figure 8.6c shows these radiation belts a few Earth radii above the surface. An indication of the panic induced by *Sputnik* in the USA, and the relief when their own satellite programme was successful, was the presence of Van Allen on the cover of *Time Magazine* the week after the announcement of his discovery. NASA was formed in 1958 and designated a civilian agency despite Eisenhower wishing 'space' to remain largely in the military domain.[78] IGY was a major triumph for Chapman, perhaps his greatest. It changed the way that science was carried out.

The final decade

We now return to the development of the Chapman–Ferraro model. In 1957 Chapman proposed that the outward extension of the hot solar corona into interplanetary space was a static, stratified atmosphere originating in a corona with a temperature of 1 MK.[79] The temperature structure of this extended corona was determined by thermal conduction, which led to a temperature of 0.2 MK at the Earth. Thus, the boundary in the Chapman–Ferraro model would be between this static, extended corona and the Earth's magnetic field. There was no mention of an interplanetary magnetic field.

Motivated by observations of extended cometary tails, in 1958 Eugene Parker (1927–2022) at the University of Chicago proposed a different scenario, namely that the Sun's corona extends into space as a dynamic highly supersonic wind, the solar wind with a velocity of hundreds of kilometres per second.[80] Parker excluded Chapman's static solution since it predicted a pressure at large distances incompatible with those inferred for the interstellar medium. The story of the initial rejection of this paper by the *Astrophysical Journal* referees and subsequent editorial decision by S. Chandrasekhar to proceed with publication in that journal is well known.[81] Parker also deduced that the solar magnetic field filled the interplanetary medium: the interplanetary magnetic field (IMF). His idea created controversy that was only resolved by the measurement of a supersonic wind and its magnetic field by spacecraft. Thus, the Chapman–Ferraro model was modified again to have a highly distorted boundary between the magnetized wind and the terrestrial magnetic field, called the magnetopause. Tommy Gold named the cavity in space the 'magnetosphere'.[82]

The second important paper was published by Jim Dungey in 1961.[83] This work was not entirely new, though the author's later claim that the inspiration arose 'sitting at a café in Montparnasse' certainly accounted for many aspects. Its genesis lay in Dungey's PhD study with Fred Hoyle in the 1940s,[84] and as early as 1948 Dungey had realized that the IMF could play a crucial role in geomagnetic activity. The delay in publishing his work seems in part to be due to the rejection by *Monthly Notices of the Royal Astronomical Society* referees of associated work.[85] By 1960, the idea of the IMF filling the heliosphere was more palatable.

Dungey uses the dynamics of magnetic field lines in a plasma (as developed by Alfvén) to deduce how the solar and terrestrial magnetic fields interact with each other. In a highly conducting plasma, the magnetic field and plasma are said to be 'frozen' to each other, so that when the solar wind and the IMF encounter the terrestrial magnetic field, solar wind plasma cannot enter the magnetosphere, the wind and IMF slide around the magnetosphere, and the magnetopause is largely impermeable, though small-scale plasma instabilities may arise. An exception occurs in regions of strong current, such as can exist at the magnetopause, when (for simplicity) the solar wind and terrestrial fields are anti-parallel: i.e. a southward-pointing IMF (Figure 8.6a). Then the field line connectivity can change, so that plasma associated with the solar wind finds itself on magnetic field lines rooted in the Earth. Thus, solar wind mass and energy can access the magnetosphere in a process known as 'magnetic reconnection' and the magnetopause is then permeable. A similar process occurs downstream of the Earth, in the magnetotail.

Know as the 'Dungey cycle', this work was of the highest importance and has concrete predictions: downstream there should be both northward and southward magnetic fields, and solar plasma should be seen inside the magnetosphere, for example, energetic electrons and highly ionized states of ions (e.g. solar O^{5+} rather than terrestrial O^+), as well as a pattern of convection in the ionosphere. All have been confirmed. A more modern representation of the Dungey model is shown in Figure 8.6b. Note the survival of the Chapman–Ferraro boundary, albeit modified in nature, on the dayside. While Dungey's paper began to resolve many of the arguments of the past thirty years, acceptance of his ideas was slow.[86] While others put some more flesh on the skeleton,[87] it was not until spacecraft data became available, first slowly in the 1960s, and then in large quantities after 1978 that Dungey's work was widely accepted.

Figure 8.6c shows a contemporary sketch of the entire magnetosphere. The 'magnetopause current' is due to the Chapman–Ferraro cavity, the ring current, responsible for the main phase of the magnetic storm, was discovered by Van Allan, and the field-aligned currents were proposed by Birkeland.

Chapman was 70 in 1958 and continued as a productive scientist until his death in 1970, though the available space precludes an extensive discussion of his work in these years. A change in his circumstances in the 1960s was his renewed interaction with young scientists. S.I. Akasofu (1930–), who arrived in Alaska at the end of 1958 and has remained there, was the primary collaborator. Peter Kendall (1934–2013) was another, a student of Ferraro's, based at Bedford College, London and then Sheffield, but making regular visits to HAO and Alaska. One does though have the impression that things were happening too quickly for Chapman to keep up. He seemed slow to catch on to the importance of a southward IMF in instigating geomagnetic activity and a review paper given in late 1961 seems to be sitting on several fences, especially in the context of Parker's work.[88] However, by the time of the posthumous Akasofu and Chapman volume,[89] both Parker and Dungey's work receive due recognition. There was also an insistence on persisting with overly simple current models when the wider community had moved on to more sophisticated and realistic models.[90] And, as we have mentioned, the discovery of Birkeland currents in 1966 should finally have ruled out his approach. There was also a fair bit of tidying up old topics such as atmospheric and lunar tides, or invited introductions or summaries of conferences. On the other hand there are gems like the work with Peter Kendall on x-point collapse, which showed that Chapman had lost none of his mathematical

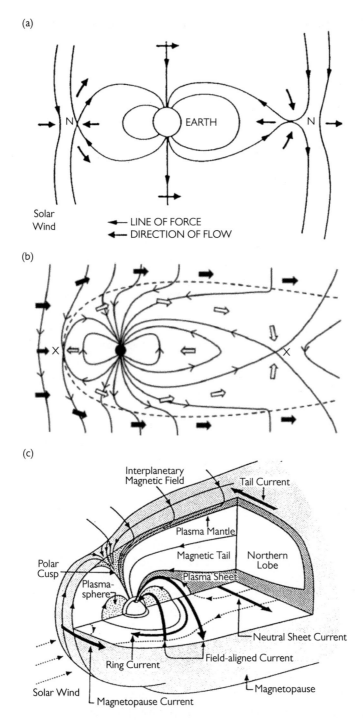

(a)

Solar
Wind

◄— LINE OF FORCE
◄— DIRECTION OF FLOW

(b)

(c)

Interplanetary
Magnetic Field

Tail Current

Plasma Mantle

Magnetic Tail

Northern
Lobe

Polar
Cusp

Plasma Sheet

Plasma-
sphere

Neutral Sheet Current

Field-aligned Current

Ring Current

Solar Wind

Magnetopause

Magnetopause Current

Figure 8.6 (a) A sketch of magnetic field lines showing the interplanetary magnetic field (IMF), magnetic reconnection when the IMF is southward (labelled N), and the formation of a tail on the nightside with a second reconnection site. (b) A more modern two-dimensional projection of Dungey's work with field lines (solid) and plasma flow (arrows). (c) A three-dimensional rendering of the magnetosphere with the various currents discussed throughout the text.

skills.[91] And he continued in the role of 'elder stateman', perhaps for longer than he should have. He worked hard right to the end.

Conclusions

Sydney Chapman was one of the great scientific figures of the twentieth century. He pioneered work into the nature of geomagnetic activity, the ionization of the upper atmosphere, and the kinetic theory of gases and atmospheric tides. His role in IGY perhaps showed him at his best. He was blessed by a long working life: he began by arguing with Jeans in 1912 and ended seeing the space age with his prediction of the magnetospheric cavity confirmed. His obituaries point to a kind and considerate colleague and friend, though tending to avoid controversy as in his arguments with Alfvén. There is also the feeling that there was a reluctance to absorb ideas that contradicted his own views.[92] But he got a lot out of life, with travel and exercise being important to him.

Oxford perhaps did not see the best of Chapman and in my opinion would have been better off appointing a much younger person: Cowling for one would have been ideal. Chapman's pre-war collaborators had moved on to other things, and the lack of university support seems petty. The relative unimportance of non-military science for both the Attlee and Churchill governments must have been discouraging. So Chapman saw greater opportunities elsewhere, in particular in the USA, and as he envisioned working past his nominal retirement age, took steps to ensure his future, although this led to lengthy absences from Oxford.

Since Chapman's death, ionospheric and magnetospheric physics have made huge advances such as the general acceptance of the open magnetosphere (Dungey) model, the identification of large solar eruptions (CMEs) as the cause of major storms, the ability to determine the origin (solar or terrestrial) of magnetospheric charged particles, and, in the past two decades, the use of multi-point spacecraft measurements to determine the three-dimensional nature of space plasmas. Figure 8.7 summarizes the present state of the link between the Sun and geomagnetic activity. The upper three panels show the magnetic field associated with the interplanetary manifestation of a large CME (called a magnetic cloud), from July 2000. The field turns sharply southward after roughly twenty hours (of order thirty hours after the solar eruption), and the 'storm' index (D_{st}) shows an immediate negative turning, peaking at ~ -300 nT. One can see the magnetic cloud as being equivalent to Chapman's stream (Figure 8.5) with the major difference that it has a magnetic field, the southward orientation of which leads to the transfer of energy to the magnetospheric cavity, as Dungey proposed.

There are also societal implications. As an example, a series of large solar eruptions between 9 and 12 March 1989 was followed by, early on March 13, a major failure of the power system in the Canadian province of Quebec. This arose due to induced ground currents associated with a large geomagnetic storm, and the effect was exacerbated by the nature of the rock which diverted the currents to long power lines, triggering circuit breakers. Expansion of the upper atmosphere caused severe difficulties with satellite tracking as well as with communications

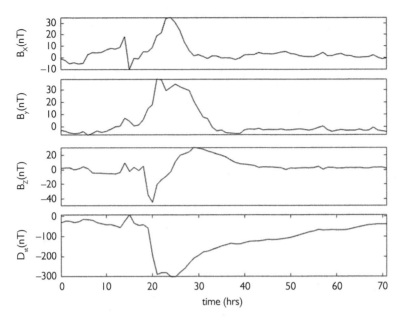

Figure 8.7 The magnetic field associated with the CME in interplanetary space and (bottom) the D_{st} index for the large geomagnetic storm associated with the July 2000 Bastille day flare. Data from 15–17 July 2000 inclusive is shown. The x- and z-components of the interplanetary magnetic field are in the Sun–Earth and north–south directions, respectively (B_z negative points south), with the y-component completing the orthogonal field vector.
[Data courtesy of NASA/Omniweb.]

systems. These effects of a geomagnetic storm had long been appreciated, but the scale of disruption in March 1989 was new. So 'space weather' was born. First in North America, then in Europe, funding became available for space weather: to understand its causes and consequences and to (if possible) predict its onset and severity. A century after Chapman's initial speculations, the field is now well established in the scientific and public conscience.

CHAPTER 9

George Temple and Albert Green

MARK MCCARTNEY

The lives of George Fredrick James Temple (1901–92) and Albert Edward Green (1912–99) spanned the twentieth century and between them they held the Sedleian Chair for nearly a quarter of that century, with Temple occupying it from 1953–68 and Green from 1968–77. Although their careers share a number of similarities, they were very different men, in both personality and their approach to scientific research.

The similarities, some quite superficial, are easily sketched: Both were born in London and spent their early lives in parts of the city that were only a couple of miles apart, Green in Kilburn and Temple in North Kensington. Neither man came from a wealthy home, and as such both had the directions of their lives set by obtaining scholarships to London schools. Temple was amongst the first intake to the (new) Ealing County School in 1913 and Green entered the (more than 200 years old) Haberdashers Aske's Hampstead School in 1924.

Both men had university chairs and were running mathematics departments in their 30s. By 1932, aged thirty-one, Temple was Professor of Mathematics at King's College, London. By 1948, aged thirty-six, Green was professor at King's College, Newcastle (then a college of Durham University). They were both at their respective 'King's Colleges' for about twenty years before moving to Oxford. Finally, both men were elected Fellows of the Royal Society in their 40s. Temple in 1943 (aged forty-two) for work on the Dirac equation, Green in 1958 (aged forty-six) for work on linear and nonlinear elasticity.

The similarities listed above are hardly surprising, indicating as they do, the relatively smooth and swift academic progressions of two extremely able men who were destined to reach the top of academia.

The differences between the men are also worth noting: Green's university education was traditional. Or at least traditional for someone who is going to end up with an FRS and an Oxford chair. He left school in 1931 with scholarships to Jesus College, Cambridge. At Cambridge he gained a first class degree, won the prestigious Smith's prize, and by 1937 had completed a PhD under the supervision of G. I. Taylor.

Mark Mccartney, *George Temple and Albert Green*. In: *Oxford's Sedleian Professors of Natural Philosophy*. Edited by: Christopher Hollings and Mark McCartney, Oxford University Press. © Mark Mccartney (2023). DOI: 10.1093/oso/9780192843210.003.0009

Temple's university education was quite different. The death of his father in 1917 meant that, as an only child, he had to leave school early to support his mother. Thus, his university education came via evening classes at Birkbeck College. By 1922 he had graduated with a first, and by 1924 had a PhD, though not through normal supervision. Rather his PhD was obtained for research on relativity completed independently while working as a steward in the Birkbeck Physics Department.

It was probably in their approach to research that they were most different. Temple published predominantly as a single author. He was also an intellectual omnivore. His research interests moved from relativity, to quantum mechanics, to fluid dynamics and aeronautics, to generalized functions, and finally to the history of mathematics and its foundations. Arguably this continual movement across a wide range of areas meant that his contribution to any one of them was more limited than it might otherwise have been.

Green's approach to research was almost the negation of Temple's. His research output was significantly higher with the majority of his papers being co-authored, in the main with a small number of long-term collaborators. Further, Green's research remained focused throughout his career within continuum mechanics (and within that in linear and nonlinear elasticity), and that focus resulted in him being considered one of the most significant contributors to the area in the twentieth century.

A final difference is to be found in their approach to Oxford life. Temple thoroughly enjoyed the variety of both university and college life, ungrudgingly taking on administrative duties and entering fully into the life of The Queen's College. In contrast, Green avoided administration and took little interest in Queen's. It was almost as if he saw his time at Oxford as an opportunity to finally devote himself fully to research without distractions.

While both were major scholars, Temple via his breadth, and Green via his depth, neither made a significant impact on the direction of mathematics within Oxford during their time as Sedleian Professor.

George Temple

According to George Temple, on both his mother and father's side of the family no one had 'been noteworthy in science, learning or public service. They were country folk who had always lived in the villages of Oxfordshire'.[1] His paternal grandfather, John, had been a shepherd, and his maternal grandfather, George, a tailor, and Clerk of Bradwell Parish. In later life John and his family lived in the neighbouring parish of Kencott and thus both sets of grandparents were living about twenty miles west of Oxford at the time of the 1861 Census of England.[2] At the age of fourteen, Temple's father, James, like his father before him, began farm work, but in 1867 he commenced what would be a forty-year career with the Great Western Railway, starting as a policeman in the passenger department at Paddington station, then working as a signalman and sergeant, and finally as an inspector, retiring in 1907 on a pension of 8 shillings and 9 pence per week.[3]

Temple's parents married late in life when his father was 51 and his mother, Fanny, was 39.[4] Thus, when their son was born on 2 September 1901 they were in their mid-fifties and mid-forties, respectively. George claimed to have little memory of his childhood, first in the London Borough of North Kensington and then, after his father retired, in Ealing. He recalls his first school (Mornington Road Board School) as 'something of a bear-garden' and was happy to move after only a year to Northfields School. There he was taught by one Ray Gilbert who[5]

> had to cope with a class of 50 boys with a hard core of some half dozen toughs, but he certainly inspired me with a love of elementary mathematics, and enabled me to win a Scholarship to Ealing County School in 1913. This did not have the cachet of Ealing Grammar School, but to me it was a paradise, and the mathematics classes of Goodall were a revelation.

Ray Gilbert at Northfields, P. D. Goodall at Ealing County School, and later C. V. Coates at Birkbeck College were mathematics 'teachers of exceptional gifts in arousing enthusiasm for the subject' whose classes Temple considered himself fortunate to have attended.[6]

Temple entered Ealing County School in its opening year as one of its first 111 pupils.[7] The scholarship which he won was a crucial one,[8] as otherwise the termly fees of £2, 2 shillings, and 6 pence[9] would have consumed a significant fraction of his father's pension and would probably have made his attendance an impossibility. That the decision to award a scholarship was a wise one was justified by the fact that by the end of the first year he was winning prizes. In 1914, at the school's first prize distribution, he won the upper third form prize for mathematics and science, and the school prize for an English essay (a prize which he got again in 1915 and 1916).[10] Over the next two years his love of the 'paradise' of the county school saw him flourish academically. In 1915 he won prize certificates in scripture, English, history and geography, and science and mathematics, and in 1916 prize certificates in languages, history and geography, and science.[11] Probably more significant than any of the school prizes was the award of a Junior County Scholarship, entitling Temple to 'education, books and an annual financial grant for a period of four years'.[12]

In 1917 Temple, now head boy, took the London University Matriculation Certificate and gained a distinction in all but one of the subjects he sat.[13] But in February 1917 his father, James, died aged 71. Scholarships and prizes would not pay household bills and Temple had to earn a wage and support his mother. The headmaster found him a post as a Junior Assistant Clerk with the Prudential Assurance Company at £70 per year. In the quote above Temple mentions the mathematics teaching of Percy Denis Goodall, a popular master who was only about 12 years older than Temple.[14] However, perhaps just as important was the chemistry mistress Alice Rebecca Rollinson, who encouraged Temple to continue his studies and apply to take evening classes at Birkbeck College, London, beginning in 1918. In 1919 the Professor of Physics, Albert Griffiths, offered him the post of part-time research assistant working on the viscosity of water at very low shear. The pay was slightly less than the Prudential, but the work much more enjoyable. He took the job, and in 1922 gained a first class in the University of London General Honours exams in pure and applied mathematics and physics. At this point Griffiths offered him a job as steward in the Physics Department at £200 per year which he gladly took, but by 1924 he moved to be a demonstrator in mathematics at Imperial College. Temple put the offer from Imperial down to the influence of Alfred North Whitehead, claiming that one of

> his last acts before he resigned from the Imperial College to become a Professor of Philosophy at Harvard was to recommend me for this post. Whitehead was very pleased with two papers which I wrote on his theories of relativity and especially attended a meeting of the Physical Society of London in order to praise my researches.[15]

In the discussion section at the end of Temple's 1923 *A generalisation of Professor Whitehead's theory of relativity*, Whitehead is recorded as stating that the paper comes from the pen of a young scientist whose work augurs a very distinguished career. The mathematics in the paper was handled in a way that showed the author to be the master and not the slave of his symbols.[16]

For such a relatively young man, Temple was already beginning to be noticed and to make his mark.

Temple's 1923 paper was the beginning of not only a set of papers on relativity, but also of his modus operandi in research whereby he would become interested in a topic, publish a set of papers, and then move on to something else.

By 1924 Temple also had a PhD which he achieved in the unusual manner of simply submitting a body of work on relativity which then went on to be published in three papers.[17] As Peter Neumann, who knew Temple when they were both fellows of The Queen's College, recalled: 'Explaining this to me over a glass of wine in 1979, he said simply "If I had done it with supervision I should have been obliged to spend two or three years over it. As I did it as an external student I was able to do it in the summer vacation".[18]

In 1928 Whitehead's successor at Imperial, Sydney Chapman (see Chapter 8), put Temple forward for an 1851 Exhibition Senior Research Fellowship. Temple was successful and spent a year at Imperial while also taking on some evening teaching at Chelsea Polytechnic, lecturing on the new quantum mechanics. In 1929 he moved to Cambridge and was accepted as a PhD student to work under the supervision of Arthur Eddington. While at Cambridge he continued his work on quantum mechanics, publishing half a dozen papers, but the offer of a readership in mathematics at Imperial College from Chapman brought him back to London in 1930 before completing a second PhD. The previous incumbent in the readership had stayed for only one year and the Rector of Imperial College, Sir Henry Tizard, made the appointment in the hope that Temple would stay for at least three. However, in 1932 the chair in mathematics at King's College, London became vacant and to his surprise not only did Sydney Chapman and Henry Tizard encourage him to apply, but he also got the job 'as a disgustingly young man, aged only 31 years'.[19]

Professor at King's

When Temple arrived at King's he had a number of problems. Unfortunately all these problems were his colleagues. The head of department, Professor A. E. Jolliffe, had not wanted Temple to get the job in the first place, with his preferred candidate being William Ferrar, who was at that stage a fellow of Hertford College, Oxford. Further, neither Jolliffe or his colleague S. A. White had any interest in research or in changing the mathematics department. For Jolliffe teaching was 'an all absorbing interest'.[20] In an uncharacteristically blunt remark Temple stated that 'Jolliffe and his colleague S. A. White hated research and all my work in changing the character of the Mathematics Department'.[21] Thus, it was only when Jolliffe retired in 1936 and J. G. Semple arrived as Professor of Pure Mathematics from Queen's University, Belfast, that changes could be made unopposed. According to Temple the honours syllabus in mathematics needed to be raised 'from the deplorably low level so vigorously maintained by Professor Harold Hilton (afterwards Simpson) of Bedford College, who eliminated from the examination papers any questions which could not be successfully attempted by the weakest woman student in his department'.[22] This does not appear to be an exaggeration, with Lilian

Button recording in her obituary of Hilton that 'As Chairman of the Board of Examiners of the B.A. and B.Sc. Special and General Degrees he often pronounced a Russian-type veto on any question he considered beyond the capabilities of his young ladies, and to the bright young men of the day he seemed a shade reactionary towards any proposals to revise the syllabus'.[23]

Temple described Semple as 'a wonderful companion, and a great friend'[24] and over the next three years they set about reforming the curriculum in conjunction with colleagues at University College to bring it up to date. They also began developing a department at King's which was to become both academically excellent and known for its informality and friendliness. All was going well, but then the war intervened. The mathematicians from King's were evacuated to Bristol University, but Temple did not join them; instead he went to the Royal Aircraft Establishment at Farnborough and stayed there for the duration of the war as a principal scientific officer.[25]

Farnborough

Temple was a familiar, if somewhat unorthodox, figure of 6 ft 1 inches cycling around the Farnborough compound wearing a deer-stalker hat.[26] That he was popular with both the other 'boffins' and the RAF pilots was shown by the fact that he was elected president of the senior staff mess,[27] but aside from his popularity he was also busy carrying out research and writing reports.[28] The work in Temple's technical reports held at Farnborough fall broadly under three categories: aerodynamics and supersonic flow; the behaviour of landing gear; and aircraft de-icing strategies. His work on supersonic aerodynamics was the most mathematical of the three and was an area of research which he would continue after the war. His work on landing gear, though more prosaic, was the problem he took most pleasure in. It related to 'shimmy', i.e. the wheel wobble that occurred on an aircraft on landing. This work was a practical and successful piece of mathematical modelling and is an example of how diverse Temple's problem solving abilities were. In tandem with his theory were experiments observing wheel behaviour during landing. Temple quipped that

> Theoretically I found that the only cure for shimmy was adequate friction to control lateral oscillations in the wheel unit, but in practise the most efficient way to suppress shimmy was found to be a summons to me to come and observe the strange and dangerous form of vibration.[29]

The final area of work at Farnborough which Temple was involved in was the de-icing of bombers. The obvious solution to the problem is to add something to prevent ice forming on the wings and tail. The three options available were to use a chemical de-icer coating; add a mechanical device to the wing (e.g. a rubber 'boot' along the leading edge of the wing which could be inflated to remove the ice); or add a heating system to prevent ice forming. All three were investigated at Farnborough, and all had weaknesses. Temple, however, was part of a team that realized that the increased mass due to the addition of a de-icing device would mean a corresponding reduction in the bomb load, resulting in an increase in the number of aircraft needed to deliver the same total number of bombs over a given month. A simple comparison of the rate of bomber loss due to military causes and the loss rate due to icing revealed that adding de-icing equipment would increase bomber losses rather than reduce them. R. V. Jones, in his history of British scientific intelligence during World War II,[30] and Clive Kilmister in his *Royal*

Figure 9.1 George Temple and his wife Dorothy in comic Dickensian mood, dressed as Mr Pickwick and Mrs Bardoll at Farnborough in the 1940s.

Society Biographical Memoir[31] both give full credit to Temple for this very nice 'back of an envelope' calculation. This full credit is, however, probably misplaced. The short report which gives the result was not written by Temple, though he was the author of a later report giving a more sophisticated analysis of the general issue of carrying safety devices.[32] Irrespective of how much credit should be given to Temple, the calculation is an excellent example of how simple analysis can show that what seems to be a wholly sensible course of action can have the opposite effect to that intended.

Back to King's

Temple stated that 'I returned to King's, somewhat reluctantly in 1945 where I remained until 1953, becoming more and more enmeshed in administrative duties in the college and in the university, and in serious danger that in time I should become Chairman of the Academic Council and, at last, Vice Chancellor of London University.'[33] This is not the only example of Temple fearing that he may be destined for a senior administrative role. He also states that he narrowly escaped being appointed Provost of The Queen's College, Oxford not once, but twice. However, none of these fates befell him, in either London or Oxford, and in particular there appears to be no evidence of him ever being in the running for the Provost's role at Queen's.

Back at King's, Temple continued to develop his diverse research interests. While continuing his research on aerodynamics, he also worked on generalized functions and set about putting Dirac's delta function (a function with properties that physicists were happy to make use of, but which mathematicians felt lacked rigour) on a more solid footing. He began what would be a long-term interest in the history of mathematics by writing an obituary of Arthur Stanley Eddington with whom he had worked at Cambridge in the late 1920s. And between 1948 and 1950 he was principal scientific advisor to the Ministry of Civil Aviation, working on new methods for air traffic control.

Clive Kilmister, who in 1950 was appointed an assistant lecturer in mathematics at King's recalled:[34]

> the very happy atmosphere [at King's] in the early 1950s. Temple's undergraduate lectures were said to fall into two categories, both of which were delivered without notes. The first happened on days when he had crossed Waterloo Bridge on the way to college by himself; they enlightened and inspired every member of the class. The second occurred when he had crossed the bridge with a colleague; then the better undergraduates had the chance of seeing a first-rate mathematician at work re-creating the subject as the lecture progressed . . . The breadth of Temple's mathematical knowledge was a constant surprise and every member of the staff benefitted from informed comment on his work. . . . The overwhelming aspects of his character where courtesy kindness and wit, so that he was much loved by all.

Though Temple seemed wary of senior administrative positions in university life, he was happy to undertake them in spheres more closely related to mathematics. He was President of the London Mathematical Society from 1951 to 1953 and acted as Treasurer, President, and Vice-President of the International Union of Theoretical and Applied Mechanics (IUTAM) in the 1950s and 1960s.

As noted by Clive Kilmister above, Temple was good humoured and good natured. He also had a ready supply of stories and anecdotes. Peter Neumann and John Kingham recalled him as 'genial, friendly, witty',[35] 'quizzical, jovial, . . . A little too worldly perhaps for a monk, a little mischievous for a Professor'.[36] It was this temperament which made him not only a pleasure to be with, but also a good committee chair, or society president. When in 1968 Temple ended his

(a) (b)

Figure 9.2 The mathematician and the monk. (a) Temple in 1947, while professor at King's College London, and (b) in the 1980s as monk at Quarr.

term as IUTAM Vice-President the German applied mathematician Henry Görtler wrote to him stating, 'We need your good humour and your elegant, if not to say graceful way of treating discussions and getting things done without pressure but with everybody feeling happy.'[37] While Temple may have been gracious he was certainly no pushover. As Kingham notes, 'He could on occasion be quietly ruthless in sabotaging the efforts of misguided reformers, but entrenched reaction earned his gentle, but effective scorn.'[38]

The Queen's College and the Sedleian Chair

Temple became Sedleian Professor and fellow of The Queen's College, Oxford in 1953. It was not, however, the first time he had considered the chair. When in 1946 he heard that the chair was to be filled after having lain vacant during the war due to the death of A. E. H. Love in 1940, Temple asked Sydney Chapman, who had been pivotal in his previous career moves, in the late 1920s and early 1930s, if he should apply. Chapman was encouraging, and even offered to write a reference, so Temple applied. After having heard nothing for six weeks he wrote to Oxford to check if his application had been received, and this inquiry was greeted with more silence. After a further six weeks it was announced in the press that none other than Sydney Chapman himself had been appointed. Presumably Temple was unaware that, as John Ball notes in Chapter 11, the electors were quite likely to simply ignore his application precisely because he had applied.

When Temple arrived at Oxford the discipline of mathematics was about to enter a phase of rapid growth, guided by the dynamism and huge energy of Charles A Coulson. Coulson had left King's College, London the year before Temple to take up the Rouse Ball Chair of Mathematics at Oxford in 1952. As part of the negotiations for Coulson's appointment he convinced the university to double the budget for secretarial support for the four professors (meaning each of them now had in effect half a secretary each) and when he arrived Coulson also oversaw the move of the Mathematical Institute from inadequate quarters in the Radcliffe Science Library to 10 Parks Road. The move provided some small lecture rooms, space for the four professors and graduate students, and Coulson insisted on incorporating a common room to encourage informal discussion. It was the first common room of its kind in any Oxford department and, as Coulson's biographer quipped, it had the remarkable effect that 'after a time the pure and applied mathematicians started to talk to each other'.[39] Later in the 1960s, as 10 Parks Road was outgrown, Coulson again took the helm and oversaw the move to the new purpose-built institute on St Giles which opened in 1966. Temple and his professorial colleagues (who in the 1950s were E. C. Titchmarch and J. H. C. Whitehead) doubtless all benefitted from Coulson's thoroughness and sheer goodwill in overseeing jobs which they, in all probability, were very happy not to do.

Figure 9.3 Three professors watching ping-pong. A table tennis match at Charles Coulson's house in 1955. From right to left, Charles Coulson (Rouse Ball Professor of Mathematics), Edward Titchmarsh (Savilian Professor of Geometry), and George Temple (Sedleian Professor of Natural Philosophy). Oxford's fourth mathematics professor, Henry Whitehead (Waynflete Professor of Pure Mathematics), is nowhere to be seen.

Temple noted that '[a]t school I had been equally proficient in Arts and Science' and on commencing studies at Birkbeck in 1918 had 'to make the first important decision in my life—whether to read Mathematics or Classics. I chose Mathematics hoping to maintain my interest in literature (in which I was successful) and guessing (correctly) that if I specialized in any Arts subject I should inevitably abandon Mathematics.'[40]

Temple did indeed maintain his interest in literature. He was clearly comfortable in a number of languages with notebooks showing him writing in French, Latin, and Greek and one fellow monk at Quarr recalling that he also had knowledge of biblical Hebrew.[41] Perhaps the most sustained example of his literary urbanity is in his 1954 inaugural lecture as Sedleian Professor, *The Classic and Romantic in Natural Philosophy*. In this lecture after suggesting 'that perhaps some light might be thrown on the concepts and methods of mathematical physics by employing the similitudes of literary and artistic criticism',[42] he states that he sees mathematics not simply as a language, but as a literature, containing within it 'almost every variety of style'.[43] For Temple, E. T. Whittaker's *History of the Theories of Aether and Electricity* was not simply two volumes of history, it was an 'epic'; the cosmological work of E. A. Milne would 'frequently rise to the heights of the lyric'; and Sir Arthur Eddington's quest for a fundamental theory was 'prophetic'.[44] He saw a text such as his predecessor Augustus Love's *A Treatise on the Mathematical Theory of Elasticity* as a classic finished work of literature or like a Greek temple constructed from inevitable analytic investigations. But for Temple the weakness of such a classic was precisely that it *was* a finished work. For him the romance in natural philosophy comes from the adventure, surprise, false starts, and struggles of discovery and creation. As an example, he turned to fluid mechanics: 'To make significant advances in this field requires the keenest physical intuition, the most powerful methods of mathematical analysis, and often the infinite capacity for arithmetic of the modern digital computing machine ... romantic fits fluid dynamics like a glove—energy, freedom, fancy, caprice—these are the authentic notes of this branch of natural philosophy'.[45] Temple saw mathematical source texts as rich, but neglected, resources for the student; elsewhere he stated: 'One of the lamentable features of contemporary university life is the decline in the discipline of the study of the mathematical classics, and the substitution of ephemeral lecture notes for the abiding companionship of a personally-acquired library, however small it may be.'[46]

Later in the same year as his inaugural lecture, Temple returned to the theme of mathematics as literature in a talk at the congress of the International Federation for Modern Languages and Literatures held in Oxford. Here he broadened his scope from fluid mechanics to aeronautics stating, 'Here all that is best in abstract analysis, in fluid dynamics, in engineering and in the spirit of the fighting services of the Crown unite and issue in some of the finest text-books and treatises which mathematicians have been privileged to study.'[47] Temple's work at Farnborough had clearly roused an enthusiasm and passion for aeronautics, even if by 1954 the subject was already on the wane in terms of his published research. Dating from the Second World War he served on the Aeronautical Research Council and various committees associated with it up until the year before his retirement from the Sedleian Chair in 1968. In addition to this, from 1954 until 1960 he served on the Academic Advisory Board of the Royal Military Academy at Shrivenham. This work was recognized by the award of a CBE in 1955.

For the modern reader the biggest surprise in Temple's inaugural address is that while he raids the rich storehouses of natural philosophy using examples from optics, special relativity, Newtonian mechanics, and continuum mechanics (across solids, liquids, and gases), he does not once mention quantum mechanics. This is all the more surprising given that between

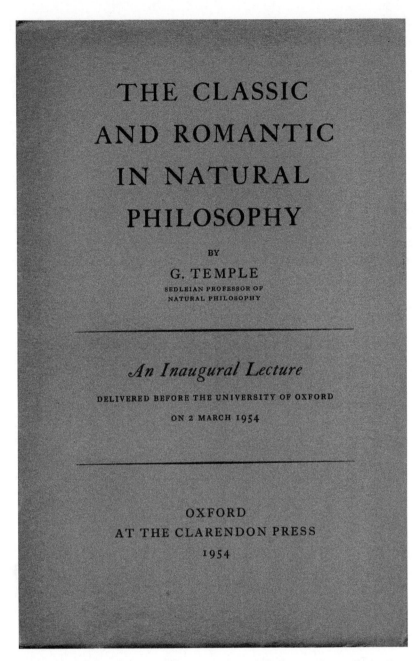

Figure 9.4 Temple's inaugural address extended the idea that mathematics is the language of nature to that it was also a literature containing 'almost every variety of style'.

1928 and 1935 he published thirteen papers and two books in the area. Indeed, one of these books was his most successful: *The General Principles of Quantum Mechanics*, a small volume which went through many editions in the *Methuen's Monographs on Physical Subjects* series. The omission from the address was not simply because Temple's interests had moved on to other things. Rather, he had decided that quantum mechanics was badly flawed, stating bluntly that he ended his work in the area in 1935 'with a short note in which I showed that the Dirac–Schrodinger theory is inherently self-contradictory'.[48]

Temple's literary interests extended to him claiming[49] that he was 'a member of the most exclusive club in Oxford—which had no name or organisation but which met every Monday in the Eagle and Child to discourse with C. S. Lewis, J. R. R. Tolkien and other great writers'.[50]

The club which 'had no name' was in fact the Inklings. Both Tolkien and Lewis would probably have been bemused at its description as an 'exclusive club', but that it contained great writers and was one of the most significant literary groups of the twentieth century is certainly true. By the time Temple became involved, the Inklings was past its 1930s and 1940s peak of regular Thursday night meetings in Lewis's rooms at Magdalen College. Indeed by the 1950s it was no longer even functioning as a literary group.[51] When Temple arrived at Oxford, Lewis was just about to move university to take up the newly founded chair of medieval and renaissance literature at Cambridge. While Lewis had moved university, he had not moved from his Oxford house and a Monday morning at the Eagle and Child pub on St Giles suited him before

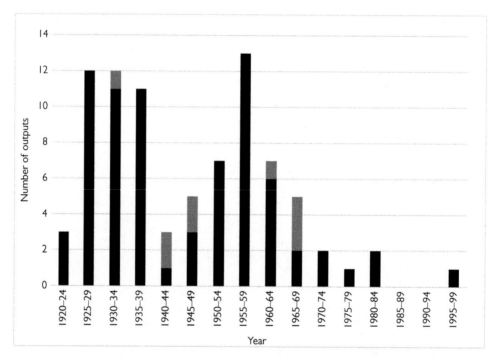

Figure 9.5 George Temple's published outputs: black, single author; grey, multiple authors. While Temple may have been gregarious in person, the majority of scholarly output (89%) was published as a single author. Hidden by this figure, however, is the sheer diversity of Temple's work, covering quantum mechanics, general relativity, fluid dynamics, the theory of functions, foundations of mathematics, and the history of mathematics. Also included are seven books: *An Introduction to Quantum Theory* (Williams & Norgate, 1931), *Rayleigh's Principle and Its Applications to Engineering* (with W. G. Bickley; OUP, 1933), *The General Principles of Quantum Theory* (Methuen, 1934); *An Introduction to Fluid Dynamics* (OUP, 1958), *Cartesian Tensors* (Methuen, 1960), *The Structure of Lebesgue Integration Theory* (OUP, 1971), and *100 Years of Mathematics* (Duckworth, 1981). Each book is recorded only once, even though a number went through several re-publications.

he got the 2:34 train across to Cambridge for his week's work.[52] Temple's introduction to the group had come from his newly acquired Oxford medical doctor, Robert 'Humphrey' Havard. Havard was also Lewis's and Tolkien's doctor, as well as being a member of the Inklings.[53]

Temple enjoyed college life: 'I greatly enjoyed the liberal and cultured atmosphere, the ancient traditions and the camaraderie of the Common Room.'[54] He was a man who appreciated fine wine, good food, convivial company, and an audience for his wit and anecdotes. Oxford agreed with him not only socially but also intellectually and his arrival in the Sedleian Chair saw a corresponding growth in his research work. His publications while at Oxford drifted towards pure mathematics, but also included two textbooks, one on fluid dynamics (1958) and another on Cartesian tensors (1960). Temple, while charming and kind, did not have a good reputation as a research supervisor. Perhaps it was because he had gained his own PhD with no formal supervision and so he did not appreciate the need to guide a student through a problem. But whatever the reason, there were a number of students who started under Temple's very light supervisory style and never finished.[55]

At Oxford he found himself taking on the roles expected of a senior academic; as a member of the General Board of Studies and of the Hebdomadal Council, and taking his turn as chairman of the Faculty of Mathematics and as examiner in the Final Honours School. He also took the opportunity to travel. During academic year 1959–60 he spent September 1959 to January 1960 at the Institute for Advanced Study at Princeton. He then moved on to the US Navy's David Taylor Model Basin, northwest of Washington, DC, for three months, where he delivered a set of research lectures on nonlinear differential equations. After Temple retired from the Sedleian Chair in 1968 his travels continued. He returned to the United Sates, spending five months at the University of Maryland before moving on to the University of Michigan for a further three. In the autumn of 1970 he visited Bulgaria as a Royal Society Visiting Professor, and in 1971 he spent six weeks as a visiting professor at the University of Aberdeen.

His career was well scattered with honours. In 1943 he was elected FRS, and as noted earlier, in 1955 he was awarded a CBE. Dublin (1961), Louvain (1966), and University of Western Ontario (1969) awarded him honorary doctorates. Finally, the year 1970 also saw him being awarded the Sylvester Medal of the Royal Society for his work on generalized functions.

Quarr Abbey

Temple had a strong Christian faith throughout his adult life. He was received into the Catholic Church in 1920 and by 1922 was professed as a Third Order Dominican (i.e. member of a lay order whose members follow Dominican spirituality while remaining in secular life). During his time at King's College and Oxford he was actively involved with the university chaplaincies and while at Farnborough during the war he became involved with the Benedictine community at Farnborough Abbey. The superior of this community, Aelred Sillem, became a close friend of Temple and supported him and his wife Dorothy through a failed pregnancy. When Sillem moved to Quarr Abbey, a Benedictine monastery on the Isle of Wight, the friendship was maintained and Temple became a frequent visitor. Thus, when Dorothy died in May 1979, after nearly forty-nine years of happy marriage, it was not entirely surprising that Temple asked to be admitted to the community at Quarr.

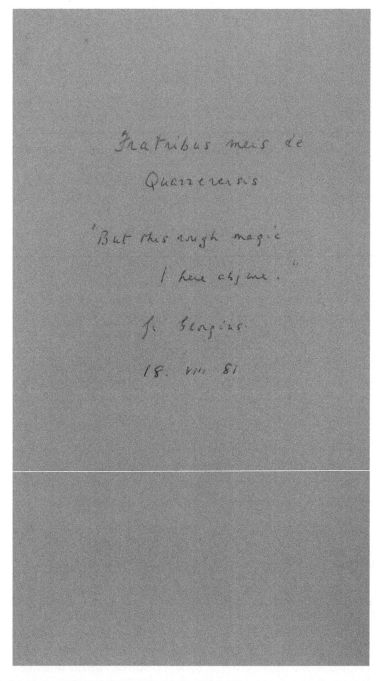

Figure 9.6 A noble, but unkept, Shakespearian promise. George Temple's inscription on the copy of *100 Years of Mathematics*, which he gave to the library of Quarr Abbey.

Temple wrapped up his affairs in Oxford,[56] and entered the Abbey in early 1980. A special dispensation was obtained from the Vatican to permit him to move directly from being a novice to taking solemn vows in 1982, with him then being ordained a priest in 1983.[57] This was an exceptionally rapid progression, but as one monk at Quarr stated, given Temple's age and experience, 'at 78 there weren't rough corners to be knocked off' and he was already very well read in both theology and philosophy. At Quarr he performed the normal duties expected of a monk, presiding at services, reading during meals, taking his turn welcoming visitors, and answering the phone at the porter's lodge, but because of his age was excused manual work.[58]

While at Quarr, Temple's *100 Years of Mathematics* was published. He inscribed the copy he gave to the Abbey library with a quotation from Prospero's speech in the last act of *The Tempest*: 'But this rough magic I here abjure'. It was a noble sentiment as he entered a new life, but not one he maintained. While at Quarr he still worked on mathematics, sending back and forth various manuscript drafts to be typed by a willing secretary at the Mathematical Institute in Oxford. The results were two papers, both stemming from an interest in the foundations of mathematics which he had developed later in life: 'Sets, Numbers and Taxa', published in 1984, and 'Fundamental Mathematical Theories', published posthumously in 1996.[59] One of his last pieces of mathematics was a note, written in 1991, to a fellow Benedictine monk on the logistic map. In it he calculated the period one and two orbits of the map and then stated, 'I am terrified of the calculation which would be necessary to examine the possibility of 4 limits!'[60] The terror, whether real or feigned, was perhaps not unreasonable for a man in his ninetieth year.

At Quarr there is significant manuscript evidence of Temple's detailed theological study, but he published on theological matters only once, and that was in 1954 long before he became a monk. And unlike his Oxford colleague Charles Coulson, who published at book length on science and religion, he published only one brief piece, again many years before Quarr, in 1929. It was a riposte to an article on Descartes whose author showed himself to have a rather confused understanding of mathematics.[61] However, Temple did speak on the links between mathematics and theology on at least two occasions. In those talks he drew parallels between the two subjects at a methodological level: both sought to argue from principles to conclusions; both seek to simplify and unify concepts across a wide landscape of ideas; and both combined science and art. Further, in considering these parallels, he returned to the ideas which he had explored in relation to his inaugural lecture at Oxford: just as natural philosophy contained both classic and romantic texts, so did theology. The systematic edifice of the *Summa* of Aquinas is the example par excellence of the classic in theology. The romantic texts are to be found in the writings of mystics such as St John of the Cross.[62]

In his last years Temple's physical health declined, but he remained mentally sharp, reading Aquinas on the Trinity a few weeks before his death. He died on 30 January 1992 and was buried on 4 February in the cemetery at Quarr.

Albert Green

Temple's paternal grandfather had been an Oxford shepherd. Green's was a factory labourer in Egham, Surrey,[63] and while Peter Chadwick's Royal Society Memoir of Green lists his father has having been an electrical engineer,[64] the truth is a little more modest. The 1911 census lists Albert Edward Green's occupation (the father and son shared the same name) as being a 'greaser' in an electrical works, with him, his wife of two years (Jane), and their first child Annie, living in a two-room property at 14 Kingsgate Rd, Kilburn, northwest London. By 1939 the family had moved a few houses along to number 22 (which was presumably bigger); Albert is listed as being a 'Bulk Supply Static Sub Station Attendant', his son (doubtless with some pride) as 'University Lecturer in Mathematics at Durham, & Research Fellow of Jesus College, Cambridge', and Jane's occupation simply as 'Unpaid Domestic Duties'.[65]

Albert Green was born in London on 11 November 1912. In 1917 he started his education at the local Kingsgate Road primary school, and in 1924 was awarded a place at the Haberdashers' Aske's Hampstead School under the London County Council Direct Grant Scheme, meaning his family did not have to pay any school fees. His mathematical ability was spotted and nurtured by the senior mathematics master Mr Oliver, a fact which in later life Green remembered with gratitude. He was a hard-working, bright, and well-organized boy and progressed smoothly through the school. In his final year he was appointed a school prefect and left at the end of July 1931 with an Open Scholarship in mathematics to Jesus College, Cambridge; a State Scholarship; and a leaving scholarship from the school.

The Jesus College which Green entered in the autumn of 1931 was one which was coming to the end of a period of post-war expansion. The number of fellows was beginning to increase for the first time since 1559 (when it had been fixed at sixteen) and various building works had been completed in the 1920s which included a block of forty-five sets of new rooms, and for the first time since the college's foundation, a set of bathrooms.[66] It was said that the aging Master, Arthur Gray, who had entered the college in 1870 and had been Master since 1912, 'knew every stone in the college, every person in its history, every book in its library'.[67] However, as far as Green's education was concerned the most important person was not the seemingly omniscient Master, but his college supervisor L. A. Pars. Pars, who Green described as a 'superb

Figure 9.7 A youthful Albert Green (back row far left) with the other prefects of 1930–1 at the Haberdashers' Aske's Hampstead School. The headmaster, Rev. F. J. Kemp, is seated in the middle.

mathematician'[68] was only fifteen years his senior, and a bachelor-don devoted to the college and to 'meticulous, lucid and inspiring teaching over the whole range of Tripos subjects'.[69] That Pars was an inspiring teacher is perhaps borne out by the fact that three of his students went on to be awarded the Smith's Prize. Green was awarded the prize in 1936 (as was Alan Turing), I. J. Good in 1940, and D. R. Taunt in 1949. Good, coincidentally, had also been a pupil at Haberdashers' Aske's school.

Pars' course of lectures on general dynamics were well attended by students and amongst the other lecturers Green experienced were Arthur Eddington, Harold Jeffreys, and Sydney Goldstein. Of Goldstein, who was only eight years Green's senior, he recalled:

> I attended many of his lectures both as an undergraduate and as a graduate student and I still have some excellent notes on Electromagnetism and Fluid Dynamics from him. He always packed a tremendous amount into lectures. One habit was to finish a lecture at 10:00 a.m. on one day in the middle of a sentence and then to begin his lecture the next day promptly at 9:00 a.m. continuing the same sentence as he walked in the door! He also disregarded physical disabilities. Occasionally he suffered from gout and would lecture seated on a bench with both feet and legs wrapped in bandages using the board above as far as he could reach.[70]

Green was one of fifty-three students placed, along with three other students from Jesus, in the first class in Part 1 of the Mathematical Tripos in 1932, meaning he was in the top 40% of his cohort.[71] After two further years of study, Cambridge finals loomed as an exacting test, with six three-hour papers being taken on the Monday to Wednesday of one week, followed by a further six on the Monday to Wednesday of the next. When the results were posted on the Senate House notice board on June 12th,[72] Green was listed as a wrangler in Part II with special credit being given for his performance is the more advanced Schedule B papers (which in the same year was renamed Part III). To give a sense of the scale of Cambridge mathematics at the time, in 1934 a total of eighty-five candidates are listed in the published results for Part II, seven being from Jesus College and thirty-eight of them listed as wranglers. Seventeen of the wranglers attempted Schedule B papers, and of those there were nine 'who in the opinion of the Moderators and Examiners deserve special credit'. Thus, Green was in the top 10% of his cohort.[73] Other wranglers that year included Alan Turing of King's College and Maurice V. Wilkes of St John's. Turing's contributions to code breaking and computing are well known. Wilkes went on to be a major driving force in early computing at Cambridge and led the team which constructed the Electronic Delay Storage Automatic Calculator (EDSAC), which ran its first code in 1949 as one of the first stored program computers.

After graduation, Jesus College awarded Green a grant allowing him to work as a PhD student under the supervision of Professor G. I. Taylor. Taylor, who is considered to be one of the most important fluid dynamicists of the twentieth century, combined abilities in mathematics and experimental science with strong physical insight and a joy in finding agreement between theory and experiment. Taylor, while an encouraging supervisor, treated his students more as independent colleagues than fledgling researchers to be closely guided. He pointed students to an area of interesting work, gave written insights in his barely legible hand, and let them get on with it. Green stated that

> G.I. always left me to do my own mathematical work, but he could always put his finger on key physical aspects. When I was talking to him he would scribble some ideas on a piece of paper in a very untidy way. I would take this valuable paper away to decipher at leisure.[74]

Even though Green claimed, like many a young research student, that 'after about 9 months I was almost in despair as I had made absolutely no progress', by 5 March 1935 his first paper, 'The Gliding of a Plate on a Stream of Finite Depth', had been submitted to the *Mathematical Proceedings of the Cambridge Philosophical Society*. It was the first of a dozen to come out of his time at Cambridge, which included further analysis of the gliding plate problem; the evolution of eddies in viscous flow; and the elastic stability of a twisted metal strip. This last topic included Green's only foray into experimental work. In the Cavendish Laboratory he set about examining the twisting of two different steel strips under increasing torque. Even though he found that his theory and experiment agreed 'fairly well' he stuck to theory thereafter.[75]

Green's clear ability was soon rewarded. In 1936 he was not only awarded a Smith's Prize, but also a research fellowship at Jesus College, and in 1937 he got his PhD. During the tenure of his research fellowship Green was also actively involved in lecturing. He taught students at Jesus, Corpus Christi, Girton, and Newnham, and also acted as scholarship examiner for these colleges. He also gave courses of lectures on hydrodynamics to Part III and postgraduate students. At the end of a glowing reference, written in June 1939 to support Green's application for the post of lecturer in mathematics at Durham University, L. A. Pars stated, 'I can recommend Dr Green unreservedly. He is a man of exemplary character and an excellent teacher, and he

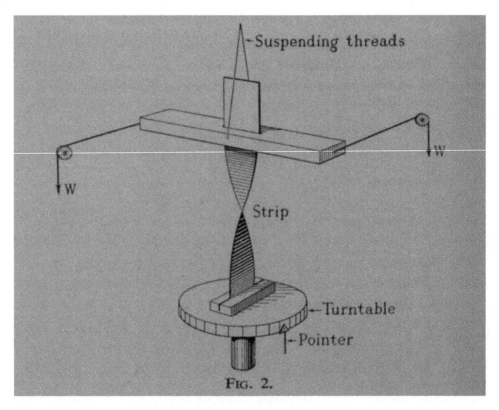

Figure 9.8 Experimenting with experiments: Albert Green's 1936 apparatus for measuring the elastic stability of a metal strip. He found that his theory and his experiment agreed 'fairly well', but apart from a follow-up paper completed later the same year, it was his last experiment.

can be relied on absolutely to carry out his duties cheerfully and conscientiously. There is no doubt that he has ability of a very high order. And he is a particularly pleasant and sympathetic colleague to work with.'[76] With a reference like that, coupled with enviable tripos results, the Smith's Prize, and a formidable list of publications, it was not surprising that in the autumn of 1939 Green moved north to take up the post at Durham.

Durham

Durham was at this stage a university of eight colleges and about 500 students and was small enough to form a community where academics, students, and domestic staff all knew each other.[77] Green briefly lived in University College until his marriage on Christmas Day 1939 to Gwendoline May Rudston. The couple had known each other since their schooldays and grew up as neighbours on the Kingsgate Road. They were married in Kilburn Methodist Church, Hampstead, honeymooned in Eastbourne, and then moved to Durham to a house on Field-house Lane, a fifteen-minute walk from University College. They were to live there for the next twenty-nine years.

As Green arrived in Durham, Percy John Heawood, the Professor of Mathematics and expert in the four-colour problem, was retiring at the age of 78. He was replaced by the promotion from reader to professor of Joseph Burchnall. Burchnall was a capable mathematician and thorough university administrator who had lost his leg, and won the Military Cross, in the First World War. As the Second World War began, Burchnall took command of a section of the Home Guard; E. F. Baxter, the only applied mathematician in the department, moved to the Meteorological Office; and Green, a pacifist, was registered as a conscientious objector 'without conditions'.[78] The loss of Baxter meant that the entire mathematics department at Durham was made up of Burchnall, Green, and a temporary lecturer. Student numbers in mathematics were small, with single figures taking the honours course, but the teaching load was heavy. Green stated:

> I was expected to lecture on any topic of the undergraduate course in either pure or applied mathematics—something which I could not do today. If present-day staff were required to do the number of lectures per week that we had to do there would be some sort of sit down strike or walk-out. It was many years before this situation changed.[79]

Two more temporary posts, filled by schoolmasters, were created in 1942 when the university started to take RAF crews in batches of 100 for six months at a time to give them training in mathematics and navigation. One of the schoolmasters, Syd Holgate, a recent Durham graduate, quickly became Green's first research student, and spent the rest of his career at the university, moving into administration after the war and then becoming the first Master of the newly formed Grey College in 1959.

Despite his heavy teaching load, Green's research carried on apace with twenty-one papers being published in the 1940s, mostly on isotropic and anisotropic materials. A simple example of an anisotropic system which Green investigated was the buckling of plywood. Pars' reference from 1939, stating Green had 'ability of a very high order', had been borne out by both the quality and quantity of his research and by the end of the decade he was recognized as one of the UK's authorities on the theory of linear elasticity.

Newcastle

In 1948 Green became professor at King's College, Newcastle upon Tyne. King's was at this stage a large college of Durham University, which in 1963 would become the University of Newcastle. The Mathematics Department was made up of Green as Professor of Applied Mathematics; W. W. Rogosinski, who had just been promoted to Professor of Pure Mathematics; and eight lecturers. Rogosinski, or Rogo to his friends, was a Polish Jew who, with the help of British mathematicians John Edensor Littlewood and G. H. Hardy, had fled Nazi Germany in 1937. Between them, Green and Rogosinski ran a department which fourished in both pure and applied mathematics and nurtured a number of young academics who went on to prominent careers.

When George Temple arrived at Oxford in the 1950s it was about to undergo a phase of rapid expansion, but this was not peculiar to Oxford or the discipline of mathematics. The post-war British Government was keen not only to expand higher education, but to specifically increase provision in science and technology. Thus, like many departments across the UK, mathematics at Newcastle grew rapidly. During the twenty years Green was there the academic staff tripled to include two professors of pure mathematics, two in applied, and one in statistics. Student numbers also grew, from single figures studying honours mathematics in the mid-1950s to around sixty by the early 1960s. Green was considered by his colleagues to be a good head of department. He was keen to give staff freedom in both teaching and research, and happy to quietly ignore aspects of wider university structures and regulations if he felt they were unhelpful. This latter pragmatism was also applied, probably to the consternation of the university authorities, when he served as Dean of the Science Faculty from 1959 to 1962. When Rogo retired from Newcastle in 1959, Green became overall head of the department and was even-handed in his support of pure and applied mathematics and statistics during the next decade as the department continued to grow, move to new buildings (where he found himself 'concerned with all the paraphernalia of architects etc'.[80]) and become part of the new University of Newcastle.

Green's administrative roles did not frustrate his progress in research. He had a string of post-graduate students who were met with regularly and expected to have their progress written up as they went along. Weekly departmental seminars, which Robin Knops, a lecturer, then reader at Newcastle between 1962 and 1971, described as 'exciting events that provoked lively and constructive discussion and occasionally some rethinking', provided regular settings to explain recent work.[81] Green also began to work increasingly in collaboration, an approach to research which would dominate his work until the end of his life. One of Green's first collaborations while at Newcastle was with Ronald Rivlin, who from 1944 to 1953 worked for the British Rubber Producers Research Association (BRPRA) based in Welwyn Garden City outside London. The work with Rivlin was prompted by Green's interest in nonlinear elasticity and the publication of exact solutions by Green and his research student R. T. Shield for the twisting of a circular cylinder and the deformation of a spherical shell. The discovery of exact solutions, a comparatively rare find in any nonlinear system, was facilitated by the assumption of the incompressibility of the solid. The results overlapped with work already done by Rivlin, who was working on a theoretical analysis of the behaviour of vulcanized rubber. Green also worked with Rivlin on the behaviour of materials with 'memory' where the stress at a point in the material depends on the deformation of the material at that point over a previous time interval.

Rivlin invited Green to become a consultant at BRPRA and a research collaboration, which began shortly before Rivlin left to take up the chair of applied mathematics at Brown University in the United States in 1953, continued until 1968. Today such a long-distance collaboration would take place via email and Teams. In the 1950s and 1960s it was via airmail, when a 6*d* stamp would take a letter from Newcastle, Co. Durham to Providence, Rhode Island in three or four days. Some of the letters between the men were as short as an email; only one or two sentences. Others were several pages long, complete with equations, blunt disagreements, and occasionally a follow-up apologizing for being too direct. At times both men grumbled about the burden of university administration; less frequently still they discussed personal life. Mostly they stuck to research.[82]

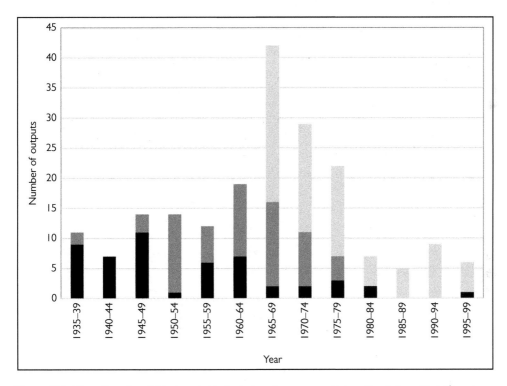

Figure 9.9 Albert Green's published work: black, single author; grey, multiple author (where Naghdi is not a co-author); light grey, multiple author (where Naghdi is a co-author). In contrast to George Temple, who mostly published as a solo author, Green worked for most of his career in collaboration with others. Of his 194 publications 143 were co-authored. By far his most significant collaborator was Paul Naghdi from the University of California, Berkeley. They published 83 joint papers between 1965 and 1996. Other significant collaborators included Norman Laws (15 joint papers) and Ronald Rivlin (14). Included in this figure are Green's two books, *Theoretical Elasticity* (with W Zerna, Clarendon Press, 1954) and *Large Elastic Deformation and Non-linear Continuum Mechanics* (with J. E. Adkins, Clarendon Press, 1960). Both went through second editions, but are only counted once.

A. J. M. Spencer, who as a young researcher worked with Rivlin and Green in the late 1950s, stated:

> Albert and Ronald were completely different personalities, but I thought they complemented each other perfectly; Ronald had the ideas and very deep physical insight, while Albert was also pretty good at the physics but in addition was an excellent technical mathematician. Ronald would always tackle a problem head on, whereas Albert tended to be more subtle. Neither of them could easily be persuaded to accept an argument, but they normally agreed in the end.[83]

Rivlin and Green also seem to have felt they worked well together. The collaboration resulted in not only fourteen papers and a number of visits by each to the other, but also with Rivlin offering Green a permanent job at Brown in 1961.[84] Green declined the offer, but he did spend two extended periods at Brown during the academic years 1955–6, and 1963–4. During the second of these visits, he spent the first semester at Brown, and then spent the second semester at the University of California, Berkeley with Paul Naghdi. This was the beginning of another, and much more substantial, transatlantic collaboration which would last for the rest of each of their academic careers and produce eighty-three papers.

(a)   (b)

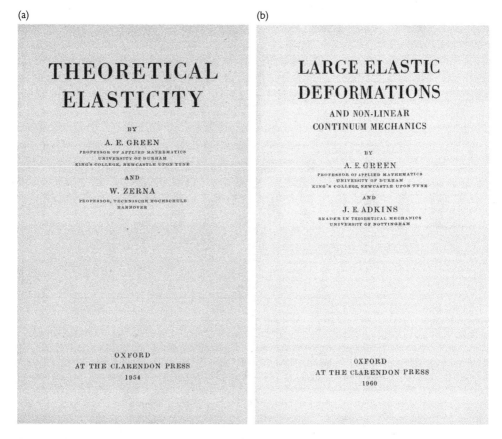

Figure 9.10 Albert Green's two co-authored books became standard reference texts in continuum mechanics, with both volumes running to second editions.

During Green's time at Newcastle he also co-authored two books. *Theoretical Elasticity* (with W. Zerna) published in 1954 gave a book-length coverage of linear and nonlinear elasticity, and *Large Elastic Deformation and Non-linear Continuum Mechanics* (with J. E. Adkins), published in 1960, formed a natural companion to it. Both became standard reference texts in the area. Election to the Royal Society came in 1958. Peter Chadwick states that 'Within the bounds of a strong natural modesty he took great pleasure in his election. There were no changes, however, in his working life.'[85] That sounds like a good summary of the man. Straightforward, modest, and not to be deflected from the research work which he thoroughly enjoyed.

Oxford

By the late 1960s Green was settled in his career and his wider life. He and Gwen were still living in the same house in Durham, the department at Newcastle was a pleasant train journey away, his career had progressed smoothly, and his research, where he was an international authority, was absorbing and agreeable. And then in 1967 came an informal approach from Oxford asking if he would be happy for his name to be considered by the electors of the Sedleian Chair. He heard nothing from Oxford for several months and then received a letter informing him that he had been appointed. He was mildly irritated by the fact that he was seemingly being told of his appointment rather than being offered it, but it was a decision which did not take long to make. Colleagues at Newcastle, while sorry to see him go, realized it was a natural step. By the summer of 1968 he was in post and Oxford awarded him a D.Sc. Perhaps not to be outdone in 1969 Durham brought him back up north and awarded him another. As part of his speech the university's Orator, W. B. Fisher, stated that Green was 'one of the most distinguished of English Applied Mathematicians of the past thirty years', but also joked that

> In a world where we are nearly all affecting to ignore or brush off the stark reality that almost everything now is an aspect of, or controlled by mathematics, it is partially reassuring to note, from the very long list of Professor Green's publications, how mild, homely and humane mathematics really can be: his repeated studies on the gliding of a plate, the theory of shells, and the intimate yet vital properties of elasticity. A paper entitled 'Nonlinear materials with memory' I take as a mathematical definition of the human species, and here also it must be admitted some uneasiness does creep in with the evidence again from publications of Professor Green's preoccupation with rods, rigid and otherwise.[86]

It may be the first and last time in print that the solution of nonlinear partial differential equations has been described as either mild or homely.

The period 1965–79 saw the years of Green's peak research output, and while this overlaps with his time as Sedleian Professor, it was the extensive collaboration with Paul Naghdi (with whom he published thirty-six papers during his time as Sedleian Professor) rather than the environment of Oxford that explains the huge productivity. Green did not build up a research group at Oxford, but he and Naghdi regularly visited each other across the Atlantic, and in between the visits there was a flow of letters and telephone calls.[87] James Casey, who was one of Naghdi's collaborators at Berkeley, noted that 'Paul and Albert were intellectually frank with one another, admired one another greatly, and had a deep, enduring friendship. They always

Figure 9.11 Albert Green (front row, far left) on his award of a D.Sc. by the University of Durham, 25 June 1969.

had a paper or two in progress, other papers planned, and there were eternally long battles with referees'.[88] As noted earlier, between 1965 and 1996 they co-authored eighty-three papers.[89]

Outside the regular supervision of the three graduate students[90] he had while Sedleian Professor, and giving lecture courses,[91] Green's engagement with Oxford and the Mathematical Institute was minimal. He served as chair of the Committee of Management of the Institute from 1973 to 1976, but according to Sir John Kingman, who was Wallis Professor of Mathematics at Oxford from 1969 to 1985, he took a decidedly hands-off approach and 'left everything to the ... Administrator Rosemary Schwerdt, and signed what was put in front of him'.[92] At one point he simply refused to become an examiner (i.e. one of the committee of academics whose job it is to set the final examination papers) with John Ockendon, an applied mathematician who has spent his career at Oxford, stating that 'how he managed to get out of that I don't know. He just was not part of the everyday workings of the maths department'.[93] His disengagement also extended to his fellowship of Queen's, with him taking little interest in college life to the extent that when Naghdi visited he would ask a colleague, Alan Day, to arrange for Naghdi to dine at Hertford College as he did not feel able to make such arrangements at Queen's.[94] None of this should be taken to imply that Green was unhappy at Oxford, nor that he was a difficult colleague. He remained, as he had always been, friendly and good natured, but his overriding concern was to make progress in his research, which was his sole focus and consumed his time. Perhaps he was simply under the misapprehension that he would be completely free of administrative duties as Sedleian Professor.

Green's research from 1968 to the end of his career continued to cover a wide range of topics within continuum mechanics, including a return to fluid dynamics, but a major focus of his work during this period was thermomechanics. This included not only investigating the interplay of elasticity, solid deformation, and heat flow, but also more formal studies of the links between continuum mechanics and the laws of thermodynamics. Formal is a keyword here. Green had always been formal in his approach to continuum mechanics, inclining towards rational mechanics, i.e. the desire to place the topic on an axiomatic basis. This inclination appears to have increased as he got older, with John Ockendon stating, 'I can remember once having my only real argument with him. He said he'd rather have a rational theory that was irrelevant to the real world than a theory that was not rational and relevant to the real world.'[95]

It was on the topic of rational mechanics that Green incurred the ire of his Oxford colleague Les Woods.[96] Woods (who incidentally had George Temple, then at King's College London, as the external examiner on his Oxford D.Phil. in 1950) was a maverick New Zealander who had worked across the areas of engineering, plasma physics, and applied mathematics, and was not afraid to speak his mind. In particular his applied science background meant that he was much more interested in mathematical models that worked (i.e. agreed with experiments) and physical assumptions that were reasonable, than with axiomatic consistency. At its simplest, and most generously interpreted, the difference between Woods and Green was a matter of emphasis. Green emphasized internal mathematical consistency, and Woods emphasized pragmatic modelling. Despite Ockendon's remark above, both men knew as well as any that good applied mathematics had to be both consistent *and* physically relevant. To an outsider this may appear arcane, but on the instigation of Woods, the disagreement made it to the pages of the *Bulletin of the Institute of Mathematics and its Applications* and became what John Ockendon described as 'the biggest row I've ever seen in Oxford applied maths'.[97] Woods described the axiomatic approach of rational mechanics as sophistry and a 'dry rot in the foundations of the subject'.[98] Green, though not explicitly singled out by Woods, was undoubtedly his target and was roused enough to respond. His reply avoided rhetoric and stuck closely to defending all but one of the particular axioms which Woods had criticized.[99]. This was then followed up by a rejoinder from Woods seeking to 'convert Professor Green to my viewpoint'.[100] He didn't. Green maintained the integrity of his approach and both in private and in print appeared unfazed by Woods' critique.[101] As he had said in his 1974 acceptance speech for the Timoshenko Medal from the American Society of Mechanical Engineers:

> We all suffer from prejudices in our everyday life and it is not surprising that this spills over into science. Some regard highly abstract mathematical presentations of work as being divorced from physics while others regard some aspects of physics as mere handwaving. I believe that there is something of value in the whole range of scientific thought. Of course, intensive discussion and argument with colleagues is sometimes a very profitable—or at least a very enjoyable exercise.[102]

While it is doubtful that Green found his disagreement with Woods 'enjoyable',[103] he seemed just as able to disagree without rancour with him in public as he was to disagree with Ronald Rivlin and Paul Naghdi when they were hammering out a joint paper in private.

Retirement

Green retired from the Sedleian Chair in 1977, and though his research output declined, it was still by any standards substantial, with thirty-six papers being published between 1978 and 1996. Virtually all of these were in collaboration with Naghdi. When Naghdi died in 1994, Green saw their last five papers through to publication and then, aged eighty-three, finally stopped. He died on the 12 August 1999, aged eighty-six.

It would be tempting to read Green's career as that of a man who, because of his devotion to research, saw it as the sole measure of an academic's worth. But he was much more circumspect, both in the value he put on wider scholarship and on the significance of an individual's work. To quote again from his 1974 Timoshenko Medal acceptance speech:

> After the 1939–45 war interest in research in university departments greatly increased and the pressure on staff, particularly younger members, is tremendous—publish or perish has almost become the watchword. I am afraid that this tends to lead to bad standards. I particularly regret that often due recognition is not given to the type of person in a university who is a true scholar but is not one to produce a large number of papers. Such a person, who often had wide knowledge and understanding, can be invaluable in a department but gets left behind in the promotion stakes. The output of scientific papers in every subject is enormous and in recent years there has been a tremendous increase in the number of journals published. It is practically impossible to keep track of every paper in a particular area of research of interest, let alone in a variety of topics. As a result some duplication of effort is inevitable. Also I guess that only a small fraction of work is ever read in a thorough way. . . . On looking back over the history of science one realises that most of us can only hope to place one small brick—if that—in the edifice—and even that may get knocked out by following generations.[104]

Fifty years on, Green's brick is still in place.

CHAPTER 10

Brooke Benjamin

TOM MULLIN AND JOHN TOLAND

Brooke Benjamin, a tall figure who always wore a suit or a blazer and tie and frequently smoked a pipe, was Oxford's seventeenth Sedleian Professor of Natural Philosophy between 1979 and 1995 (Figure 10.1). He was somewhat mysterious, softly spoken with deliberate enunciation that conveyed a mixture of modesty and authority, and with a gentle sense of humour he inspired admiration, loyalty and respect. His handwriting was exquisite, with serifs carefully placed on every capital letter and he spoke slowly, every word having been chosen with care. Because of his imposing presence, and since several of his best-known papers are signed T. Brooke Benjamin, it was sometimes thought that his surname was hyphenated, but it was not. He was Thomas Brooke Benjamin (1929–95) and everyone called him Brooke, which was his mother's maiden name.

The early years

The eldest of three children, Brooke was born half way between the two world wars, on 15 April 1929. His father, Thomas Joseph, was a lawyer in Liverpool who served as Registrar of the High Court of Justice and the County Court, and his mother, Ethel Mary (née Brooke), was an accomplished amateur pianist. They lived just north of Liverpool in Wallasey where between the ages of five and seven Brooke attended the local primary school before going away to boarding school for four years. When he returned, he enrolled as a day pupil in Wallasey Grammar School, but with the outbreak of the Second World War the family moved to Mold in North Wales. Mold was thirty-five miles south of Wallasey on the other side of the river Mersey, well away from Liverpool docks which were obvious targets for enemy aircraft.

In Mold Brooke attended the Alun School, but at the age of fourteen became ill and was diagnosed as diabetic. So when in 1943 the family returned to Wallasey and Brooke went back to the grammar school, his days were often spent at home, with teaching material and work assignments being brought to him by fellow pupils. Despite all the difficulties, he excelled in formal

Tom Mullin and John Toland, *Brooke Benjamin*. In: *Oxford's Sedleian Professors of Natural Philosophy*. Edited by: Christopher Hollings and Mark McCartney, Oxford University Press. © Oxford University Press (2023). DOI: 10.1093/oso/9780192843210.003.0010

Figure 10.1 Brooke at home in Oxford shortly after becoming Sedleian Professor.

academic work, with particular interest in the sciences, while at the same time developing a sophisticated interest in music. During his senior years at Wallasey Grammar School he built radio sets as a hobby and saved pocket money to buy a second-hand violin. In addition to playing music, he was interested in composing, notably a piano quintet which was performed at a public concert in Wallasey when he was sixteen. So when it came to deciding what to do next he was torn between music and science and it was not easy. He chose engineering.

A student at Liverpool, Yale, and Cambridge

Despite the effect of illness on his schooling, his headmaster strongly recommended him to Liverpool University from which he graduated with a first class honours degree in engineering in 1950. As an undergraduate, he conducted numerous concerts for the university music society, his own string quartet (1948) being particularly well received when reviewed by the local press, and he was one of those who worked to found the Liverpool Mozart Orchestra, still one of the UK's leading amateur orchestras.[1] Also, while at university he spoke in public on social and political causes, and developed a capacity for leadership which suggested the shape of things to come.

In 1951 Brooke was awarded a Rotary Foundation Fellowship to study electronics in the United States and went to Yale, from which he emerged as a Master of Engineering with a life-long interest in American football. At Yale he learned to treat his diabetes by self-injecting accurate doses of insulin that led to significant, long-term improvements in his general health. Indeed, in later life no one would have guessed, and only very close friends knew that he was diabetic. When he returned to the UK, on a cargo vessel via Manila and Singapore in October 1952, Cambridge University accepted him as an electronics research student in the Department of Engineering and he matriculated as a student of King's College. But fate intervened and Brooke enjoyed telling how.

One day, when passing the hydraulics laboratory in the Inglis building, quite by chance he stuck his head around the door and was fascinated when he saw a hive of activity, including G. I. Taylor's experiments which involved an unusual square-shaped wave tank and swirling flow in tubes, with A. M. Binnie (later FRS) clambering around the huge apparatus.[2] Brooke was enthralled and almost at once transferred to work there.

Supervised by Binnie, his PhD project was to study the tiny bubbles which are created by the screw of a ship as it drives through water. When these bubbles collapse, the pressure developed is so great that the metal screw becomes pitted with small cavities which erode the surface, and eventually the performance of the screw is impaired. Brooke devised ingenious experiments to study the formation and sudden collapse of these bubbles. He used electronics to initiate bubble formation, and then took remarkable high-speed photographs to capture the behaviour of individual bubbles. The sophistication of the experiment was due to his mastery of electronics, and his photographs of jets emerging from collapsing bubbles and damaging the screws were the first confirmation of this phenomenon.

Early career

On the basis of his thesis on cavitation in liquids, in 1955 Brooke was elected a fellow of King's College Cambridge and appointed to a research position in the Department of Engineering. In 1958 he became Assistant Director of Research, jointly in Engineering and in Applied Mathematics and Theoretical Physics, and spent the next twelve years in Cambridge comparing experiments with the implications of mathematical models. In 1966, only eleven years after his PhD, he was elected Fellow of the Royal Society for having shown a remarkable combination of theoretical and experimental ability in tackling a wide range of problems in fluid dynamics, and having distinguished himself both as an applied mathematician and as an engineer.

In a series of elegant papers he was deemed to have shown physical insight, originality, and clarity of thought in contributions to the theory of open-channel hydraulics (undular bores, solitary waves, and the flow due to obstacles in stream), cavitation in water and the pressure

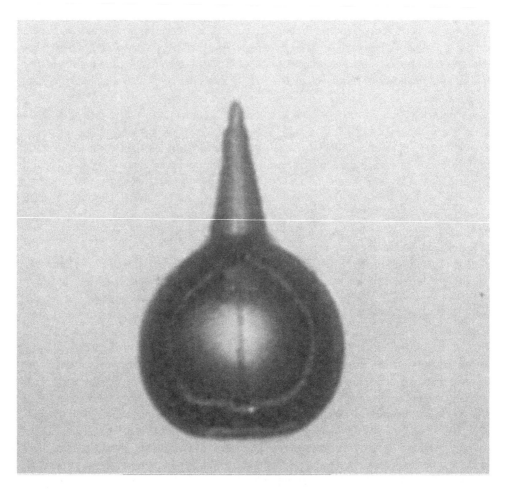

Figure 10.2 Jet formation during the high-speed collapse of a vapour cavity adjacent to a solid boundary. Formation of the jet takes place in less than two milliseconds.

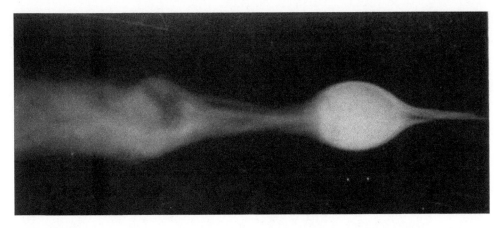

Figure 10.3 The breakdown of a vortex core in air. The core was created inside a Perspex tube using vanes, and the motion was visualized using smoke.

waves from collapsing cavities (Figure 10.2),[3] the stability of flow near a rigid wall which is either rigidly wavy or flexible and the stability of moving liquid layers on walls, vortex breakdown phenomena (Figure 10.3),[4] and the dynamics of a system of articulated pipes conveying fluid.

Scientific outlook

Since careful experiments will produce the same data if repeated in a hundred years' time, and since rigorous mathematics yields irrefutable consequences of models, Brooke believed that experiments and rigorous analysis of models are the pillars upon which the physical sciences rest.

He knew that in 1834 John Scott Russell had accidentally discovered, and subsequently studied in great experimental detail, a water wave with a single bump and no ripples (Figure 10.4), propagating at constant speed along a canal,[5] and he was aware that the discovery was controversial at the time because illustrious mathematical physicists such as G. G. Stokes and G. B. Airy were unable to explain his findings, which Russell called the Great Wave of Translation, by existing theory.[6] Brooke also knew that by the mid-twentieth century a rich mathematical theory had been developed to explain Russell's wave of translation, and that similar phenomena had been found to play key roles in many branches of the physical and life sciences, technology, and medicine. Nowadays these waves are collectively called solitons and the literature about them is vast.

Russell's discovery of the wave of translation confounded mathematical theory of the time because it was not fully appreciated that the observed phenomenon was a subtle consequence of the equations governing water-waves being nonlinear. Brooke was extremely interested in the role of nonlinearity in many different contexts.

In experiments which Brooke suggested to his mature PhD student, Jim Feir, he was surprised to find that it is almost impossible to create periodic water waves of moderate amplitude:

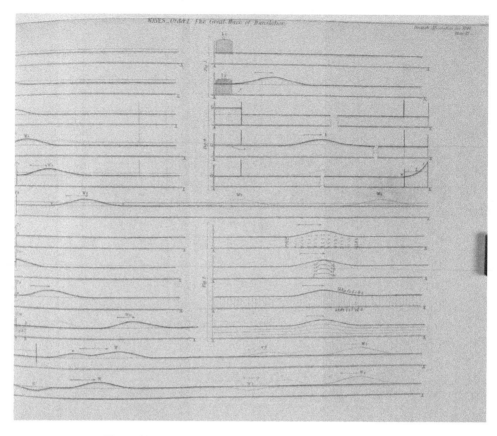

Figure 10.4 Russell's illustrations of his Great Wave of Translation.

no matter how carefully the experiment had been set up, the waves always disintegrated as they rolled along the canal, as in Figure 10.5.

After recognizing that the experimental difficulties encountered are of physical significance, Brooke went on to explain them mathematically in what is now commonly known as the Benjamin–Feir theory of side-band instability.[7] In it the observed disintegration of periodic waves is attributed to nonlinearity and the transfer of energy to waves of nearby frequencies. Incidentally, the notion of side-bands would have been familiar to a radio enthusiast such as Brooke, since side-bands in radio communication theory are bands of frequencies, higher or lower than the carrier frequency, which arise as a result of modulation.

In their celebrated paper, Benjamin and Feir wrote

> The present findings are perhaps most striking when considered as an epilogue to the famous controversy about the existence of water waves of permanent form. [. . .] Until now, however, it has apparently not been suspected that for waves on deep water the equilibrium so long in question is in fact unstable.[8]

Here the 'famous controversy' refers to the fact since the 1920s, when the existence of solutions of the water-wave equations had been established for periodic waves of very small amplitude,[9]

Figure 10.5 The experiments were performed in a large wave tank facility at the National Physical Laboratory, Feltham. The top photograph shows well-defined waves close to the wave maker and the bottom shows how they have disintegrated 60 m away.

little progress had been made and it was difficult to see how the equations could be used to predict the variety of waves seen on water surfaces. But this was an old problem. Already in 1845 Samuel Earnshaw had complained that a hundred years after Euler had derived the equations of fluid motion he was not aware of any fluid motion which had been explained by them, except for some very simple and uninteresting cases, and not for want of effort but because of the 'peculiarly rebellious character of the equations themselves which resist every attack'.[10]

The wave of translation was a case where the experimental evidence eclipsed what was then current theory. In his seminal report to the meeting of the British Association for the Advancement of Science in York in 1844, Russell said bitterly that 'the greater part of the investigations of M. Poisson and of M. Cauchy under the name of wave theory, are rather to be regarded as mathematical exercises than as physical investigations', and he blatantly challenged mathematicians to account for his wave of translation:

> Having ascertained that no one had succeeded in predicting the phenomenon which I have ventured to call the wave of translation [. . .] it remained to the mathematician to predict the discovery after it had happened, i.e. to give an *à priori* demonstration *à posteriori*.[11]

Despite this, the Euler equations for water waves remained essentially unexamined for more than a hundred years. However, Brooke was aware that by the middle of the twentieth century significant progress was being made and he was particularly impressed by Krasovskii's 1961 existence theory for water waves with any maximum inclination to the horizontal less than 30°.[12] To Brooke it was obvious that this theory of existence of solutions 'in the large' of a nonlinear equation, which was based on abstract topological methods, could not be replicated using perturbation methods familiar in engineering studies of solutions close to equilibrium, and it was striking that Krasovskii had proved the existence of waves of moderate amplitude, which he (Brooke) knew from Benjamin–Feir instability theory were unstable. Brooke was equally aware that the topological methods recently used by Krasovskii had been important in mathematical fluid mechanics since the seminal work of Leray in 1934.[13] However, since the abstract approaches of Leray and Krasovskii were not being taught in the UK, Brooke feared that it was losing ground in theoretical aspects of a subject where it had once been pre-eminent. So he set about studying the Russian literature, such as Krasnosel'skii, Ladyzhenskaya, and Vainberg, in translation.[14] His aim was to adapt their methods to other problems in hydrodynamics where nonlinear theory was either inadequate or nonexistent.

Essex and the Fluid Mechanics Research Institute

At about the same time, the Mathematics Committee of the UK Science Research Council, spurred on by Michael Atiyah,[15] had become concerned that the increasing volume of research in partial differential equations worldwide, especially in France, Russia, Scandinavia, and the USA, meant that the UK was falling behind. So it set up a national scheme to incentivize fundamental theoretical research in this important area of mathematics, which includes the theory of nonlinear equations that arise in fluid mechanics. As part of this initiative, in 1970 Brooke was appointed to a professorship at the University of Essex where he set up the Fluid Mechanics Research Institute (FMRI). Its purpose was to focus on what he later described as the alliance of practical and analytical insights in nonlinear problems of fluid mechanics.[16]

The Science Research Council grant for the institute provided for five post-doctoral research assistants, a laboratory technician, and a number of senior visitors, each of whom would spend a sabbatical year at the institute. John Mahony, on leave from the University of Western Australia, was the first of these visitors,[17] and the young research assistants included theoreticians Ron Smith and Jerry Bona, and experimentalists Barney Barnard and Bill Pritchard. Although the goal was to develop mathematical theory capable of treating questions about existence, uniqueness, and stability of solutions of complicated nonlinear equations and systems, a key component of the institute was the laboratory. In it there were several wave-tanks, equipment for creating carefully stratified fluids, a pond with an island at its centre for the study of edge waves, and various apparatus for studying rotating viscous fluids such as golden syrup. Crucial to the success of the institute were the prolonged coffee sessions in an unusually shaped, small triangular room adjacent to the laboratory. Brooke sometimes led the conversation with vivid descriptions of experiments he had conceived, and often tried out in his kitchen or bathroom, at home. But everyone, from research students to senior visitors, had their say in what were sometimes lively scientific discussions of equations and inequalities, or the design of physical apparatus. It was a very stimulating environment.

Because of his interest in stability of waves, Brooke saw the need for a proof of the existence of solutions to the initial-value problem of a seemingly simple equation modelling waves on shallow water introduced in 1895 by Korteweg and De Vries,[18]

$$\varphi_t + \varphi_x + \varphi\varphi_x + \varphi_{xxx} = 0, \qquad x \in R, t > 0, \tag{10.1}$$

but none existed. However, discussions in the triangular coffee room led Benjamin, Bona, and Mahony to develop a satisfactory theory for a different, but formally equivalent, nonlinear equation for dispersive waves, now known as the BBM equation,

$$\psi_t + \psi_x + \psi\psi_x - \psi_{xxt} = 0, \qquad x \in R, t > 0. \tag{10.2}$$

This turned out to be a key ingredient in a theory, subsequently developed by Bona and Smith,[20] that gave an affirmative answer to the original question about the initial-value problem for the Kortewig–De Vries equation. Around the same time Benjamin and Bona settled the question of stability of solitary wave solutions of (10.1) and (10.2), and the institute was off to a flying start.[21] Later, two young colleagues at FMRI published the first rigorous mathematical account, based on Euler's equations, of the large amplitude waves of translation that had led Russell to his challenge, to find an 'à priori demonstration à posteriori'.[22] At the same time Brooke sought to cultivate the French tradition, which went back to Leray, Hadamard, Poincaré, and beyond, of using abstract mathematics in problems of mechanics and physics. He wanted everyone in the institute to adopt this approach, and he urged everyone to learn to speak French!

Jerry Bona, who had just finished his PhD at Harvard under the supervision of Garrett Birkhoff, made an immediate impact on the mathematical culture of the institute. His thesis, entitled *Cauchy Problems with Random Initial Values in the Space of Tempered Distributions* (1971), meant that Jerry had the right mathematical background for the challenges and opportunities offered by FMRI. Moreover, he recognized the emerging role of computing machines when comparing experimental data with the numerically calculated solutions of model equations, and he understood the necessity of relating numerical methods to theoretical studies of the differential equations which they solved if there was to be reliable comparison of theory with experiment, as Brooke envisaged. An example is the rigorous comparisons of Eqs (10.1) and (10.2).[23] After Bona left Essex for the University of Chicago and subsequently Penn State, his influence continued as he frequently returned to work with Brooke.

Among key experimentalists were Bill Pritchard, an Australian whose PhD (1968) at Cambridge was supervised by Brooke; B. J. S. (Barney) Barnard, who had been recruited as a senior lecturer from Civil Engineering at Queen's University Belfast; and John C. Scott, who did experiments on the effect of monomolecular surface layers of polymers on wave motion. Curiously, although Brooke was involved in the entire range of experiments undertaken at Essex, he published only one paper based on experiments carried out in the laboratory.[24] However, this work led to a cover story with John Scott in *Nature*.[25]

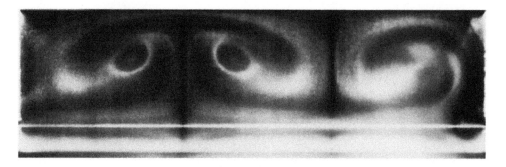

Figure 10.6 Flow visualisation of the cross-section of an anomalous, steady, three-cell flow formed in viscous liquid in the gap between rotating concentric cylinders. N.B. The view has been rotated so that the rotating inner cylinder is at the bottom of the image, the left and right end-walls and the outer cylinder at the top are all stationary.

But Brooke also published experimental work that he carried out at home. For example, his laboratory notebooks show how, after making painstakingly careful measurements, he developed theory to explain Taylor–Couette flow in an apparatus small enough to fit on the dining room table.[26] Fortuitously, the small scale of the experiment meant that he observed for the first time novel flows that had been overlooked for more than fifty years. Brooke was the first to publish the anomalous flow in Figure 10.6.

Extramural activity at Essex

In contrast to Cambridge, the University of Essex was a very political place. There were protest meetings, a rent strike, a student sit-in at the Vice Chancellor's office, and students and lecturers organized a blockade of deliveries to the university. There were posters for obscure political parties and graffiti everywhere, and the Member of Parliament thought it should be turned into an agricultural college. Yet under the leadership of its first Vice Chancellor, Albert Sloman, who had delivered the 1963 BBC Reith lectures 'A University in the Making', the protesters were treated humanely and Brooke instantly became their hero when he went to Colchester police station after a demonstration, with cigarettes for students who had been held there overnight.

At Essex University he typically arrived in his office at about 10:30 a.m. because his practice was to work into the small hours, and he used to say that he never had a really good idea until around three in the morning. In addition to his science, he wrote poetry for his own enjoyment, and maintained his interest in music by playing the violin and piano at home and conducting the university choir. A member of that choir was Natalia Court, a linguist who helped him learn French and to whom he was soon to be engaged. In addition, he was a life-long steam-engine enthusiast, and his interest in cars led him to drive a rather grand Austin Princess. With this range of cultural, scientific, and personal interests he was something of a renaissance man, but he had other commitments too.

He chaired the Science Research Council Mathematics Committee and the Mathematical Education Committee of the Royal Society. Moreover, as Vice President of the Institute of Mathematics and its Applications (IMA) when Prince Philip was its President, he greatly enjoyed the council meetings in Buckingham Palace. (When teased by colleagues he just smiled and said, 'A word in the right ear goes a long way.')

So when Albert Green retired, the Sedleian Professorship of Natural Philosophy became vacant and Brooke, who had all the right credentials, was an obvious candidate.

The Sedleian Chair and the Clarendon Laboratory

In due course he was elected to the chair and to the associated fellowship of The Queen's College, and he and his family moved to Oxford in 1979.

In Oxford, Brooke and his students and collaborators continued to investigate the mathematical structure of water-wave equations, studied the classical fluid-dynamics theory of Kelvin–Helmholtz instability,[27] and predicted the existence of a new type of solitary wave, yet another example, this time on interfaces, of the wave of translation discovered by John Scott Russell in 1834.[28]

At the same time he pursued his interests in experimental research with a new assistant, Tom Mullin, in the Clarendon Laboratory of the Physics Department. In addition to continuing Brooke's work on Taylor–Couette flow,[29] together they began experimenting with water drops suspended on soap films and with the buckling of thin viscous layers of viscous fluids under shear (Figure 10.7).[30]

Figure 10.7 The view from above of buckling under shear of a 1-mm layer of very viscous oil, approximately 10,000 more viscous than the water, floating on a deep layer of water. Buckling waves first appear close to a disc, the rotation of which provides shear.

They also encouraged several of Brooke's students to do small experiments in Tom's lab. The research discussions Brooke led in the Clarendon tearoom which he frequented, and visiting international experimentalists such as Al Ellis from Caltech, who did experiments in the Clarendon on the propagation of bubbles in sound fields, were important for Brooke's work, and that of his collaborators, in Oxford.[31]

The focus of Brooke's mathematical research in Oxford was the weekly group meetings on Friday afternoons. Held in Brooke's office they typically lasted three or more hours, with lively debate involving students, staff, and visitors somewhat reminiscent of the discussions in the triangular tearoom at FMRI. In the background was his programme of testing subtle mathematical models against data from sensitive experiments. The programme contributed to theoretical and experimental studies of gravity currents, waves on thin films, shallow water waves, Hamiltonian structures and conservation laws, impulse and flow force, and bifurcation and stability theory for inviscid flows.

He continued to play the violin and the piano at home and continued to appreciate musical performances. Alongside his scientific programme at Oxford, he had a growing involvement in science education policy. As chair of the Royal Society Mathematical Education Committee between 1979 and 1985 he wrote several letters to the *Times Higher Educational Supplement*,[32] and published an article in the *Oxford Review of Education*.[33]

Legacy and honours

Brooke's vision for FMRI in Essex was significantly ahead of his time. Nowadays there are similar institutes all over the world, but in 1970 they were rare. Not long after his move to Oxford, some of the remaining research group in Essex moved to Pennsylvania State University and established a similar institute, where sometime later Brooke became an adjunct professor.[34] Also much ahead of his time, he urged governments publicly to recognize the material contribution of UK universities to technology, culture, and society. Finally, in the UK, this idea has taken seed, and universities are required on a regular basis to assess the 'impact' beyond academia of their scholarly activity.

Brooke enjoyed public debate and was the founding President of the UK's National Conference of University Professors.[35] He deplored mathematical exclusivity and campaigned successfully to rename the British Theoretical Mechanics Colloquia as the British Applied Mathematics Colloquia, thereby broadening its relevance, for example by including mathematical biology. Fittingly, in 1991, the first meeting under the new name was held under his leadership in Oxford.

Brooke's scientific distinction and position as Sedleian Professor of Natural Philosophy enabled him to maintain close contact with international colleagues. He was a strong supporter of international science and admired in particular the International Centre for Theoretical Physics in Trieste for its support of third-world young scientists.[36] In the 1960s he had been a regular visitor to the United States, particularly the Institute of Geophysics and Planetary Physics in California, where he had done some of the work which established his international reputation. In 1966 he received the Lewis F. Moody award of the American Society of Mechanical Engineering. In the 1980s he spent sabbaticals at the Universities of Wisconsin and Houston, during which he visited colleagues in Chicago and elsewhere. But he returned most often to Pennsylvania State University where he was an adjunct professor, and renewed his enthusiasm for American football by following the Nittany Lions, their very successful football team. Penn State honoured him with a major international symposium on his sixtieth birthday.

He maintained close contacts with colleagues in France because of his admiration of their leading role in tackling fundamental nonlinear problems in mechanics using abstract methods. At symposia in France he got great satisfaction from lecturing in the French which he had learned, with support from his wife Natalia, as part of the culture he created in Essex. On 30 March 1992 he was elected *associé étranger de'l Académie des Sciences, section des sciences mécaniques* (a foreign member of the French Academy of Sciences in the mechanics section):[37]

> Benjamin (Thomas, Brooke) 15 avril 1929 à Wallasey, Grande Bretagne—16 août 1995 à Oxford, Grande Bretagne É lu associé étranger le 30 mars 1992 (section des sciences mécaniques).

Also in 1992 he was awarded the Royal Society's Bakerian Medal and delivered the Bakerian Lecture 'On the Mystery of Vortex Breakdown'.[38] In addition to the honours already mentioned, he was awarded the 1969 William Hopkins Prize of the Cambridge Philosophical Society,[39] received honorary degrees from the universities of Bath (1989) (Figure 10.8), Brunel (1991), and Liverpool (1993), and in 1994 was granted the rare distinction of honorary membership of the American Society of Mechanical Engineers.[40] Annually since 2007, the Brooke Benjamin Lecture in Fluid Dynamics with distinguished international speakers is held in his honour at the Oxford Mathematical Institute,[41] and biannually the T. Brooke Benjamin Prize

(a) (b)

Figure 10.8 Brooke aged 28 in 1957, and aged 60 when he received an honorary doctorate from the University of Bath.

in Nonlinear Waves is awarded by Society for Industrial and Applied Mathematics (SIAM), to a mid-career researcher for outstanding work in the theory of nonlinear waves.[42]

Epitaph

Towards the end of his life, Brooke confided that he took comfort from feeling that despite his many responsibilities he had made no real enemies. Everyone around Brooke had the privilege of sharing the insight of a deep thinker whose ideas continue to stimulate mathematical developments.[43]

CHAPTER 11

An Interview with John Ball

MARK MCCARTNEY

Early life

You were born in Farnham, Surrey in 1948. Tell me about your early home life and school education.

My father was a civil engineer in Farnham. He was also the director of a water company and was connected with another engineering firm in Kent. He was in some sense an intellectual. He had done well at Cambridge and had an academic mind. My brother, father, and I all went to St John's College Cambridge. Like my father, my elder brother did engineering at Cambridge, and so I suppose I thought that I would do something different to avoid comparison.

My mother was a housewife and her father was a doctor who had been born in Stornoway, and who as a medic won the Military Cross in the Boer War.

I can remember doing maths at preparatory school with a teacher called Mr Sorrell, and being very interested in geometry and constructions with rulers and compasses. I remember him at the board giving the example of 5, plus something hidden under his hand, equals 6 and saying, 'What is under my hand?' The class said, '1', and he said, 'Let's call it x.' I thought that was very nice and it was, I suppose, my first encounter with abstract mathematics. Later I attended Mill Hill, an independent school about 9 miles northwest of the centre of London. The teaching in maths was pretty good. In fact there was one young teacher there who had apparently been top of his year in maths at Oxford. I studied maths and physics at A level. In upper school I entered an essay competition with a topic something like 'In science the most important thing is to ask the right question.' I suppose that stayed with me. I know lots and lots of brilliant mathematicians, but you can be less brilliant and work on the right problem, and that makes a fantastic difference.

Mill Hill was founded in the early nineteenth century by nonconformists. Francis Crick was an old boy and when he was awarded the Nobel Prize, instead of getting a whole day off we just got half a day, because Crick was a humanist!

Mark Mccartney, *An Interview with John Ball*. In: *Oxford's Sedleian Professors of Natural Philosophy*. Edited by: Christopher Hollings and Mark McCartney, Oxford University Press. © Mark Mccartney (2023). DOI: 10.1093/oso/9780192843210.003.0011

When I was probably 14 or 15, I ground a telescope mirror for an 8-inch reflector telescope in my bedroom, following instructions from a book. I took two thick disks of glass, fixed the bottom piece and attached a handle using pitch to the top piece (which became the mirror) and ground the disks together with finer and finer carborundum powder to get a spherical surface. Then it was polished using optical rouge. To change it from a spherical to a parabolic mirror I used Foucault's test with a projector and pinhole, which amazingly allows you to see a relief map of the mirror surface. I suppose it wasn't brilliantly done, but it was sent away to be silvered. The telescope had to be motor-driven and so I bought the motors for it and my father erected a concrete base in the garden. I was able to look at planets and galaxies. I think at the same time I got interested in Bode's Law, which states that each planet in the Solar System should be approximately twice as far from the Sun as the one before. I wrote to an astronomer, Michael Ovenden, who was very kind in replying to me. I also read a 1913 paper by Mary Blagg, who gave a more complicated form of Bode's Law. I was very interested in it at the time and really worked a lot on it. In some sense it was a first piece of research.

Before university you spent a year working in the Mathematical Services Department of the British Aircraft Corporation in Weybridge, Surrey. How did this come about, and what sort of work did you do?

I'm not sure how that happened. My brother had worked there, and that probably gave me the idea. Weybridge was on the railway line from Farnham, where we lived, to Waterloo, so it was a very easy commute. It was a very interesting time because I was there exactly during the first major transition in computing. So, for example, I did what they called 'ghosters'. I would be there alone during the night looking after the computers, which were running what was called a weights programme, keeping track of the weights of different components of each aircraft. At that stage the main memory store for the computers was on a rotating magnetic drum. The computers would suffer from parity failures, and I can't quite remember how it was done now, but you would in some sense correct the parity failure and then set the computer going again. It involved some machine code. While I was there they moved to ICL1905s, which were housed in a huge room full of what we viewed as super-fast computers, though of course today there's probably more computing power on our wrist.

I learned programming in FORTRAN and worked on what were called mishmash charts. When an aircraft goes down the runway Civil Air Regulations say that if there's an engine failure you have to be able to either stop or take off. There is a critical engine revs, and if you have an engine failure before that you slam on the brakes, but after that you take off. All sorts of information goes into that calculation; the altitude, number of passengers, slope of the runway, whether you have to clear a tree or building, etc. This was done by charts of intersecting curves and we were trying to computerize it. It was said that the only time engine failure had occurred exactly at the critical revs was in a Vickers Viscount. The pilot had slammed on the brakes and the aircraft had stopped with its wheels exactly on the end of the runway.

It was very interesting working in a company, both socially and technically. I got to do some really amazing stuff. I don't think I would have ever done engineering at university, but I could have ended up in industry doing some sort of research. I'm not sure I would have been a very good manager, but then I don't know, as I've ended up being a manager in some situations, so maybe would have been reasonable!

Cambridge

In 1966 you entered St John's College, Cambridge with an Open Exhibition to study mathematics. What was Cambridge and college life like in the 1960s? What was the mathematics curriculum like at that point?

It was, first of all, all men. One of the things that struck me was that I was one of the two best people at school in maths, but of course when I went to Cambridge I was by no means the best. That particular year there were some really fantastic mathematicians at St John's. The speed of the course was an incredible shock to me and the mathematics was exciting, but completely different from anything I had encountered. I remember in the first term we did special relativity. A teacher at school, Mr Wormell, had told me I should do some reading before I went to Cambridge, but I didn't take any notice because I had always been good at maths and didn't think I needed to do that. But it would have been a help actually!

There were some very good courses and some very bad courses. I was not very good at doing exams under time pressure and didn't do so well in terms of exams. I'm not a particularly fast thinker on my feet. I think I was also sort of nervous and stressed and this didn't help. I know a large number of the very best mathematicians in the world and lots of other mathematicians too. There is the sort of person you can sit next to in a seminar who will always be able to understand the seminar and explain it to you, whereas I may have understood almost nothing. Such people are incredibly valuable, but may not do very original stuff themselves. The attributes that are important for being successful in research are not necessarily the same ones that get you a first in mathematics in Cambridge, which I didn't get.

There were exams at the end of each year, Part 1A, Part 1B, and Part 2. Part 2 was quite specialized. Unfortunately, I chose some wrong courses in Part 2. I did a measure and integral course taught by Ben Garling from St John's. It was very good. Earlier as a tutor he had taught me epsilons and deltas so I really was taught basic analysis very well. But I also chose quantum theory, which I probably shouldn't have done. We had Dirac's book. I don't know why one would give Dirac's book to undergraduates! In it, the world motors on according to Schrödinger's equation until you make a measurement, and then something quite different happens. I asked my advisor why, say, this chair wasn't measuring everything all the time. He said that I should leave thinking about questions like that to Dirac! And then I heard Dirac giving his farewell lecture as Lucasian Professor, and he said he thought quantum theory was in big trouble and it would require some advanced mathematical logic to solve the measurement problem, which made me feel better. If you're a thinking person you cannot understand quantum mechanics. We have great authorities like Roger Penrose who recognize this, but at some point people will typically say that it is an incredibly successful theory and you've just got to accept it and move on. Which means you have to suppress all these doubts, and so when somebody else comes to ask you about quantum mechanics you say the same thing and you feel kind of guilty!

I really regret not having done functional analysis and topology in Part 2 because these are things I have had to use, and it is so much more difficult having to learn new subjects when you're older. I must have done some algebra courses because that was my first idea of what to do after Cambridge.

There was a cultural division between pure and applied maths at Cambridge which was in the long term damaging to Cambridge mathematics. We were taught these two subjects, pure maths and applied maths, and certainly I never got the impression that there was some kind of link between them!

As I said, some of the other students were really very, very good. Shaun Bullett and Tim Goodman both had academic careers, as did Peter Johnstone, who never left St John's. The brightest guy was perhaps Mike Christie, who was just sensationally good. He was eventually supervised by Ian Cassells, who had a reputation of setting his students extremely difficult problems. I think Mike got his PhD and then went into military research, but I lost touch with him.

Do you have memories of any particular lecturers or tutors during your time as an undergraduate?
Frank Smithies taught me complex analysis and George Reid, who became Mayor of Cambridge, was very good on algebra. Raymond Lyttleton, who was an astronomer, was always worried about whether you were wearing gowns or not. He tended to get phone calls from NASA in the middle of supervisions, which always seemed a little bit suspicious!

Dennis Sciama was an excellent lecturer. I remember him telling us about a brilliant student called Roger Penrose who had just done a calculation to show that due to relativity you could just see the other side of the moon.

But I think the main people who taught me were the students at St John's. It was really an intellectual community in a way, which I doubt happens so much now. But maybe I'm wrong.

In 1969 after graduating from Cambridge you moved from one of the oldest universities in the UK to one of the newest, the University of Sussex. What were your initial experiences at Sussex?
My first idea had been to do research in algebra. I remember having an excellent conversation with Coulson in Oxford. I didn't have the exam results to get accepted to do research at Oxford, and he said I would have had to do a one-year examined course. I thought, well I screwed up in my exams at Cambridge so I would probably screw up the Oxford exams as well. This was actually a piece of good luck. I took a very easy option, which was that there was a school teacher of mine called Michael Semple who had left Mill Hill and gone to Sussex as a lecturer in the Applied Sciences Department. I knew I wanted to do research and that by going to Sussex I could get three years funding to do it. Incidentally there is an amazing connection with Queen's College, Oxford. Many years later I got a note in my mailbox at Queen's from Semple, because he had later become a vicar in one of the parishes owned by the college.

Even though your D.Phil. was in mechanical engineering your supervisor, David E. Edmunds, was a mathematician. How did this come about? Were there academics (either faculty members or visitors) who had a significant influence on your time at Sussex?
I was very lucky to be helped by Mike Alford, who was Semple's research student. Semple had effectively abandoned Mike, who didn't want the same thing to happen to me. He intervened on my behalf and after some months I managed to get transferred to the Maths Department.

It so happened that at Sussex at that time there was a differential equations programme with money from the SRC which David Edmunds had got, and so there were fantastic people who were visiting the Maths Department. For example, there was Guido Stampacchia, who was a very good Italian specialist on partial differential equations and a very kind man. Mike Alford and I learned that he was giving a course, but only after he had already given one lecture. We went to the second lecture and it was amazing. At the end we were saying to each other how good it was, and that it was a pity we had missed the first lecture, and Stampacchia overheard us and asked if we would like him to give us the first lecture again. Of course we didn't let him do it.

Robin Knops, who eventually gave me my first job at Heriot-Watt, came to give a seminar and he put me in contact with Stuart Antman. I remember being in Mike Alford's Mini taking Antman and another well-known mathematician, Avron Douglis, to the railway station in Brighton, and he said, 'I hear from Robin that you need a research project. What I suggest is that you look at this paper by Dickey. This is what it does . . . and this is what I suggest you do with it . . .'. Up to this point I had had no conception that somebody could know about research at that level. I had been working on my own, and Semple had very little understanding of how research should be done. It was an absolute revelation to me. So that is where my project came from. Thus, when I started working with David Edmunds I already had a project, and so it was more a question of him listening to me, giving encouragement, and putting me right on some things. For example, I had to read a book by Jacques-Louis Lions, which was in French and was a bible for people at the time. There was a footnote which was some incredibly abstract thing about distributions and some sort of complicated function space. I said to David I couldn't understand this, and he came back with a bit of paper with his very, very small handwriting which got to the conclusion in eight lines. He said that I shouldn't feel bad about it, as it really was quite difficult.

One other thing I did at Sussex was to sing in a fantastic choir called the Brighton Festival Chorus. The choirmaster László Heltay was incredibly talented and had been a student of the composer Zoltán Kodály. We worked with some great conductors and sang at the Stravinsky Memorial Concert in London which was conducted by Leonard Bernstein. Later when I came to Edinburgh I joined the Edinburgh Festival Chorus and sang with it for about three years.

Post-doctoral work

Between 1972 and 1974 you worked as a post-doc for a year at Brown University in America and then at Heriot-Watt in Scotland. What did you work on during this time, and who were the people that influenced you most?
I had an SRC research fellowship which provided me with money to go to Brown. Brown was the obvious place to go because of the way my research had gone and because there were people there who were experts. I was married to my first wife at the time. The year at Brown was very good for my mathematics, but not very good for my marriage. I already knew about the problem of existence of minimizers in elasticity from a lecture that Stuart Antman had given. At Brown Constantine Dafermos told me about all the difficulties of this problem, before he travelled back to Greece, leaving me to some really solitary work. Brown has a most fantastic library. Nowadays you would do things differently, but then it was an incredible resource, and in the library I went back to read the early papers. There was a professor at Brown called Jack Pipkin and I was thinking a lot about a particular kind of inequality and whether it was the right one to use, and he was able to tell me very quickly that it was completely wrong. That was very important. It was a very hot summer and the key technical idea that I had came when I was lying on my bed simply trying to keep cool! It turned out that it had already been found by a Russian, Yuri Reshetnyak, but he wasn't working on elasticity and so didn't realize its importance for that subject.

Heriot-Watt

In 1974 you were appointed to a lectureship at Heriot-Watt. By 1982 you had risen through the ranks to the level of Professor, and you remained at Heriot-Watt until 1996. How did the department, and your own research interests and collaborations, evolve over this period?

Robin Knops created the modern department at Heriot-Watt. He had come from Newcastle and had been appointed by a committee chaired by Ian Sneddon. Robin had been a student of Rodney Hill, who was very strong and did very important work on the theory of plasticity. Robin was faced with the huge problem of trying to make a department which was not active in research research-active. As you can imagine personality difficulties ensued. There was already some research going on. There were people like Aubrey Truman, who later went to Swansea and was a mathematical physicist, and Ken Brown, who did good research in differential equations. I represented a different kind of area which was very close to Robin's heart. We got money from the Science and Engineering Research Council and all sorts of top people came to Edinburgh, including Dick James and Jerry Marsden, both of whom became important collaborators, and others such as Constantine Dafermos and Craig Evans who became close mathematical and personal friends.

I ended up with some amazing students and post-docs. For example, Stefan Müller came on a German scholarship during his undergraduate career and then came back to do a PhD with me. And then there was Vladimír Šverák, who was an absolutely brilliant post-doc. Others went on to become professors at leading universities. Müller and Šverák were there together. It was quite intimidating for me, actually. They were sensational and in your lifetime you will never get people like that more than once. But they were at Heriot-Watt, not at Oxford or Cambridge. In the UK there would be an institutional prejudice and such top people would not end up with somebody at Heriot-Watt. But they were coming from Germany and Prague and they were looking at me as a person to work with. Today big is beautiful. We have Centres for Doctoral Training and I have bought into that model. It gives a lot of opportunities to those working in such centres. But these people's careers were not damaged by being at a small university, with maybe not so brilliant library facilities. Being in a small group with lively people was great for them.

The department at Heriot-Watt became more and more research-oriented. An early appointment was that of Jack Carr, who became a collaborator and great friend. We used to discuss mathematics while going to football games. Later Oliver Penrose was appointed, and added a big new dimension to the department. I was extremely well treated by the university administration and at the end had a research only position. It was really quite a difficult decision to leave. I still think that Edinburgh is a nicer place to live than Oxford!

At the time there was a very good physics department at Heriot-Watt and there were other good departments too. And then of course there was the community in Scotland and the Edinburgh Maths Society. I remember once giving an Edinburgh Maths Society talk in, I think, Strathclyde or Glasgow, which was called 'What is compensated compactness?' Compensated compactness is a generalization of the ideas which I used to prove existence in elasticity. One of the people in the audience was Arthur Erdélyi, who was professor of maths at Edinburgh. He was always very good in these seminars and would ask all sorts of interesting questions. At

Figure 11.1 John Ball in Berkeley in 1983.

the end of my seminar he put up his hand and said, 'But what is compensated compactness?', which was a terrible failure! After this seminar I was at a conference, and flying back from it the person in the seat in front of me had his copy of *The Times* open, and I could see the obituary of Erdélyi. So that must have been the last question he asked in a seminar.

The other important thing about Edinburgh was the founding of the International Centre for Mathematical Sciences (ICMS). When I was first in Edinburgh, Edinburgh University looked down on Heriot-Watt. But then when Elmer Rees became professor at Edinburgh in 1979 that began to change. When there was to be the creation of a National Mathematics Research Institute, John Coates from Cambridge was President of the London Mathematical Society (LMS) and we were all invited to a meeting in the Isle of Thorns in Sussex to discuss its formation. But we weren't given the information that the institute was going to be in Cambridge! Adrian Smith, who was then Chair of the Mathematics Committee of the Science and Engineering Research Council, and is now President of the Royal Society, was also at this meeting and he was very tough. He said that if there was going to be a National Research Institute, then there had to be a national competition for it. This was not, I think, what was planned. Elmer and I worked very much together on writing the proposal which was eventually instrumental in founding the ICMS. That really changed the relationship between Heriot-Watt and Edinburgh. And now we also have the Maxwell Institute, which combines the mathematics research of the two universities and has been a great success.

The Sedleian Chair

In 1996 you moved to the Sedleian Chair. How did you find the process of the move between Oxford and Heriot-Watt?
Well first of all, I didn't apply! I had the impression that that would be the kiss of death. In fact, Ron Mitchell, who was professor of numerical analysis at Dundee, told me that he had been on the selection panel for a chair at Oxford and that the chair of the panel had opened the discussion with 'Well I take it, gentlemen, that we will not consider anyone who has applied.' But then at some point I was asked to apply. Michael Berry, who was on the panel, told me that their first choice had been Vladimir Arnold.

It was an incredible process. One day I got a letter saying that I had been elected and to fill in a medical certificate and so on and so forth. There was no sort of question about whether I was going to accept it or not. However, there was a long negotiation and I took about six months before I accepted. The reason was because I was asked to take a cut in salary. I thought of trying to find some other college, but I was told that it would require an act of Parliament to change the college.

By this stage I had offers to go to the US but I had more or less rejected the idea. I think if I'd applied for the Directorship of the Newton Institute I might have been a strong candidate. So I did feel that somehow there was a choice between Oxford and Cambridge.

Before moving to Oxford I had married Sedhar, a Tibetan refugee and actress, and we already had two children, Kesang and Tenzin, with a third, Palden, born later. It was a wrench

Figure 11.2 At the International Congress of Mathematicians in Madrid, 2006 with family and King Juan Carlos of Spain. Left to right: wife Sedhar, John, King Juan Carlos, children Tenzin, Palden and Kesang.

for Sedhar to leave all her friends in Edinburgh. She has been incredibly supportive during my career.

One of the duties of the Sedleian Professor at Queen's is to be the keeper of the orrery, a mechanical model of the Solar System made around 1765 that is kept in the library. I got to turn its handle at a few ceremonial occasions, but fundamentally it's a very boring thing to do or witness. I got it maintained once; there are not very many people in the world who are experts in maintaining orreries, and it was very interesting to see all the gears underneath. The handle can be turned only by the Sedleian Professor and the Patroness or Patron of the College. When I was there it was the Queen Mother. When the Queen Mother turned 99 I didn't necessarily have the confidence that she would make it to 100, so wrote a letter to her for posterity congratulating her on her 99th birthday and mentioning this connection between her and me. She didn't reply directly, but I got a very nontrivial reply from an equerry which actually named the exact day when the Queen Mother had last turned the handle.

Aspects of your research work follow in a tradition of some of the other holders of the Sedleian Chair in the twentieth century, namely Augustus Love, Albert Green, and, to a lesser extent, Brooke Benjamin. What do you see as their most significant contributions to applied mathematics?

Love wrote the bible on linear elasticity so there is a connection there. But there is a closer connection between my work and that of Albert Green's because he was interested in nonlinear continuum mechanics and wrote a book on nonlinear elasticity.

I met Albert Green at Oxford and went to visit him at an old people's home. His style was very different from mine. He used what I thought was awful notation with lots of indices, and I hate indices as I can't see what's going on. He was not in any way an analyst, but he had a rigorous view of mechanics which I appreciated. Peter Neumann once told me that one of his students came to a tutorial very impressed by a lecture she had been to by Green, in which he had written down one equation that had covered the entire blackboard.

My predecessor had been Brooke Benjamin. Brooke rated me at some level and had been very helpful to me with a paper I wrote on cavitation. Brooke's thesis was on cavitation in fluids and my paper was on cavitation in solids. It was published in the *Philosophical Transactions of the Royal Society*. It was a good paper actually! Initially it got not exactly rejected, but nearly, and Brooke fought very hard to get the paper accepted.

There is a much closer philosophical link with Brooke Benjamin because he was very interested in applying analysis, and French analysis in particular, to fluid mechanics. That's why he left Cambridge, because he felt there was no appreciation at Cambridge for that kind of work. In some sense I suffered a bit from that. I remember I gave a talk at a conference on my work on existence and elasticity. There is a particular model of elasticity called the neo-Hookean model in which the energy function is the square of the deformation gradient, but the power two is not large enough to get the estimates for existence. At the conference I said it makes one wonder whether you do have the existence of minimizers in this neo-Hookean case; this has still not being resolved many, many years later. At the end of the talk someone put up their hand from the audience and said, 'Am I correct in understanding that the reason you don't think there is existence in the neo-Hookean case is because you can't prove it?' This was my first encounter with a truly aggressive question!

On the subject of French analysis, I spent two wonderful sabbaticals in Paris at the invitation of my great friend Francois Murat. These years abroad, together with others in Berkeley and the Institute for Advanced Study in Princeton, were unforgettable and mathematically inspiring.

I'm a great believer in talking to experimentalists directly, without intermediaries. It's hard work because there are language barriers to break down though. But experimentalists are always very happy if someone takes an interest in what they're doing and you get problems that way which you won't get in any other way.

Twice in my life I've been involved in moving into a new area, once with materials science and then with liquid crystals. And now I work a little bit on computer vision, so I may find a similar thing there. Despite the fact that my doctorate is in mechanical engineering that's really a lie, and in particular I didn't learn any materials science as an undergraduate. There's something that attracts you into a particular area and you've got to be humble, but also have a little bit of confidence that you know something that could be helpful. But you've got to be prepared for some pushback, which would typically be that you're ignoring this, that, or the other important thing.

The fact is that different fields have a different understanding of what understanding means. A mathematician may feel that if you have a model and you really understand exactly what this model predicts, then that can be very useful. Whereas if between the model and you are numerical calculations, or approximations whose validity you can't really assess, then that creates a slight fog in the understanding. Materials scientists may say that they understand the problem when they understand the physical processes which are at play, but in some sense that can be the beginning of where a mathematician takes over. In liquid crystals it is a bit different and because of the topology of singularities there's been a greater mathematical tradition there. So I found in that area there was much less pushback and now there's no pushback at all.

Wider issues

How much of academic life and success in research do you feel is a result of luck and serendipity?
My case is probably a bit unusual because I didn't have the exam results to propel me naturally through the system. The choice of problems you study, whom you happen to meet, and their personalities, can be hugely important.

You never know the effect that lectures you give can have on other people. It can change people's lives. I remember I gave a lecture and there was a now very well-known mathematician who heard it, and it changed his research direction. It meant he married somebody he wouldn't otherwise have married because of the place he went to and so on. Small events can result in large changes.

During your career you have held a number of significant roles within a wide range of scientific bodies, including being President of the Edinburgh Mathematical Society (1989–90), President of the London Mathematical Society (1996–98 and in an interim role in 2009), and President of the International Mathematical Union (2003–6), and you are currently President of the Royal Society of Edinburgh. What have been the highs (and lows) of serving is such roles?

Figure 11.3 Presenting the inaugural Gauss Prize to Kiyosi Itô in 2006. Itô was too ill to travel to the ICM in Madrid and so John travelled to Kyoto to make the presentation. Also in the picture is Itô's daughter, Junko.

I was President of the LMS twice. The first time it was very satisfying, because we bought a new building for the society in Russell Square. The second time when I was President was when I was trying to sort out the mess of the failed merger with the Institute of Mathematics and its Applications (IMA), and it was pretty much a nightmare. The IMU was also exciting, especially as regards Perelman and the Fields Medal. This is a story which is now related in Wikipedia, various books, and even a novel! After returning from the International Congress in Madrid I went to have my hair cut, and the barber asked me what I did, and I told him I was a mathematician. He said, 'What about that Russian who turned down a medal?' It was really an event that touched the general public in unexpected ways.

In general in such administrative positions you may sometimes have to take really difficult, tough, and unpleasant decisions, but if you make a beneficial change to an institution, that can be a really long-lasting thing.

Your career in applied mathematics has lasted over fifty years. What are the most significant changes you have seen over that period?
There was a period when analysis and rigorous methods in partial differential equations became part of applied mathematics, and that happened in my lifetime in a way that had not happened before. There were people like Jean Leray who in 1934 proved existence for the Navier–Stokes equations. The great British fluid mechanics community thirty years later would have known nothing about what Leray had done.

Stefan Müller related a joke told by one of my former students, Nick Owen, at a seminar he was giving in Bonn. Halfway through the lecture Nick had to clean the board. And he said, 'You know in Cambridge they have a bucket of water by the board. It's for two reasons. First, so you can clean the board and second to prove that there exists a smooth solution to the Navier–Stokes equations!' I felt I had taught him well.

What do you think are likely to be the most significant changes in applied mathematics (both in terms of what research directions are pursued and how research is likely to be carried out) over the next fifty years?

There's a revolution with AI and machine learning that's already changing the way science is done. It's a bit scary and while I see its great potential, I also worry about it being so different. What will the position of traditional science be when the understanding is inside some computer but not outside it, so to speak? Traditional applied mathematics will remain important, but there will be changes. Young people will now have to understand these kinds of methods and incorporate them into what they do. It will affect pure mathematics too in my opinion. I'm quite sure that in thirty or so years, maybe much sooner, if you write a paper in rigorous mathematics you'll be writing it in some higher order language that will enable the proof to be checked; and of course if computers can check the proofs they can find the proofs. This is tied up with deep philosophical and scientific problems such as the nature of consciousness and so on, and whether computers can be conscious.

Machine learning is a very sophisticated form of curve fitting. You take the data and see some pattern in it, and predict what happens with different values of the parameters. But there are cases, and I expect it maybe is not so uncommon, that the very interesting behaviour is in a very small area of parameter space. Machine learning is not in general going to find you that sort of thing. As an example, I did some work with a collaborator in engineering. When you have an alloy that is a mixture of certain elements in certain proportions, it turns out that there are very special alloy compositions that produce special geometric microstructures, which don't generally occur. I always regarded those special compositions as being nongeneric and therefore not of interest. But my collaborator Dick James took the opposite point of view and said, 'Well let's make a material which satisfies these conditions and see what happens.' So he had more faith in mathematics than me. And you get fantastic, spectacular, things that happen in these materials, such as a reduction in thermal hysteresis from about 50°C to 2°C. That's the sort of thing that machine learning is never going to do, unless machines take over rational thought, because it's not in the data and it's not a natural extrapolation from the data.

The more quantitative a subject becomes, the more open it is to mathematization. If you look at the life sciences, things are incredibly complicated. If you go to the doctor and say, 'I've got such and such symptoms, and what are you going to do about it?', that's a sort of classic formulation of a control problem. You're asking the doctor to prescribe a control in terms of pills or surgery or whatever to steer your symptoms to an optimal situation. So why is it that doctors don't use control theory? It's because they don't know what the equations are. So you can imagine that mathematics will become more and more important in the life sciences. Another area is modelling human behaviour, whether that be a pandemic or the behaviour of crowds or how people come to opinions and so on. There are all sorts of things like that which I think will become increasingly important areas of research.

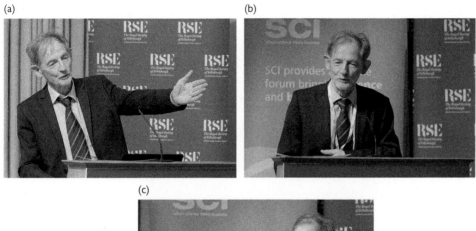

Figure 11.4 John in the chair as President of the Royal Society of Edinburgh, 2022.

FURTHER READING, ACKNOWLEDGEMENTS, NOTES, AND REFERENCES

CHAPTER I

Further reading

For a prior overview of the foundation and early years of the Sedleian Professorship (which has been drawn upon here), see Kenneth Dewhurst, *Thomas Willis's Oxford lectures* (Oxford: Sandford Publications, 1980), 37–49. On the more general Oxford context, see John Fauvel, Raymond Flood, and Robin Wilson (eds), *Oxford Figures: Eight Centuries of the Mathematical Sciences*, 2nd ed. (Oxford: Oxford University Press, 2013). On the parallel development of Oxford's Savilian Professorship of Geometry, see Robin Wilson (ed.), *Oxford's Savilian Professors of Geometry: The First 400 Years* (Oxford: Oxford University Press, 2022). Details of many of the individuals mentioned in this chapter may be found in the published register of Oxford alumni, *Alumni Oxonienses*.

Acknowledgements

I am grateful to Margaret Pelling for helpful suggestions and references when this project was in its very earliest stages, and in particular to her and to Peter Elmer for supplying information from the database of the Early Modern Practitioners Project (https://practitioners.exeter.ac.uk/), then based at the University of Exeter under the direction of Jonathan Barry and Peter Elmer. Thanks are due for the help offered by Simon Bailey and Faye McLeod (Oxford University Archives), Diana Smith and Steven Archer (Trinity College, Cambridge), and the staff of the Waddesdon Archive. I am also grateful to Michael Riordan and Amy Ebrey of The Queen's College Archive—most particularly to Mike for sharing his speculations on the origins of the tradition linking the Sedleian Professor and the college's orrery (see note 153, this chapter). I am particularly grateful to Mark McCartney, Nigel Aston, Christopher Stray, and Philip Beeley for their useful comments on an early draft of this chapter. Finally, special thanks must go to Steve Martin for making me so welcome on a visit to St Nicholas's Church, Southfleet in September 2022.

Notes and references

1. Other chairs founded around this time include White's Professorship of Moral Philosophy (1621, now the Sekyra and White's Professorship), the Camden Professorship of History (1622, now explicitly Ancient History), and the Laudian Professorship of Arabic (1636). On the general background to these foundations and others, see Mordechai Feingold, 'Patrons and Professors: The

Origins and Motives for the Endowment of University Chairs—In Particular the Laudian Professorship of Arabic', in *Brill's Studies in Intellectual History*, ed. G. A. Russell, vol. 47 (Leiden, New York, Köln: E. J. Brill, 1994), 109–27.

2. We note that for the first three centuries or more of the Sedleian post, there does not appear to have been any great consistency in university documents as to the precise title given to its incumbent: he appears as Reader and Professor, and very occasionally (in earlier years) as Lecturer, with little to suggest that the titles were systematically applied. Only in the nineteenth century did he become consistently a Professor (see the comments on the 1860 retitling of some university positions in Robert Fox, 'The University Museum and Oxford Science, 1850–1880', in *The History of the University of Oxford*, vol. VII : *Nineteenth-Century Oxford, Part 2*, ed. M. G. Brock and M. C. Curthoys (Oxford: Clarendon Press, 2000), 641–93, at p. 674). In the interests of simplicity, we gloss over this unevenness and refer to him throughout as 'Professor'.

3. The description given here of Oxford teaching *c.*1600 is of course vastly simplified. For a more detailed account, see Modechai Feingold, 'The Humanities', in *The History of the University of Oxford*, vol. IV : *Seventeenth-Century Oxford*, ed. Nicholas Tyacke (Oxford: Oxford University Press, 1997), 211–357; and Modechai Feingold, 'The Mathematical Sciences and New Philosophies', in *History of the University of Oxford*, vol. IV, 359–448. For the more general context of mathematics and natural philosophy in the early seventeenth century, see Mordechai Feingold, *The Mathematicians' Apprenticeship: Science, Universities and Society in England 1560–1640* (Cambridge: Cambridge University Press, 1984).

4. William Poole, 'Sir Henry Savile and the Early Professors', in *Oxford's Savilian Professors of Geometry*, 2–27, at p. 4; William Poole and Christopher Skelton-Foord (eds), *Geometry and Astronomy in New College, Oxford: On the Quatercentenary of the Savilian Professorships 1619–2019* (Oxford: New College Library & Archives, 2019), 10.

5. Robert G. Frank Jr., *Harvey and the Oxford Physiologists: A Study of Scientific Ideas* (Berkeley; Los Angeles; London: University of California Press, 1980), 45–8; Feingold, 'Patrons and Professors', p. 112. A Regius Professorship of Physic (now titled the Regius Professorship of Medicine) had existed in Oxford since the 1540s (Robert G. Frank Jr., 'Medicine', in *History of the University of Oxford*, vol. IV, 505–58, at p. 516).

6. See: Feingold, 'Patrons and Professors'.

7. On Savile, see R. D. Goulding, 'Savile, Sir Henry (1549–1622)', *Oxford Dictionary of National Biography*, https://doi.org/10.1093/ref:odnb/24737; Poole, 'Sir Henry Savile and the Early Professors'; Poole and Skelton-Foord, *Geometry and Astronomy in New College*, ch. 5.

8. Feingold, 'Patrons and Professors', p. 111.

9. See, for example, Robert Goulding, *Defending Hypatia: Ramus, Savile, and the Renaissance Rediscovery of Mathematical History*, Archimedes: New Studies in the History of Science and Technology, vol. 25 (Springer, 2010).

10. A general history of the Sedley family, drawing upon a range of sources, may be found in Chapter 1 of V. de Sola Pinto, *Sir Charles Sedley 1639–1701: A Study in the Life and Literature of the Restoration* (London: Constable & Company, 1927).

11. A later manor house survives on the site (https://historicengland.org.uk/listing/the-list/list-entry/1099222). It should not be confused with a different Scadbury Manor which stands near Orpington, around 12 miles west of Southfleet. On Southfleet, see Edward Hasted, *The History and Topographical Survey of the County of Kent: Volume 2* (Canterbury: W. Bristow, 1797), 421–40 (British History Online: http://www.british-history.ac.uk/survey-kent/vol2/pp421-440, accessed 1 September 2022); see also John Newman, *Kent: West and the Weald*, The Buildings of England (New Haven; London: Yale University Press, 2012), 557–9.

12. De Sola Pinto, *Sir Charles Sedley*, p. 18; Hasted, *History and Topographical Survey of the County of Kent*.

13. In the south aisle of St Nicholas's Church in Southfleet there stands a marble tomb whose surviving inscriptions indicate that it contains not only the remains of John and Anne Sedley, but also those of John's grandparents: the John Sedley who was an exchequer auditor, and his wife Elizabeth. Most of the tomb's memorial brasses have been stolen over the centuries, but some remain, including two featuring small effigies of children: four boys and two girls. Some accounts of the church brasses associate these children with (now missing) effigies of John and Elizabeth, although they abut a plaque commemorating John and Anne, so may in fact be the children of the latter couple. If this is the case, however, we do not know the name of the fourth son, nor those of either of the daughters. Rubbings of the brasses, including some since lost, appear in W. D. Belcher, *Kentish Brasses*, vol. 1 (London: Sprague & Co., 1888), 104.

14. De Sola Pinto, *Sir Charles Sedley*, p. 20. On Aylesford, see Edward Hasted, *The History and Topographical Survey of the County of Kent*: Volume 4 (Canterbury: W. Bristow, 1798), 416–47 (British History Online http://www.british-history.ac.uk/survey-kent/vol4/pp416-447, accessed 1 September 2022); see also Newman, *Kent*, pp. 105–11. It is not clear why John should have inherited the family lands, since both *Alumni Oxonienses* and the inscription on his monument in Southfleet (see note 16) indicate that he was younger than William.

15. A community of Carmelite Friars reoccupied the priory in 1949: https://www.thefriars.org.uk/

16. William commissioned an elaborate alabaster and marble wall monument for his brother which still stands in the south aisle in St Nicholas's Church in Southfleet, adjacent to the tomb of their parents and great-grandparents (see note 13). The monument features an armoured reclining effigy of John beneath an arch between marble columns. The family coat of arms appears above the arch, and beneath it a plaque carries a Latin funeral ode to John by William ('Guilielmus Sedley, eques et baronettus, mærens posuit': 'William Sedley, knight and baronet, grieving, placed this').

17. There is an unsubstantiated claim that William's great-grandfather, the John Sedley who had worked at the exchequer of Henry VIII, was educated at Exeter College, Oxford (De Sola Pinto, *Sir Charles Sedley*, p. 18). William's brother John matriculated at Hart Hall in the same year as William, but he does not appear to have completed a degree (*Alumni Oxonienses*).

18. Unless stated otherwise, the details here about William Sedley are drawn from *Alumni Oxonienses* and Cockayne's *Complete Baronetage*, vol. I: *1611–1625* (Exeter: William Pollard, 1900).

19. He would later be treasurer at Lincoln's Inn from 1608.

20. Kent History and Library Centre, Maidstone: Kent Quarter Session Rolls (1595–1605) (Q/SR). In September 1603, for example, one Thomas Rayneshawe was bound over 'as a result of speaking opprobrious words about William Sedley, esq., connected with the execution of his office' (Q/SR/4/m.7).

21. Kent History and Library Centre, Maidstone: Sevenoaks Library Collections, Middleton (Willoughby) Manuscripts, Official Papers, Militia Correspondence, U1000/3/O5/24, U1000/3/O5/25, U1000/3/O5/31.

22. In July 1616, for example, he was part of a commission, which also included the Archbishop of Canterbury and the Lord Admiral, among others, that was empowered to investigate the misemployment of monies that had been bequeathed for charitable uses in Kent (*Calendar of state papers, domestic series, of the reign of James I: preserved in the State Paper Department of Her Majesty's Public Record Office*, 5 vols (London: Longman, Brown, Green, Longmans, & Roberts, 1857–72), vol. 2, p. 387).

23. William Camden, *Britain, or A chorographicall description of the most flourishing kingdomes, England, Scotland, and Ireland, and the ilands adioyning, out of the depth of antiquitie beautified vvith mappes of the severall shires of England: vvritten first in Latine by William Camden Clarenceux K. of A. Translated newly into English by Philémon Holland Doctour in Physick: finally, revised, amended,*

and enlarged with sundry additions by the said author (London: [Printed at Eliot's Court Press] impensis Georgii Bishop & Ioannis Norton, 1610), p. 332.

24. Newman, *Kent*, p. 110.

25. *John Owen's Latine epigrams, Englished by Tho. Harvey, Gent.* (London: Printed by Robert White, for Nevil Simmons [. . .], and Thomas Sawbridge [. . .], 1677), book II, epigram 45. Epigrams 4 and 99 of book II are also addressed to Sedley, and no. 5 to his son John. On John Owen, see D. K. Money, 'Owen, John [Joannes Audoenus] (1563/4–1622?)', *Oxford Dictionary of National Biography*, https://doi.org/10.1093/ref:odnb/21013.

26. Newman, *Kent*, p. 558; Peter Clark, *English Provincial Society from the Reformation to the Revolution: Religion, Politics and Society in Kent 1500–1640* (Hassocks, Sussex: The Harvester Press, 1977), 193; Ralph Penniston Taylor, The Sir John Sedley Charity (Wymondham: Sir John Sedley Educational Foundation, 1999); https://www.sedleys.kent.sch.uk/.

27. They certainly lived in Oxford at the same time: when Sedley was an undergraduate, Savile was already a fellow of Merton, and was active within the wider university, but there is no reason to believe that they would have met at this time.

28. *Calendar of state papers, domestic series, of the reign of James I*, vol. 2, p. 169.

29. The offer came from Robert Sidney (1653–1626), Viscount Lisle (later Earl of Leicester), for his son Sir Robert (1595–1677) to marry Elizabeth Savile (*Calendar of state papers, domestic series, of the reign of James I*, vol. 2, p. 175). John and Elizabeth Sedley would go on to have nine children together; the best documented of them is the poet Sir Charles Sedley (1639–1701) (De Sola Pinto, *Sir Charles Sedley*). His descendants include Sir Francis Galton (1822–1911). The poet Edmund Waller (1606–87) penned an epitaph for Elizabeth Sedley, 'the learned Savil's heir': 'All that her father knew or got, // His art, his wealth, fell to her lot [. . .]' (G. Thorn Drury (ed.), *The Poems of Edmund Waller*, 2 vols (London: George Routledge & Sons, 1893), vol. 2, pp. 114–15).

30. Feingold, 'Patrons and Professors', p. 111.

31. The National Archives, Kew: Will of Sir William Sedley of Aylesford, Kent, https://discovery.nationalarchives.gov.uk/details/r/D907623.

32. There is some small doubt about this last statement. In his will, Sedley instructed that he should be buried in the church at Southfleet. This is no reason to believe that this didn't happen, but at the same time there is no record of the burial, and no obvious monument. It is possible that Sedley was interred in the tomb already occupied by his parents and great-grandparents (see note 13), but this is speculation.

33. Payment of the annual sum continued into the nineteenth century, when it was allowed to lapse (Kent History and Library Centre, Maidstone: Records of Holy Trinity and Other Charities, Aylesford (1600–1952), Correspondence relating to Sedley Charity, Ch26/E11). The nature of Sedley's connection to some of the other parishes named, such as Navestock in Essex, is unknown.

34. Cambridge Alumni Database: https://venn.lib.cam.ac.uk/.

35. Trinity College, Cambridge: *Memoriale Collegii Trinitatis*, Manuscript R.17.8, https://mss-cat.trin.cam.ac.uk/Manuscript/R.17.8, f. 103: 'hic quanqum istius Amplitudinem Collegii nunquam oculis delibauerit, sed celebritatem quondam eius promis (ut videtur) auribus exceperit: tamen ad bibliothecam hanc optimorum Scriptorum monumentis instruendam centenas libras (pro sua Nobili Munificentia) Testamento Legauit'.

36. On Sedley's bequest to Trinity, see also Phillip Gaskell, *Trinity College Library: The First 150 Years* (Cambridge, UK: Cambridge University Press, 1980).

37. Understandably, this extract from the will occurs more than once in the university archives (SEP/D/18 and WP/α/60/2), and may also be found, in updated spelling, on p. 284 of vol. I of G. R. M. Ward (tr.), *Oxford University Statutes*, 2 vols (London: William Pickering, 1845–51).

38. These were Nathaniel Brent, Savile's successor as Warden of Merton; John Keeling of the Inner Temple; and the university registrar, Thomas French.

39. C. Oscar Moreton, *Waddesdon and Over Winchendon, being a slight sketch of the history of two parishes in the county of Buckingham* (London: Society for Promoting Christian Knowledge, 1929), pp. 50–2. The farmhouse no longer stands, but its site is known: https://heritageportal. buckinghamshire.gov.uk/Monument/MBC6418.

40. Oxford University Archives: NEP/supra/Reg N, Register of Convocation, 1615–28, p. 104v. The principle of fining professors who did not deliver their lectures was still part of the statutes governing the Sedleian Chair in the nineteenth century, but there is no evidence that it was ever enforced—not even during the eighteenth century when the successive professors almost certainly did not teach any natural philosophy.

41. Mordechai Feingold, 'The Universities and the Scientific Revolution: The Case of England', in *New Trends in the History of Science: Proceedings of a Conference Held at the University of Utrecht*, ed. R. P. W. Visser, H. J. M. Bos, Lodewijk C. Palm, and H. A. M. Snelders, Nieuwe Nederlandse Bijdragen tot de Geschiedenis der Geneeskunde en der Natuurwetenschappen, vol. 30 (Amsterdam: Rodopi, 1989), 29–48, at p. 39.

42. C. L. Kingsford, rev. Sarah Bakewell, 'Lapworth, Edward (1574–1636)', *Oxford Dictionary of National Biography*, https://doi.org/10.1093/ref:odnb/16066. Details concerning Lapworth may also be found in the database of the Exeter University Early Modern Practitioners Project (https:// practitioners.exeter.ac.uk/).

43. The academic halls were educational institutions within the university that had their origins in the properties that emerged as officially sanctioned accommodation for students. The main difference between the halls and the colleges was that whereas the colleges are governed by their fellows and are founded on secure financial endowments, the halls were directed by their principals and were financially more precarious. Most halls were eventually taken over and absorbed into nearby colleges. See Christopher Hibbert and Edward Hibbert (ed.), *The Encyclopædia of Oxford* (London: Macmillan, 1988), 3–6.

44. R. G. Lewis, 'The Linacre Lectureships Subsequent to Their Foundation', in *Essays on the Life and Work of Thomas Linacre, c. 1460–1524*, ed. Francis Maddison, Margaret Pelling, and Charles Webster (Oxford: Clarendon Press, 1977), 223–64 at p. 240.

45. Feingold, *Mathematicians' Apprenticeship*, p. 65.

46. Lewis, 'The Linacre Lectureships Subsequent to Their Foundation', p. 240.

47. Thomas Guidott, *A discourse of Bathe, and the hot waters there: also some enquiries into the nature of the water of St. Vincent's rock, near Bristol, and that of Castle-Cary: to which is added a century of observations, more fully declaring the nature, property, and distinction of the Baths: with an account of the lives, and character, of the physicians of Bathe* (London: Printed for Henry Brome [. . .], 1676), p. 168. Guidott also hints at Lapworth's activities as a poet: a laudatory Latin verse by Lapworth appears, for example, at the beginning of Edward Jorden's *A discourse of natural bathes, and mineral waters: Wherein, the original of fountains in general is declared. The nature and difference of minerals, with examples of particular bathes. The generation of minerals in the earth, from whence both the actual heat of bathes, and their virtues proceed. By what means mineral waters are to be discover'd. And lastly, of the nature, and uses of bathes, but especially, of our bathes at Bathe in Somerset-shire*, 4th ed. (London: For George Sandbridge, at the Bible on Ludgate-Hill, Thomas Salmon, Bookseller in Bathe, 1673). On Lapworth's poetic bent, see also Jeffrey Powers-Beck, 'George Herbert and Edward Lapworth: On Prince Charles's Return from Spain', *Notes and Queries*, December 1994, 515–17.

48. Kenneth Fincham, 'Oxford and early Stuart polity', in *History of the University of Oxford, vol. IV*, 179–210.

49. Ward, *Oxford University Statutes*, vol. I, p. 22.

50. Feingold, 'Patrons and Professors', p. 119.

51. See the comments in Frank, 'Medicine', pp. 525–6.

52. Feingold, 'Mathematical Sciences and New Philosophies', p. 389.

53. Occasionally spelt 'Edwardes'. For biographical material, see *Alumni Oxonienses*; C. J. Robinson, rev. S. E. Mealor, 'Edwards, John (b. 1600, d. late 1650s)', *Oxford Dictionary of National Biography*, https://doi.org/10.1093/ref:odnb/8543; Andrew Hegarty, *A Biographical Register of St. John's College, Oxford, 1555–1660*, Oxford Historical Society, New Series, vol. 43 (Woodbridge: The Boydell Press, 2011), pp. 48–9; Exeter University Early Modern Practitioners Project (https://practitioners.exeter.ac.uk/).

54. Feingold, *Mathematicians' Apprenticeship*, p. 108.

55. Montagu Burrows, *The register of the visitors of the University of Oxford, from A.D. 1647 to A.D. 1658* (London: Printed for the Camden Society, 1881), pp. 127, 240.

56. *Alumni Oxonienses*; William Dunn Macray, *A register of the members of St. Mary Magdalen College, Oxford, from the foundation of the college*, new series, vol. IV: Fellows, 1648–1712 (London: Henry Frowde, 1904), pp. 61–4; John Ward, *The lives of the professors of Gresham College: To which is prefixed the life of the founder, Sir Thomas Gresham, with an appendix, consisting of orations, lectures, and letters, writen [sic] by the professors, with other papers serving to illustrate the lives* (London, 1740), pp. 245–7.

57. On the Oxford Experimental Philosophy Club, see Charles Webster, *The Great Instauration: Science, Medicine and Reform 1626–1660*, 2nd ed., Studies in the History of Medicine, vol. 5 (Bern: Peter Lang, 2002), pp. 153–74. The Magdalen Hall mentioned here should not to be confused with Magdalen College: Magdalen Hall was an academic hall (see note 43) that subsequently became Hertford College.

58. Both Magdalen Hall and Lincoln College were known for their Puritan leanings: A. Clark, *Lincoln*, University of Oxford College Histories (London: F. E. Robinson, 1898), p. 94.

59. Ward, *Lives of the Professors of Gresham College*, p. 246.

60. Macray, *Register of the members of [. . .] Magdalen College*, p. 62.

61. F. J. Varley (ed.), *The Restoration Visitation of the University of Oxford and Its Colleges*, Camden Miscellany, vol. XVIII (London: Royal Historical Society, 1948), pp. 9–10.

62. Crosse's Latin epitaph is reproduced in Macray, *Register of the members of [. . .] Magdalen College*, p. 64.

63. Oxford University Archives: Register of Convocation 1659–71 (cited in Dewhurst, *Willis's Oxford Lectures*, p. 38).

64. Robert L. Martensen, 'Willis, Thomas (1625–1675)', *Oxford Dictionary of National Biography*, https://doi.org/10.1093/ref:odnb/29587; Alastair Compston, *All manner of ingenuity and industry: A bio-bibliography of Thomas Willis 1621–1675* (Oxford: Oxford University Press, 2021).

65. Frank, *Harvey and the Oxford Physiologists*, p. 179.

66. A detailed treatment of these surviving lecture notes may be found in Dewhurst, *Willis's Oxford Lectures*.

67. A. J. Turner, 'Millington, Sir Thomas (1628–1704)', *Oxford Dictionary of National Biography*, https://doi.org/10.1093/ref:odnb/18764.

68. Burrows, *Register of the visitors of the University of Oxford*, p. 241.

69. Matthew Steggle, 'Bathurst, Ralph (1619/20–1704)', *Oxford Dictionary of National Biography*, https://doi.org/10.1093/ref:odnb/1699. Bathurst outlined the reasons for electing Millington in a letter to the Duke of Ormond, Chancellor of the university, in November 1675 (Thomas Warton, *The life and literary remains of Ralph Bathurst, M.D., Dean of Wells, and President of Trinity College in Oxford* (London, 1761), pp. 138–9).

70. Scott Mandelbrote, 'Morison, Robert (1620–1683)', *Oxford Dictionary of National Biography*, https://doi.org/10.1093/ref:odnb/19275.

71. A manuscript copy of Millington's inaugural lecture survives in Cambridge University Library (MS Add.8861/2).

72. Robert Plot, *A natural history of Oxford-shire, being an essay toward the natural history of England* (Oxford: Sheldonian Theatre, 1677), p. 122.

73. John Henry, 'Keill, John (1671–1721)', *Oxford Dictionary of National Biography*, https://doi.org/10.1093/ref:odnb/15256. Keill's appointment as Millington's deputy was made on the basis of his scientific credentials—in particular, his 1698 critique of Thomas Burnet's theory of the Earth (Feingold, 'Mathematical Sciences and New Philosophies', p. 434).

74. This text eventually appeared in English as *An introduction to natural philosophy: or philosophical lectures read in the University of Oxford, anno dom. 1700. To which are added the demonstrations of Monsieur Huygens's theorems, concerning the centrifugal force and circular motion* (London, 1720).

75. Allan Chapman, 'Oxford's Newtonian School', in *Oxford Figures*, 167–80.

76. A. H. T. Robb-Smith, 'The Life and Times of Dr Richard Frewin (1681–1761): Medicine in Oxford in the Eighteenth Century', 15th Gideon de Laune lecture, Worshipful Society of Apothecaries of London (1972), https://practitioners.exeter.ac.uk/wp-content/uploads/2014/11/Frewin.pdf (accessed 20 September 2022).

77. Evidence of Keill's continued teaching between 1704 and 1712 is found in the lecture notes kept by a student, John Ivory, who matriculated at Christ Church in 1707 (Cambridge University Library: MS.Add.9317).

78. Allan Chapman and Christopher Hollings, 'A Century of Astronomers: From Halley to Rigaud', in *Oxford's Savilian Professors of Geometry*, 55–91.

79. *Alumni Oxonienses*.

80. Thomas Hearne, *Remarks and collections of Thomas Hearne*, vol. I: *July 4, 1705–March 19, 1707* (Oxford: Printed for the Oxford Historical Society at the Clarendon Press, 1885), pp. 85, 188.

81. Thomas Hearne, *Remarks and collections of Thomas Hearne*, vol. VI: *Jan. 1, 1717–May 8, 1719* (Oxford: Printed for the Oxford Historical Society at the Clarendon Press, 1902), p. 339.

82. J. R. Bloxam (ed.), *Magdalen College and James II, 1686–1688: A series of documents* (Oxford: Printed for the Oxford Historical Society at the Clarendon Press, 1886).

83. Thomas Hearne, *Remarks and collections of Thomas Hearne*, vol. VII: *May 9, 1719–Sept. 22, 1722* (Oxford: Printed for the Oxford Historical Society at the Clarendon Press, 1906), p. 101.

84. Hearne alleged that the reason was that Bertie was in debt to All Souls (ibid.), but there does not appear to be any corroborating evidence for this claim.

85. *Alumni Oxonienses*.

86. *Alumni Oxonienses*; William Hunt, rev. S. J. Skedd, 'Browne, Joseph (1700–1767)', *Oxford Dictionary of National Biography*, https://doi.org/10.1093/ref:odnb/3685.

87. William Hutchinson, *The history of the county of Cumberland* [. . .], 2 vols (Carlisle: F. Jollie, 1794), vol. 1, pp. 426–7; Margaret Evans (ed.), *Letters of Richard Radcliffe and John James of Queen's College, Oxford, 1755–83* (Oxford: Printed for the Oxford Historical Society at the Clarendon Press, 1888); Timothy L. S. Sprigge (ed.), *The Correspondence of Jeremy Bentham*, vol. 1: *1752–76* (London: UCL Press, 2017), letter 61.

88. A minute of Wheeler's election to the Sedleian Professorship is preserved in the university archives: WP/β/7/1. On Wheeler, see *Alumni Oxonienses* and the *Electronic Enlightenment Biographical Dictionary*, https://doi.org/10.13051/ee:bio/wheelbenja026060.

89. The notes taken by an anonymous student at lectures given in 1772 are preserved in the Wellcome Collection (MS 4990, https://wellcomecollection.org/works/zk7ukvvv).

90. These papers relating to Wheeler are preserved among Routh's papers at Magdalen: PR30/1/C3/8. The lectures were later published as *The theological lectures of the late Rev. Benjamin Wheeler, D.D. Canon of Christ Church, and Regius Professor of Divinity in the University of Oxford. With a prefatory sketch of his life and character, by Thomas Horne, D.D. late Fellow of Trinity College in the same university*, vol. 1 (Oxford: At the University Press, for J. Parker, 1819).

91. The subject is treated at length in Nigel Aston, *Enlightened Oxford: The University and the Cultural and Political Life of Eighteenth-Century Britain and Beyond* (Oxford: Oxford University Press, 2023), ch. 5. We focus here on teaching organized within the central university, but there are also hints of (experimental) natural philosophy being taught in colleges: in a letter of March 1762, for example, Jeremy Bentham wrote to his father that 'Mr. Jefferson [Bentham's tutor at Queen's] began Natural Philosophy with us yesterday, but whether we shall improve much or no I can't tell, as he has no apparatus' (Sprigge, *Correspondence of Jeremy Bentham*, vol. 1, letter 40). Bentham also attended the centrally organized lectures of Nathaniel Bliss (1700–64), Savilian Professor of Geometry, who lectured on experimental philosophy in succession to Bradley (see below in the main text) for a short time in the early 1760s, though not entirely successfully, for Bentham observed that 'Mr. Bliss seems to be a very good sort of a Man, but I doubt is not very well qualified for his Office, in the practical Way I mean, for he is oblig'd to make excuses for almost every Experiment they not succeeding according to expectation' (ibid., letter 46). A few years later, Bentham would also attend Thomas Hornsby's lectures on natural philosophy: 'he has it seems made several Improvements and additions to this course, which will render it completer than any of the former ones, either of his own or his predecessors' (ibid., letter 71).

92. Anita Guerrini, 'Gregory, David (1659–1708)', *Oxford Dictionary of National Biography*, https://doi.org/10.1093/ref:odnb/11456; A. V. Simcock, 'Whiteside, John (bap. 1679, d. 1729)', *Oxford Dictionary of National Biography*, https://doi.org/10.1093/ref:odnb/38124; Patricia Fara, 'Desaguliers, John Theophilus (1683–1744)', *Oxford Dictionary of National Biography*, https://doi.org/10.1093/ref:odnb/7539.

93. Chapman, 'Oxford's Newtonian School'. On Bliss's lectures, see also note 91.

94. Robert Fox, Graeme Gooday, and Tony Simcock, 'Physics in Oxford: Problems and Perspectives', in *Physics in Oxford 1839–1939: Laboratories, Learning, and College Life*, ed. Robert Fox and Graeme Gooday (Oxford: Oxford University Press, 2005), 1–23 at p. 1.

95. A. V. Simcock, 'Walker, Robert (1801–1865)', *Oxford Dictionary of National Biography*, https://doi.org/10.1093/ref:odnb/38098.

96. Allan Chapman, 'Thomas Hornsby and the Radcliffe Observatory', in *Oxford Figures*, 203–20.

97. G. L'E. Turner, 'The Physical Sciences', in *The History of the University of Oxford*, vol. V: *The Eighteenth Century*, ed. L. S. Sutherland and L. G. Mitchell (Oxford: Clarendon Press, 1986), 659–81.

98. This is certainly the trajectory that would later be emphasized by Baden Powell in his *History of natural philosophy, from the earliest periods to the present time* (London: Longman etc., 1834). By the second half of the nineteenth century, when William Thomson and Peter Guthrie Tait prepared their *Treatise on natural philosophy* (Oxford: Oxford University Press, 1867), the subject had become heavily mathematical. We note that the latter book was seen through the press by the Sedleian Professor Bartholomew Price in his role as Secretary to the Delegates of the University Press (see Chapter 6).

99. C. Webster, 'The Medical Faculty and the Physic Garden', in *History of the University of Oxford*, vol. V, 683–723.

100. Peter J. T. Morris, 'The Eighteenth Century: Chemistry Allied to Anatomy', in *Chemistry at Oxford: A History from 1600 to 2005*, ed. R. J. P. Williams, John S. Rowlinson, and Allan Chapman (Cambridge: Royal Society of Chemistry, 2009), 52–78.

101. The link, established during Hornsby's tenure, between the Savilian Professorship of Astronomy and the Radcliffe Observership also persisted for a time; see Chapman, 'Hornsby and the Radcliffe Observatory'.

102. Chapman and Hollings, 'A Century of Astronomers'. Upon Robertson's death, his successor in the Geometry Chair, Stephen Peter Rigaud (1774–1839), would similarly switch to the Astronomy

Chair with the associated Readership of Experimental Philosophy. At Rigaud's death, however, the connection between the two posts was broken: Fox et al., 'Physics in Oxford', p. 1.

103. Oxford University Archives: WP/β/7/2.

104. Ward, *Oxford University Statutes*, vol. II, p. 236.

105. M. G. Brock, 'The Oxford of Peel and Gladstone', in: *The History of the University of Oxford*, vol. VI: *Nineteenth-Century Oxford, Part 1*, ed. M. G. Brock and M. C. Curthoys (Oxford: Clarendon Press, 1997), 7–71. On nineteenth-century Oxford examinations generally, see M. C. Curthoys, 'The Examination System', in *History of the University of Oxford*, vol. VI, 339–74. On mathematics specifically, see Keith Hannabuss, 'Mathematics', in *History of the University of Oxford*, vol. VII, 433–55. See also Arthur Engel, 'Emerging Concepts of the Academic Profession at Oxford 1800–1854', in *The University in Society*, vol. I: *Oxford and Cambridge from the 14th to the Early 19th Century*, ed. Lawrence Stone (Princeton, NJ: Princeton University Press, 1974), 305–51.

106. Edmund Venables, rev. M. C. Curthoys, 'Parsons, John (1761–1819)', *Oxford Dictionary of National Biography*, https://doi.org/10.1093/ref:odnb/21466; S. J. Skedd, 'Eveleigh, John (1748–1814)', *Oxford Dictionary of National Biography*, https://doi.org/10.1093/ref:odnb/8994; W. R. Ward, 'Jackson, Cyril (1746–1819)', *Oxford Dictionary of National Biography*, https://doi.org/10.1093/ref:odnb/14522.

107. K. C. Turpin, 'The Ascendancy of Oriel', in *History of the University of Oxford*, vol. VI, 183–92.

108. Between 1807 and 1850, only around 10–15% of candidates sat for honours in mathematics (Curthoys, 'Examination System', p. 352). For a very brief history of the mathematics honour school, see E. B. Elliott, 'Honour School of Mathematics and Physics', in *English Education Exhibition: Accounts of the Honour Schools & Faculties of the Univ., &c.*, 1899 (Oxford: Bodleian Library, G. A. Oxon b. 41).

109. On mathematics in Cambridge, see W. W. Rouse Ball, *A History of the Study of Mathematics at Cambridge* (Cambridge, UK: Cambridge University Press, 1889); Andrew Warwick, *Masters of Theory: Cambridge and the Rise of Mathematical Physics* (Chicago: University of Chicago Press, 2003); Alex D. D. Craik, *Mr Hopkins' Men: Cambridge Reform and British Mathematics in the 19th Century* (London: Springer, 2007); Tony Crilly, 'Cambridge: The Rise and Fall of the Mathematical Tripos', in *Mathematics in Victorian Britain*, ed. Raymond Flood, Adrian C. Rice, and Robin Wilson (Oxford: Oxford University Press, 2011), 17–32.

110. Brock, 'Oxford of Peel and Gladstone', p. 13.

111. Ibid., p. 16. For further background, see F. Sherwood Taylor, 'The Teaching of Science at Oxford in the Nineteenth Century', *Annals of Science* 8(1) (1952) 82–112.

112. Pietro Corsi, 'Powell, Baden (1796–1860)', *Oxford Dictionary of National Biography*, https://doi.org/10.1093/ref:odnb/22642.

113. Baden Powell, *The present state and future prospects of mathematical and physical studies in the University of Oxford, considered in a lecture* (Oxford: W. Baxter, 1832).

114. As is often the way of these things, Powell's views were countered in one anonymously published pamphlet (A Master of Arts, *A short criticism of a lecture published by the Savilian Professor of Geometry* (Oxford: W. Baxter, 1832)), and reinforced, in a slightly more temperate tone, in another (Philomath: Oxoniensis, *A few words in favour of Professor Powell and the sciences, as connected with certain education remarks*, Oxford, 1832).

115. Hannabuss, 'Mathematics', p. 446.

116. [B. Powell], *A short treatise on the principles of the differential and integral calculus* (Oxford: University Press, 1830).

117. Since there was otherwise little compulsion to be present, the attendance figures for nineteenth-century scientific lectures in Oxford have subsequently been taken as a measure of the popularity of individual lecturers: N. A. Rupke, 'Oxford's Scientific Awakening and the Role of Geology', in *History of the University of Oxford*, vol. VI, 543–62 at p. 546.

118. On the motivation for the formation of the Commission, see *Hansard* HL Deb. 13 June 1850, vol. 111, cc. 1146–58 ('Universities Commission'), https://api.parliament.uk/historic-hansard/lords/1850/jun/13/universities-commission (accessed 17th October 2022).

119. See instead W. R. Ward, *Victorian Oxford* (London: Frank Cass & Co., 1965); L. W. B. Brockliss, *The University of Oxford: A History* (Oxford: Oxford University Press, 2016), ch. 9.

120. *Report of Her Majesty's Commissioners appointed to inquire into the State, Discipline, Studies, and Revenues of the University and Colleges of Oxford: together with the Evidence, and an Appendix* (London: W. Clowes and Sons for Her Majesty's Stationery Office, 1852).

121. J. H. C. Leach, 'Jeune, Francis (1806–1868)', *Oxford Dictionary of National Biography*, https://doi.org/10.1093/ref:odnb/14806. See also Charles Edward Mallet, *A History of the University of Oxford*, 3 vols (London: Methuen, 1924–7), vol. 3, p. 415.

122. *Report of Her Majesty's Commissioners*, Evidence, 59–67.

123. Ibid., Evidence, p. 65.

124. Ibid., Evidence, p. 189.

125. For a summary of the Commission's recommendations, see 'Oxford University Commission', *The Literary Gazette: A weekly journal of literature, science, and the fine arts*, 29 May 1852, 449–51.

126. Engel, 'Emerging Concepts of the Academic Profession at Oxford', p. 333. For the Act itself, see: Oxford University Act 1854 (17 & 18 Vict. c. 81), https://www.legislation.gov.uk/ukpga/Vict/17-18/81/contents (accessed 30 October 2022).

127. Ward, *Victorian Oxford*; Christopher Harvie, 'Reform and Expansion, 1854–1871', in *History of the University of Oxford*, vol. VI, 697–730.

128. See, for example, a student handbook of 1881, which indicates that the Savilian Professor of Geometry will provide instruction in pure mathematics, and the Sedleian Professor in applied (*The student's handbook to the university and colleges of Oxford*, 6th ed. (Oxford: Clarendon Press, 1881), p. 50).

129. Curthoys, 'Examination System', pp. 352–3.

130. These discussions are reflected in the minutes of the university's Hebdomadal Council; see, in particular, Oxford University Archives: HC1/2/1.

131. 'University intelligence', *The Times*, no. 22002, Thursday 15 March 1855, 5f. The Sedleian Professor would similarly find himself an *ex officio* member of a number of other electoral boards, including that of the Reader (later Professor) of Experimental Philosophy (*The historical register of the University of Oxford, being a supplement to the Oxford University Calendar, with an alphabetical record of university honours and distinctions, completed to the end of Trinity Term, 1888* (Oxford: Clarendon Press, 1888), pp. 61, 69).

132. 'The new professorial statute' [signed 'A member of Congregation'], Oxford, 1857; two copies of the pamphlet survive in the Bodleian Library under shelfmarks G. A. Oxon b. 29 and G. A. Oxon b. 73 (156).

133. Oxford University Archives: NW/17/2, University Professorships, 1858. The latter features a copy of the text of the statute, as does the pamphlet cited in note 132.

134. 'University intelligence', *The Times*, no. 22002, Thursday 15 March 1855, 5f.

135. The original proposal seems to have been to replace the President of Magdalen and the Warden of All Souls by the Astronomer Royal and the President of the Royal Society outright, but this was rejected by the university council on 25 February 1857; the revised arrangement whereby the two colleges' heads would alternate in their participation was approved instead a month later (Oxford University Archives: HC1/2/1, Hebdomadal Council—Signed Council Minutes, 1854–66, pp. 171, 176). But see note 136.

136. As the involvement of Queen's with the Sedleian Chair increased (see below in the main text), a further person nominated by the college was added to the electoral board (Oxford University Archives: NW/17/56, Sedleian Professorship of Natural Philosophy, 1882). Other changes were the

inclusion of the Professor of Experimental Philosophy and of another elector appointed by the university's Hebdomadal Council; after more than 250 years, the President of Magdalen and the Warden of All Souls were quietly dropped as electors.

137. It appears to have been a letter from the Provost of Queen's to the university's Hebdomadal Council that prompted the reconsideration of the board of electors to the Sedleian Chair in the first place (Oxford University Archives: HC1/2/1, Hebdomadal Council—Signed Council Minutes, 1854–66, p. 170, 25 February 1857).

138. Universities of Oxford and Cambridge Act 1877 (40 & 41 Vict. c. 48), https://www.legislation.gov.uk/ukpga/Vict/40-41/48/contents (accessed 30 October 2022).

139. The connection between Queen's and the Sedleian Chair is dealt with only very briefly in the histories of the college: John Richard Magrath, *The Queen's College*, 2 vols (Oxford: Clarendon Press, 1921), vol. II, pp. 183, 202; R. H. Hodgkin, *Six Centuries of an Oxford College: A History of the Queen's College, 1340–1940* (Oxford: Basil Blackwell, 1949), p. 186.

140. Oxford University Archives: HC1/2/1, Hebdomadal Council—Signed Council Minutes, 1854–66, p. 150 (7 November 1856).

141. *Report of Her Majesty's Commissioners*, Report, pp. 180–1.

142. The Queen's College, Oxford: College Register L, 1809–1862, 26 June 1851.

143. Ibid., 5 January 1854.

144. Ibid., 21 June 1855.

145. E. I. Carlyle, rev. H. C. G. Matthew, 'Thomson, William (1819–1890)', *Oxford Dictionary of National Biography*, https://doi.org/10.1093/ref:odnb/27330.

146. *Statutes of Queen's College, Oxford, as approved by the Provost and Fellows, submitted for the approval of the commissioners* ([Oxford], December 1878); *Report of committee [of Queen's college] (Mr. Armstrong, Mr. Grose) on [the college's contribution to the stipend of the] Sedleian professorship* [Oxford, 1900] (Bodleian Library: G. A. Oxon c. 153); Oxford University Archives: Sedleian Professorship of Natural Philosophy, note 5 (*c.*1980).

147. Waddesdon Archive, Buckinghamshire: Chancellor, Masters & Scholars of the University of Oxford to the Baron F. J. de Rothschild—Conveyance of a messuage and lands called 'Philosophy Farm' at Waddesdon, Bucks, 14 May 1875, accession number: 88.2005.105.

148. Oxford University Archives: NW/4/3/21, Sedleian Professorship and Waddesdon estate, 1876.

149. E. A. Milne, rev. Julia Tompson, 'Love, Augustus Edward Hough (1863–1940)', *Oxford Dictionary of National Biography*, https://doi.org/10.1093/ref:odnb/34603.

150. See note 132.

151. Price was one of the first people to be elected to such a position at Queen's (The Queen's College, Oxford: College Register M 1862–1873, 11 June 1868).

152. Although the Sedleian job advertisement prior to Love's appointment mentions the contribution by Queen's to the professor's salary, it does not explicitly say that the professor will be a member of the college (*Oxford University Gazette*, no. 943, vol. XXIX, Friday 20 January 1899, p. 238). Quite how to fit Love into the college structure was a further tedious detail that needed to be resolved, for the number and nature of fellowships was strictly regulated by the college statutes and tradition. In the first instance, he was elected an honorary member of the Senior Common Room, which gave him access to college facilities and the right to take meals in college, but which conferred little status (The Queen's College, Oxford: Governing Body Minute Book 1883–1902, p. 298, 17 March 1899). A little later, he was made an honorary fellow (The Queen's College, Oxford: Governing Body Minute Book 1903–1914, p. 48, 9 November 1904).

153. Indeed, the connection has been in place long enough for local traditions to have developed: aside from the college's patroness (a role filled by queens consort), only the Sedleian Professor is permitted to operate the orrery located in the college's Upper Library. This custom probably stems from George Temple's time as Sedleian Professor (1953–68) when he oversaw the repair of the

orrery in 1958 (The Queen's College, Oxford: Governing Body Minute Book, 1957–1961, p. 99, 21 June 1958).

154. Oxford University Archives: UR6/PHS/1, file 1, Correspondence relating to Sedleian Chair 1933–1963.

155. Ibid., ff. 1, 29, 31.

156. Ibid., f. 40.

157. Ibid., f. 23 (E. A. Milne to D. Veale, University Registrar, 21st June 1944).

158. Ibid.

159. Ibid., f. 29.

160. Royal Society, London: FS/7/4/1/1, Simon Papers, 'The future of physics, Oxford University', 1945.

161. See, for example: J. D. Bernal, 'A new industrial revolution', *The Observer*, no. 8030, 22 April 1945, p. 4; *Report on the needs of research in fundamental science after the war* (London: Royal Society, 1945). Copies of both of these items are included in the file cited in note 160.

162. See, for example, the letters of the physicist Francis Simon (1893–1956) contained in the file cited in note 160.

163. A particularly detailed picture of the appointment process may be reconstructed from materials in the archives of Oxford University (UR6/PHS/1, file 1) and of the Royal Society (FS/7/4/1/1).

164. Again by Francis Simon (see note 162), who encouraged the physicist Sybren Ruurds de Groot (1916–94) to apply (Royal Society, London: FS/7/2/179, Simon Papers, Correspondence with S. R. de Groot, 1953).

165. 'Sedleian Professor of Natural Philosophy' (Part of Council Regulations 24 of 2002: Regulations for Academic and Other Posts), https://governance.admin.ox.ac.uk/legislation/sedleian-professor-of-natural-philosophy (accessed 6 November 2022).

166. Job advertisement preserved at https://data.ox.ac.uk/doc/vacancy/131761.html (accessed 6 November 2022).

CHAPTER 2

Further reading

Hansruedi Isler, *Thomas Willis: Ein Wegbereiter der Modernen Medizin 1621–1675* (Stuttgart: Wissenschaftliche Verlagsgesellschaft, 1965); translated in English as *Thomas Willis 1621–1675: Doctor and Scientist* (New York; London: Hafner, 1968).

Kenneth Dewhurst, *Thomas Willis's Oxford Lectures* (Oxford: Sandford Publications, 1980).

Kenneth Dewhurst, *Willis's Oxford Casebook (1650–1652)* (Oxford: Sandford Publications, 1981).

J. Trevor Hughes, *Thomas Willis 1621–1675: His Life and Work* (London: Royal Society of Medicine Services, 1991; reprinted with minor amendments, Oxford: Rimes House, 2009).

Carl Zimmer, *Soul Made Flesh: The Discovery of the Brain, and How It Changed the World* (London: Heinemann, 2004).

Alastair Compston, *All Manner of Ingenuity and Industry: A Bio-bibliography of Dr Thomas Willis 1621–1675* (Oxford: Oxford University Press, 2021).

Notes and references

1. [John Aubrey], *'Brief lives', chiefly of contemporaries*, ed. Andrew Clark, 2 vols (Oxford: Clarendon Press, 1898), vol. 2, pp. 302–4; John Aubrey, *Brief lives with an apparatus for the lives of our English mathematical writers*, ed. Kate Bennett, 2 vols (Oxford: Oxford University Press, 2015), vol. 1, pp. 550–1, with notes on the edited entry, vol. 2, pp. 1477–81; [Anthony Wood], *Athenæ Oxoniensis etc.*, 2 vols (London: Printed for Tho. Bennet at the Half-Moon in S. Pauls Churchyard, 1690/2), vol. 1, pp. 402–3, entries 364–5; *Athenæ Oxoniensis etc.*, 2nd ed., 2 vols (London: Printed for R. Knaplock, D. Midwinter, and J. Tonson, 1721), vol. 2, pp. 549–52, entry 408; and *Athenæ Oxoniensis etc.*, 4

vols (London: Printed for F. C. and J. Rivington [...]; Oxford: J Parker, 1817 [1813–20]), vol. 3, pp. 1048–53, entry 550.

2. John Fell, *Pharmaceutice rationalis, sive diatriba de medicamentorum operationibus in humano corpore, pars secunda* (Oxford: E theatro Sheldoniano, 1675), pp. [12–17]; translated [Samuel Pordage: 1633–91] in *Pharmaceutice rationalis or an exercitation of the operations of medicines in humane bodies. The second part* (London: Printed for T. Dring, C. Harper, and J. Leigh in Fleetstreet [...], 1679), pp. [6–8].

3. The plaque identifying the house where Thomas Willis lived is on 4 Merton Street. Julian Reid, archivist to Corpus Christi and Merton Colleges, confirms that Beam Hall originally also referred to 3 Merton Street where Willis lived, number 4 forming the kitchen and service area. The sites were bought by Gilbert de Biham (dates unknown), Chancellor of the University, in 1246 and 1248. They were rebuilt and purchased by Corpus Christi College in 1553, from whom Willis leased the buildings.

4. Thomas Willis, *Pharmaceutice rationalis* etc., *pars secunda* (Oxford: E. Theatro Sheldoniano, 1675), pp. [5–11]: translated in *Pharmaceutice rationalis* etc. (1679), pp. 3–5.

5. [Browne Willis], 'Willis (Thomas)', in [John Bernard, Thomas Birch, John Lockman, and other hands], *A general dictionary, historical and critical* etc., 10 vols (London: Printed by James Bettenham, for G. Strahan, J. Clarke, T. Hatchet in Cornhill [...], 1741), vol. X, pp. 171–5. Although unsigned, everything suggests that the entry is by Browne Willis, as confirmed in his correspondence with Philip Morant (British Library, Add MS 37222, ff. 114–15, ff. 127–8, and f. 139).

6. *De Febribus*, pp. [9–15], in Thomas Willis, *Diatribæ duæ medico-philosophicæ* etc. (London: Typis Tho. Roycroft, impensis Jo. Martin, Ja. Allestry, & Tho. Dicas [...], 1659); translated in *The remaining medical works of that famous and renowned physician Dr Thomas Willis* etc. (London: Printed for T. Dring, C. Harper, J. Leigh, and S. Martyn, 1681), pp. 53–5.

7. Thirteen letters are held at St John's College library, Oxford; three are written by Willis, two in English and one in Latin. Other correspondence is held in the British library (Add MS 34727, Sloane MS 810, f. 229, and Sloane MS 1513, f. 1).

8. Wellcome Library 788A: inscribed 'Liber Manuscriptus Clarissimi Doctor Thome Willis Medici Insign [...] Two of these MSS of Dr Willis Lent Dr Chace abt. 5 years agoe and not returned 1737'. One of at least three known to have been kept in his lifetime, this was purchased in December 1961 from Mr Richard Hatchwell (1927–2009), a well-known dealer in antiquarian books. It came from the library at Llantarth Court near Abergavenny in South Wales, when the building passed from private ownership to become a boys' preparatory school. Almost certainly, the manuscript was originally in the possession of Browne Willis. The fate of the other two is unknown.

9. Dewhurst, *Willis's Oxford Casebook*.

10. [William Petty], *The Petty papers. Some unpublished writings of Sir William Petty* etc., 2 vols (London: Constable & Co.; Boston/New York: Houghton Mifflin Co., 1927), vol. 2, pp. 155, 157–67; Richard Watkins, *Newes from the dead* etc. (Oxford: Printed by Leonard Lichfield [and H. Hall], for Tho. Robinson, 1651); reprinted in: Joseph Morgan, *Phoenix Britannicus* etc. (London: printed for the compiler [...], 1732), vol. 1 (all printed), pp. 233–48.

11. Bishop Burnet, *History of his own time: from restoration of King Charles the Second to the treaty of peace at Utrecht in the reign of Queen Anne* (London: Henry G. Bohn, 1857), p. 154. The infected Duke of York and his wife passed on the venereal disorder to several of their children. The two Queens, Mary and Anne, survived but the story went that their younger half-brother, James the Old Pretender (1688–1766: sometime Prince of Wales), being unaffected, 'could neither be his [Duke of York], nor be born of any wife, with whom he had long lived'.

12. There is no consensus on the dates at which meetings of the Oxford Experimental Philosophical Club were first held in Oxford. A detailed description is given by Robert Frank, *Harvey and the*

Oxford Physiologists: A Study of Scientific Ideas (Berkeley; London: University of California Press, 1980), pp. 53–5, and table 3, pp. 63–89, with a full list of participants throughout the entire period.

13. Thomas Sprat, *The history of the Royal-Society of London for the improving of natural knowledge* (London: Printed by T[homas]. R[oycroft]. for J. Martyn [...], 1667); see pp. 53, 55 for mention of Willis.

14. Robert Frank, 'Thomas Willis and his circle: brain and mind in seventeenth-century medicine', in *The Languages of Psyche: Mind and Body in Enlightenment Thought*, ed. G. S. Rousseau (Berkeley: University of California Press, 1990), pp. 107–46, at p. 119.

15. Frank, *Harvey and the Oxford Physiologists*, p. 165, ch. 7, n. 5; Frank, 'Thomas Willis and his Circle', p. 115, n. 32.

16. See note 8: annotations on rear free-endpaper and paste down.

17. See Royal Society Archives: Journal Book, volume 1 (28 November 1660–11 November 1663), JBO/1, p. 2 for the initial mention of Willis; and Thomas Birch, *The history of the Royal Society of London for improving of natural knowledge* etc., 4 vols (London: Printed for A. Millar in the Strand, 1756–7), vol. 1, pp. 4, 15, 54, 332; vol. 2, p. 201; vol. 3, pp. 95, 242.

18. John Wallis, *A defence of the Royal Society, and the Philosophical transactions* etc. (London: Printed by T[homas] S[nowden] for Thomas Moore, 1678); John Wallis, 'Account of some passages of his own life' etc., appendix to *Peter Langtoft's Chronicle*, ed. Thomas Hearne, 2 vols (Oxford: Printed at the Theater, 1725), vol. 1, pp. clxi–clxiv; Harold Hartley, *The Royal Society: Its Origins and Founders* (London: Royal Society, 1960), pp. 1–39; Charles Webster, 'The Origins of the Royal Society', *History of Science* 6 (1967), 106–28.

19. Sir Archibald Geikie, *Annals of the Royal Society Club: The record of a London dining- club in the eighteenth & nineteenth centuries* (London: Macmillan, 1917).

20. The following are dedications to the third edition of *Diatribæ duæ medico-philosophicæ*, etc. (London: Typis Tho. Roycroft, Impensis Jo. Martin, Ja. Allestry, & Tho. Dicas, 1662); *Cerebri anatome: cui accessit nervorum descriptio et usus* (London: Typis Tho. Roycroft, impensis Jo. Martyn & Ja. Allestry, 1664).; and *De anima brutorum* etc. (Oxford: E Theatro Sheldoniano, 1672); translations in *The remaining medical works*, pp. [5–6] and 51–2; and *Two discourses concerning the soul of brutes* etc. (London: Printed for Thomas Dring, Ch. Harper, John Leigh, 1683), pp. [3–4].

21. Robert L. Martensen, *The brain takes shape: an early history*, Oxford; New York: Oxford University Press, 2004 p. 112.

22. Dewhurst, *Thomas Willis's Oxford lectures*; based on John Locke's commonplace book (British Library, Ms Locke, ff. 1–68), and material located in Robert Boyle's papers, library of the Royal Society (Boyle papers, vol. XIX, ff. 1–35). Frank, 'Thomas Willis and His Circle', p. 122, n. 58, explains that it was he who, in 1969, identified Lower as the author of information on Willis's lectures in the Boyle papers.

23. Robert Plot, *The natural history of Oxford-shire: being an essay toward the natural history of England* (Oxford: Printed at the Theater, 1677), pp. 301–7, paragraphs numbered 215–34.

24. Dewhurst, *Thomas Willis's Oxford Lectures*, pp. 138–50.

25. Ibid., pp. 72–87.

26. Ibid., pp. 96–113.

27. Book 1. 1659, containing treatises: 1. *De fermentatione*; 2. *De febribus*; and 3. *De urinis*. Book 2. 1664, containing treatises: 4. *Cerebri anatome*; and 5. *Nervorum descriptio et usus*. Book 3. 1667, containing treatises: 6. *Pathologiae cerebri*; and 7. *De scorbuto*. Book 4. 1670, containing treatises: 8. *Affectionum hystericæ et hypochondriacæ*; 9. *De sanguinis accensione*; and 10. *De motu musculari*. Book 5. 1672, containing treatise: 11. *De anima brutorum*. Book 6a. 1674, containing treatise: 12. *Pharmaceutice rationalis* (part 1); with Book 6b. 1675, containing treatise: 13. *Pharmaceutice rationalis* (part 2). Book 7. 1691, containing treatise: 14. *A plain and easie method for preserving those that are well from the infection of the plague*. Book 8. 1676, *Opera omnia* or *Opera medica et physica*

containing treatises: 1–13 (14 present only in the 1694 edition). Book 9. 1679, 1681, 1683, 1684, English translations under various titles containing treatises 12, 13, 7 (1679); 1–6, 9, 10 (1681); 11 (1683); 1–7, 9–13 (1684). Book 10. 1685, 1701, Extractions in English from treatises 2, 6, 7, 11–13. See Compston, *All Manner of Ingenuity and Industry*, ch. 4, pp. 211–30; chs. 5–14, pp. 231–558. To date, 103 different editions, issues, or states of these books published between 1659 and 1721 have been identified.

28. Martyn Ould, *Printing at the University Press, Oxford 1660–1780*, vol. 1, Seaton: The Old School Press, 2015; and Vivienne Larminie, 'The Fell era 1658–1686', in *The History of Oxford University Press*, Volume 1: *Beginnings to 1780*, ed. Ian Gadd (Oxford: Oxford University Press, 2013), pp. 79–104.

29. Compston, *All manner of ingenuity and industry*, ch. 3, pp. 125–210.

30. [Thomas Willis], *The anatomy of the brain and nerves*, ed. William Feindel, 2 vols (Montreal: McGill University Press, 1965), vol. 1, p. 22.

31. See Philip Oldfield, *A bibliography of Thomas Willis's De anima brutorum* (typescript, 1991), p. 5 n. 3, who also mentions a catalogue issued by Dawson's of Pall Mall, 'The Royal Society of London for the promotion of natural knowledge 1660–1960: a tercentenary tribute London' (1961) in which #1801 claims that Wren engraved the plates for *De anima brutorum*.

32. See Antonio Clericuzio, *Elements, Principles and Corpuscles: A Study of Atomism and Chemistry in the Seventeenth Century* (Dordrecht; London: Kluwer Academic Publishers, 2000), pp. 75–101 for discussion of chemistry and atomism in England (1600–60), and the work of Thomas Willis on corpuscular theory, at p. 100; and Compston, *All Manner of Ingenuity and Industry*, ch. 15, pp. 559–613.

33. Steven Shapin and Simon Schaffer, *Leviathan and the Air Pump: Hobbes, Boyle, and the Experimental Life* (Princeton: Princeton University Press, 1985).

34. Thomas Willis, *Of feavers* (1681), p. 131; Charles Creighton, *A history of epidemics in Britain*, 2 vols (Cambridge: The University Press, 1891–4), vol. 1, pp. 559–60; J. F. C. Shrewsbury, *A History of Bubonic Plague in the British Isles* (Cambridge: Cambridge University Press, 1970), suggests that Dr Sayer was 'naturally antipathetic to fleas' and so remained immune to the plague, probably dying of 'the old Irish "bed-sickness" (typhus fever) brought on by sleeping naked with the friend in whom he had induced vomiting and sweating, thus spraying the bedroom with infected droplets'.

35. *De motu musculari* (1670), p. 79; Edwin Clarke and C. D. O'Malley, *The Human Brain and Spinal Cord: A Historical Study Illustrated by Writings from Antiquity to the Twentieth Century* (Berkeley: University of California Press, 1968), pp. 333–5, citing *Cerebri anatome* (1664), pp. 136–7, and *De anima brutorum* (1672), part 1, pp. 104–5; and Compston, *All Manner of Ingenuity and Industry*, ch. 16, pp. 614–62 (with page numbers amended to those in the original editions).

36. S. T. Soemmering, *De basi encephali et originibus nervorum cranio egredientum libri quinque* (Göttingen, 1778), p. 184; C. Wilbur Rucker, 'History of the Numbering of the Cranial Nerves', *Mayo Clinic Proceedings* 41 (1966), 453–61.

37. For example, Gabrielle Falloppio, *Observationes anatomicæ* (Venice, 1561); Johan Vesling, *Syntagma anatomicum, locis plurimis auctum, emendatum, novisque iconibus diligenter exornatum* (Padua, 1647); Johann Wepfer, *Observationes anatomicæ ex cadaveribus eorum quos sustulit apoplexia* (Schaffhausen, 1658).

38. *Cerebri anatome* (1664), pp. 95–6; translated in *The remaining medical works* etc., pp. 82–3; Sir Charles Symonds, 'The Circle of Willis', *British Medical Journal* 1 (1955), 119–24.

39. In dealing with muscle mechanics, *De motu musculari* (1670) leaned upon William Croone's *De ratione motus musculorum* (London, 1667), printed anonymously and added to continental editions of *Cerebri anatome* between 1664 and 1667, leading to a Dutch translation of Croone's essay erroneously attributed to Willis: see Compston, *All Manner of Ingenuity and Industry*, ch. 6, pp. 263–7, 314–17.

40.	Claire Crignon, 'How Animals May Help Us Understand Men', in *Human and Animal Cognition in Early Modern Philosophy and Medicine*, ed. Roberto Lo Presti and Stefanie Buchenau (Pittsburgh: University of Pittsburgh Press, 2017), pp. 173–85; Compston, *All Manner of Ingenuity and Industry*, ch. 17, pp. 663–730.

41.	William G. Lennox, 'Thomas Willis on Narcolepsy', *Archives of Neurology and Psychiatry* 41 (1939), 348–51; *Two discourses concerning the soul of brutes* (1683), p. 134. The poet Epimenides (fl. 600 BC), who slept in a cave for 57 years, is remembered for the paradox 'all Cretans are liars' which turns back and forth around the paradox that, as a Cretan himself, he might be lying in which case 'All Cretans are honest', and so on.

42.	*Pathologiæ cerebri, et nervosi generis specimen: in quo agitur de morbis convulsivis, et de scorbuto* (Oxford: Excudebat Guil. Hall [...], 1667), pp. 90, 124–5; translated in *The remaining medical works* [*An essay on the pathology of the brain*, etc.], pp. 48, 68–9. In describing what are now designated Tourette's syndrome and torsion dystonia, Willis is making an important point that the movements are complex and could not be fabricated, distinguishing between organic and (so-called) functional neurological disease.

43.	Macdonald Critchley, 'The malady of Anne, Countess Conway: A Case for Commentary', *King's College Hospital Gazette* 17 (1937), 44–9; reprinted in Macdonald Critchley, *The Black Hole and Other Essays* (London: Pitman Medical, 1964), pp. 91–7; Compston, *All Manner of Ingenuity and Industry*, pp. 704–7.

44.	*De anima brutorum* (1672), p. 389; translated in *Two discourses concerning the soul of brutes*, p. 160.

45.	Leonard G Guthrie, '"Myasthenia Gravis" in the Seventeenth Century', *Lancet* 1 (1903), 330–1; Sir Geoffrey Keynes, 'The History of Myasthenia Gravis', *Medical History* 5 (1961), 313–26, reprinted with some modifications as Geoffrey Keynes, 'The History of Myasthenia Gravis', in *Myasthenia Gravis*, ed. Raymond Greene (London: Heinemann Medical, 1969), pp. 1–13; J. Trevor Hughes. 'The Early History of Myasthenia Gravis', *Neuromuscular Disorders* 15 (2005), 878–86; Compston, *All Manner of Ingenuity and Industry*, pp. 716–18.

46.	Paul Cranefield, 'A Seventeenth Century View of Mental Deficiency and Schizophrenia; Thomas Willis on Stupidity or Foolishness', *Bulletin of the History of Medicine* 35 (1961), 291–316, with reprinting of chapter XIII from *Two discourses concerning the soul of brutes*; J. Vinchon and J. Vie, 'Un maître de la neuropsychiatrie au XVIIe siècle: Thomas Willis (1662–1675)', *Annales Médico-Psychologiques* 2 (1928), 108–44; Compston, *All Manner of Ingenuity and Industry*, pp. 719–26.

47.	*Pharmaceutice rationalis* (1675), part 1, p. 164; translated in *Pharmaceutice rationalis* (1679), part 1, p. 79; Compston, *All Manner of Ingenuity and Industry*, ch. 18, pp. 731–83.

48.	Marcelli Malpighi, *De externo tactus organo anatomica observatio*, Naples, 1665, pp. 1–45; *Pharmaceutice rationalis* (1675), part 2, p. 382; translated in *Pharmaceutice rationalis* (1679), part 2, p. 161. Willis rarely provided citations and the mention of Malpighi's rete mucosum is no exception.

49.	*Cerebri anatome* (1664), pp. [10–14]; translated in *The remaining medical works* [*The anatomy of the brain*], pp. 53–4.

50.	Kenneth Dewhurst, *Richard Lower's Vindicatio: A Defence of the Experimental Method* (Oxford: Sandford Publications, 1983).

51.	Frederick Slare, *Experiments and observations upon oriental and other Bezoar stones [...] to which is annex'd a vindication of sugars against the charge of Dr Willis, other physicians, and common predujices* etc. (London: Printed for Tim. Goodwin, 1715); Compston, *All manner of Ingenuity and Industry*, pp. 736–9.

52.	[Anon], *Discours de Monsieur Stenon sur l'anatomie du cerveau a Messieurs de l'assemblée, qui se fait chez Monsieur Thevenot* (Paris: chez Robert de Ninuille [...], 1669); [Nicolas Steno] *A dissertation on the anatomy of the brain by Nicholas Steno read in the assembly held in M. Thevenot's house in the year 1665* (Copenhagen, 1950), pp. 11–12, 23–6 in French; and 8–9, 17–18 in English.

53. See note 38.

54. Sir Charles Sherrington, *Man on His Nature* (Cambridge: Cambridge University Press, 1940), pp. 189, 245.

CHAPTER 3

Further reading

Nigel Aston, *Enlightened Oxford. The The University and the cultural and political life of eighteenth-century Britain and beyond*, Oxford: Oxford University Press, 2023.

L. W. B. Brockliss, 'Science, the universities, and other public spaces', in *The Cambridge History of Science*, Vol. 4. *Eighteenth-century science* (ed. Roy Porter), Cambridge: Cambridge University Press, 2003, 44–86.

John Fauvel, 'Eight centuries of mathematical traditions', in *Oxford Figures: Eight Centuries of the Mathematical Sciences* (eds. John Fauvel, Raymond Flood and Robin Wilson), 2nd ed., Oxford: Oxford University Press, 2013, 3–34.

Mordechai Feingold, *The Newtonian Moment. Isaac Newton and the making of modern culture*, New York: New York Public Library; Oxford: Oxford University Press, 2004.

G. L'E. Turner, 'The Physical Sciences', in *The History of the University of Oxford*, Vol. 5: *The Eighteenth Century* (eds. L.S. Sutherland and L.G. Mitchell), Oxford: Oxford University Press, 1986, 659–81.

Jeffrey R. Wigelsworth, *All Souls College, Oxford in the Early Eighteenth Century: Piety, Political Imposition, and Legacy of the Glorious Revolution*, Leiden: Brill, 2018.

Notes and references

1. For some helpful thoughts on pre-disciplinarity see G. S. Rousseau, *Enlightenment Borders: Pre- and Post-modern Discourses: Medical, Scientific* (Manchester: Manchester University Press, 1991), pp. 217–18.

2. John Friesen, 'Christ Church Oxford, the Ancients-Moderns Controversy, and the Promotion of Newton in Post-Revolutionary England', *History of Universities* 23(1) (2008), 33–66 at pp. 37, 53.

3. On the terms *science* and *natural philosophy* see A. Cunningham, 'Getting the Game Right: Some Plain Words on the Identity and Invention of Science', *Studies in History and Philosophy of Science* 19 (1988), 365–89.

4. G. L'E. Turner, 'The Physical Sciences', in *The History of the University of Oxford*, Vol. 5: *The Eighteenth Century*, ed. L. S. Sutherland and L. G. Mitchell (Oxford: Oxford University Press, 1986), pp. 659–81 at p. 670. John Fauvel's observation that the Sedleian Professors in the eighteenth century shared 'in promoting Newtonian science alongside the Savilian professors' is, alas, also wide of the mark: John Fauvel, 'Eight Centuries of Mathematical Traditions', in *Oxford Figures: Eight Centuries of the Mathematical Sciences*, ed. John Fauvel, Raymond Flood, and Robin Wilson, 2nd ed. (Oxford: Oxford University Press, 2013), pp. 3–34 at p. 8.

5. B. W. Young, *Religion and Enlightenment in Eighteenth-Century England: Theological Debate from Locke to Burke* (Oxford: Oxford University Press, 1998); Robert H. Hurlbutt, *Hume, Newton, and the Design Argument*, rev. ed. (Lincoln, NE: University of Nebraska Press, 1985); James Dybikowski, 'Natural Religion', in *Encyclopedia of the Enlightenment*, ed. Alan Charles Kors, 4 vols (Oxford: Oxford University Press, 2003), vol. III, pp. 142–50.

6. N. Aston, 'Hutchinsonians (act. *c*.1724–*c*.1770)', *Oxford Dictionary of National Biography* (2008), https://doi.org/10.1093/ref:odnb/59223.

7. J. E. McGuire, 'Boyle's Conception of Nature', *Journal of the History of Ideas* 33 (1972), 523–42.

8. Elaine Chalus, *Elite Women in English Political Life, c. 1754–1790* (Oxford: Oxford University Press, 2005), p. 107. For step models of patronage such as the 'ladder of patronage' and 'pagoda of patronage' see L. P. Curtis, *Anglican Moods of the Eighteenth Century* ([Hamden, CN]: Archon Books,

1966). S. N. Eisenstadt and L. Roniger, *Patrons, Clients and Friends: Interpersonal Relations and the Structure of Trust in Society* (Cambridge, UK: Cambridge University Press, 1984) set out several sociologically informed comparative studies of interpersonal relationships and their connections to the institutional matrices within which they develop. See also C. Clapham (ed.), *Private Patronage and Public Power. Political Clientelism in the Modern State* (London: Pinter, 1982).

9. The University and patronage is a central theme of my forthcoming *Enlightened Oxford: The University and the Cultural and Political Life of Eighteenth-Century Britain and Beyond* (Oxford: Oxford University Press, 2023).

10. John Gutch, *The History and antiquities of the University of Oxford*, 2 vols (Oxford, 1792–6), vol. II, pp. 869–70. That provision still applied in the early nineteenth century. See *Oxford University Calendar* (Oxford: Printed by J. Munday, 1810), p. 25.

11. For his career see A. J. Turner, 'Millington, Sir Thomas (1628–1704)', *Oxford Dictionary of National Biography* (2004), https://doi.org/10.1093/ref:odnb/18764. Millington was one of Willis's protégés but there is no evidence that he anointed him as his successor in the Sedleian Chair. Archbishop Gilbert Sheldon, who was behind Willis's election as Sedleian Professor in 1660, is likely to have been influential in the choice of Millington to replace him. Both had been fellows of All Souls.

12. For Millington's work on the observation of sexuality in plants in relation to the comparable pioneering studies of Nehemiah Grew (1641–1712), Secretary to the Royal Society and author of *The Anatomy of Plants*, 4 vols (London, 1682), see Robert Gunther, *Early Science in Oxford*, 14 vols (Oxford, 1923–45), vol. III, pp. 209–10; vol. XI, p. 159.

13. Campbell R. Hone, *The Life of Dr. John Radcliffe 1652–1714* (London: Faber and Faber, 1950), p. 50.

14. For his subsequent career, see John Henry, 'Keill, John (1671–1721)', *Oxford Dictionary of National Biography*, https://doi.org/10.1093/ref:odnb/15256. He succeeded Gregory as Savilian Professor of Astronomy at the second attempt in 1712. The *Introductio ad Veram Physicam* (Eng. trans., 1720) lectures, in which he attacked Thomas Burnet's *Theory of the Earth* (1698) for its Cartesian shortcomings, were originally published in 1701. See R. T. Gunther, *Early Science in Oxford*, vol. II (Oxford: Printed for the subscribers, 1923), pp. 197–8; David B. Wilson, *Seeking Nature's Logic: Natural Philosophy in the Scottish Enlightenment* (University Park, PA: Pennsylvania State University Press, 2009), pp. 35, 50–1. Keill's text has been called 'arguably [one of] the first popular expositions of Newtonian philosophy ever published' (J. B. Shank, *The Newton Wars and the Beginning of the French Enlightenment* (Chicago: University of Chicago Press, 2008), p. 182). See also D. Kubrin, 'John Keill', in *Dictionary of Scientific Biography*, ed. Charles C. Gillespie, 16 vols (New York: Charles Scribner and Sons, 1970–80), vol. VII, pp. 275–7; Derek Gjertsen, *The Newton Handbook* (London; New York: Routledge & Kegan Paul, 1986), pp. 284–6.

15. Quoted in Mordechai Feingold, *The Newtonian Moment: Isaac Newton and the Making of Modern Culture* (New York: New York Public Library; Oxford: Oxford University Press, 2004), p. 42.

16. Wilson, *Seeking Nature's Logic*, p. 53.

17. John Keill, *An Introduction to Natural Philosophy* (London, 1720), pp. x, 1–7.

18. L. W. B. Brockliss, 'Science, the Universities, and Other Public Spaces', in *The Cambridge History of Science*, Vol. 4: *Eighteenth-Century Science*, ed. Roy Porter (Cambridge, UK: Cambridge University Press, 2003), pp. 44–86, at pp. 79–80. Aristotelianism retained an honoured place on the margins of the Oxford curriculum in the early eighteenth century, by no means all its insights considered outmoded. See C. Mercer, 'The Vitality and Importance of Early Modern Aristotelianism', in *The Rise of Modern Philosophy. The Tension between the New and Traditional Philosophies from Machiavelli to Leibniz*, ed. T. Sorell (Oxford: Clarendon Press, 1993), pp. 33–67 at p. 54.

19. For an overview of science teaching at Oxford, see Laurence Brockliss, 'L'enseignement des sciences naturelles dans les universités britanniques: réforme ou stagnation', in *Universités et Institutions Universitaires Européennes au XVIIIe siécle. Entre Modernisation et Tradition*, ed. François

Cadilhon, Jean Mondot, and Jacques Verger (Bordeaux: Presses universitaires de Bordeaux, 1999), pp. 61–79, esp. 69–73.

20. Thomas Hearne, *Reliquiae Hearnianae*, ed. J. Buchanan-Brown, rev. ed. (London: Centaur, 1966 (originally 1857)), pp. 237–8, 1 Sept 1721, quoted in David Money, *The English Horace: Anthony Alsop and the Tradition of British Latin Verse* (Oxford: Oxford University Press, 1998), pp. 159, 162–3.

21. J. R. Bloxam, *A register of the members of St Mary Magdalen College, Oxford*, 7 vols (Oxford: William Graham, 1853–81), vol. V, pp. 247–9; R. Darwall-Smith, 'The Monks of Magdalen, 1688–1854', in *Magdalen College Oxford. A History*, ed. L. W. B. Brockliss (Oxford: Magdalen College, 2008), pp. 253–386 at p. 261. Hearne intimated that Bayley was a secret Jacobite: *Remarks and collections of Thomas Hearne*, ed. C. E. Doble, R. W. Rannie, and H. E. Salter, 11 vols (Oxford: Printed for the Oxford Historical Society at the Clarendon Press, 1885–1921), vol. I, p. 283, 15 Aug 1706.

22. For Delaune see W. C. Costin, *The History of St John's College Oxford 1598–1860* (Oxford: Clarendon Press for the Oxford Historical Society, 1958), pp. 165–70.

23. For the leading role of Magdalen College in resisting the Catholic intrusion of James II see L. W. B. Brockliss, G. Harriss, and A. Macintyre, *Magdalen College and the Crown: Essays for the Tercentenary of the Restoration of the College 1688* (Oxford: Magdalen College, 1988).

24. John Clarke, 'Warden Gardiner, All Souls, and the Church, c.1688–1760', in *All Souls under the Ancien Régime. Politics, learning, and the arts, c.1650–1800*, ed. S. J. D. Green and P. Horden (Oxford: Oxford University Press, 2007), pp. 197–213. Gardiner and his travails at All Souls are central to Jeffrey R. Wigelsworth's *All Souls College, Oxford in the Early Eighteenth Century: Piety, Political Imposition, and Legacy of the Glorious Revolution* (Leiden: Brill, 2018).

25. For Tenison and All Souls see Edward Carpenter, *Thomas Tenison Archbishop of Canterbury: His Life and Times* (London: SPCK, 1948), pp. 272–84. Hearne referred to 'the great projudice' that Tenison had done to the university for being 'the main instrument' to bring in Fayrer ('a Fellow all guts and no brains'): Hearne, *Remarks and collections*, vol. I, p. 85.

26. He was the eldest of 16 children. For his Oxford progress, see https://www.british-history.ac.uk/alumni-oxon/1500-1714/pp480-509, and also Bloxam, *Register*, vol. VI, p. 4. Fayrer's father, also James, was rector of Sulhamstead Bannister and Sulhamstead Abbots, Berkshire from 1652. He had been a fellow of Queen's (1649), where he had matriculated as a plebian in 1638. The Fayrers were originally from Crosby Ravensworth, Westmorland, but became associated with Berkshire.

27. For Halley's exceptionally varied career see Allan Chapman, 'Edmond Halley', in Fauvel, Flood, and Wilson (eds), *Oxford Figures*, pp. 140–64; Alan Cook, *Edmond Halley: Charting the Heavens and the Seas* (Oxford: Clarendon Press, 1998).

28. He was rector of St Martin's Church (Carfax), Oxford, from 1693 until his death: C. J. H. Fletcher, *History of the Church and Parish of St Martin, Oxford* (Oxford: Blackwell, 1896). He was also briefly Rector of Appleton, near Didcot, March 1710–March 1711, a Magdalen living: 'Parishes: Appleton', in *A History of the County of Berkshire*, vol. 4, ed. William Page and P. H. Ditchfield (London: St Catherine Press, 1924), pp. 335–41; *British History Online*: http://www.british-history.ac.uk/vch/berks/vol4/pp335-341 (accessed 28 May 2022). He quit Appleton to return to residence in Magdalen to, according to Hearne, 'do just nothing but eat the Founder's bread' (Hearne, *Remarks and collections*, 22 Sept 1717).

29. Nicholas Keene, 'John Ernest Grabe, Biblical Learning and Religious Controversy in Early Eighteenth-Century England', *Journal of Ecclesiastical History* 58 (2007), 656–74; G. Thomann, 'John Ernest Grabe (1666–1711): Lutheran syncretist and Anglican patristic scholar', *Journal of Ecclesiastical History* 43 (1992), 414–27.

30. Hearne, *Remarks and collections*, vol. I, p. 188, 19 Feb 1706. Hearne's verdict on Fayrer on his death in February 1720 was damning: 'He was a very proud, haughty man, of no learning, and therefore

altogether unfit for the Natural Philosophy lecture.' See also W. D. Macray, *A register of the members of St. Mary Magdalen College, Oxford*, [. . .] *New Series*, 8 vols (London: Henry Frowde, 1894–1915), vol. IV, pp. 131–2.

31. Hearne, *Remarks and collections*, 23 Feb. 1719/20. Details of West's career are scanty. He was probably vicar of Weare, Somerset, from 1705 until his death in 1712. He graduated at Oriel College in 1688: https://theclergydatabase.org.uk/jsp/search/index.jsp

32. He was admitted to Doctors' Commons, 1 Dec1712: G. D. Squibb, *Doctors' Commons: A History of the College of Advocates and Doctors of Law* (Oxford: Clarendon Press, 1977), p. 188.

33. *Historical Manuscripts Commission* [hereafter HMC], *Portland MSS*, 10 vols (London, 1891–1913), vol. VII, pp. 115, 139; Linda Colley, *In Defiance of Oligarchy the Tory Party, 1714–1760* (Cambridge, UK: Cambridge University Press, 1982), p. 181. Abingdon's influence in Oxford urban elections was also declining despite his being High Steward and Recorder of the Borough: C. L. S. Linnell (ed.), *The Diaries of Thomas Wilson, DD, 1731–37 and 1750: Son of Bishop Wilson of Sodor & Man* (London: SPCK, 1964), 6 Dec 1732, p. 82.

34. His Oxford career is summarized in J. Mordaunt Crook, *Brasenose: The Biography of an Oxford College* (Oxford: Oxford University Press, 2008), pp. 123–41. See also R. W. Jeffery, 'An Oxford Don Two Hundred Years Ago', unpub. MS, Brasenose College Archives, MPP 56F4/10.

35. Hearne reported on Harwar's death in 1722 that 'He was a Man that seldon appear'd abroad in the University, nor did any University Duty, being a quiet Man' (Hearne, *Remarks and collections*, vol. VII, p. 83). See also Darwall-Smith, 'The Monks of Magdalen', pp. 261–2; Bloxam, *Register*, vol. VI, pp. 9–19; Macray, *Register*, vol. IV, pp. 128–9.

36. All Souls College, Acta, 4 Nov. 1721.

37. All Souls College, Warden's Manuscripts, 6.

38. There is no evidence that he had promised to take Orders (as he would have been required to do) before the election, and this may have been a consideration for Wake: John Clarke, 'Warden Niblett and the Mortmain Bill', in Green and Horden (eds), *All Souls under the Ancien Régime*, pp. 217–32 at p. 219.

39. Courtenay also presented him to the livings of Wolborough (1739) and Honiton (1740): Thomas Cann Hughes, 'Some Notes on Rectors of Honiton since the Commonwealth', *Reports and Transactions of the Devonshire Association* 30 (1898), 127–31 at p. 128.

40. W. H. Wilkin, 'The Rectors of Honiton, 1505–1907', *Transactions of the Devonshire Association* 69 (1937), 403–10. I am grateful to Robin Darwall-Smith for elucidating Bertie's career. He points out that all the defeated candidates for the Wardenship between 1702 and 1793 had left the college within three years of their defeat.

41. J. R. Magrath, *The Queen's College*, 2 vols (Oxford: Clarendon Press, 1921), vol. II, p. 314; Joseph Foster, *Alumni Oxonienses 1715–1886*, 4 vols (London, 1888), vol. II, p. 175. The latter mistakenly gives 1731 as the date of his election to a Queen's fellowship. BA 1721, MA, 1724.

42. *HMC, Weston Underwood MSS* (London, 1885), pp. 487, 490; http://www.history ofparliamentonline.org/volume/1715-1754/member/butler-edward-1686-1745

43. John Jones, *Balliol College. A History, 1263–1939* (Oxford: Oxford University Press, 1988), p. 159. The Berties supplied four younger sons to fellowships in All Souls between the 1680s and the 1840s: John Davis, 'Founder's Kin', in Green and Horden (eds), *All Souls under the Ancien Régime*, pp. 233–67 at pp. 257–8.

44. See Victoria Huxley, *Jane Austen and Adlestrop: Her Other Family* (Moreton in Marsh: Windrush, 2013).

45. https://www.british-history.ac.uk/alumni-oxon/1500-1714/pp1050-1083. Not 1726 as given here.

46. William Hunt, rev. S. J. Skedd, 'Browne, Joseph (1700–1767)', *Oxford Dictionary of National Biography*, https://doi.org/10.1093/ref:odnb/3685. He resigned Bramshott in 1758: Gerald Aylmer (ed.), *Hereford Cathedral: A History* (London: Hambledon Press, 2000), pp. 128–9, 132–3, 640. He was

mainly absent from Hereford from 1760 because of his Oxford responsibilities and his general infirmity.

47. Beauclerk was described by one who knew him as 'a most egregious blockhead': Edmund Pyle to Dr Thomas Kerrich, n.d. [1748], in Albert Hartshorne (ed.), *Memoirs of a Royal Chaplain, 1729–1763* (London, 1905), p. 157. Browne was later pictured as the bishop's 'nurse—makes his sermons & charges &c & c'. Same to same, 10 Jan 1760, in ibid., p. 320.

48. Stephen Green to Provost Smith, 27 Nov 1741, in Queen's College, MS 456. Some of Browne's correspondence with Smith is in Queen's College, MS 473.

49. His only known work was an edition of the poems of Urban VIII (Oxford, 1726).

50. It was Browne who led a delegation from the university to present a congratulatory address to George III on his accession in 1760: G. Roberson, 'Oxford in the Eighteenth Century. First Series', in *Studies in Oxford History Chiefly in the Eighteenth Century*, ed. C. L. Stanier (Oxford: Printed for the Oxford Historical Society at the Clarendon Press, 1901), pp. 269–343 at p. 283.

51. 'The Provost is continued Vice-Can. For the fourth year, a proof of his being a most excellent magistrate: indeed, all ranks and parties are pleased with him': *Letters of Richard Radcliffe and John James, of Queen's College, Oxford, 1755–1783*, ed. Margaret Evans (Oxford: Printed for the Oxford Historical Society at the Clarendon Press, 1888), pp. 18, 19, 22.

52. Nicholas A. Hans, *New Trends in Education in the Eighteenth Century* (London: Routledge & Kegan Paul, 1951), p. 52.

53. Thomas Horne (ed.), *The Theological Lectures of the late Rev. Benjamin Wheeler, DD*, vol. 1 (Oxford: University Press, 1819), p. v.

54. Richard Polwhele, *Traditions and Recollections* (London: John Nichols, 1826), p. 91.

55. Wheeler to Thomas Warton, 17 Aug 1761, 28 Sept 1761, in David Fairer (ed.), *The Correspondence of Thomas Warton* (Athens, GA: University of Georgia Press, 1995), pp. 102, 104.

56. Scott Mandelbrote, 'Lowth, Robert (1710–1787)', *Oxford Dictionary of National Biography* (2008), https://doi.org/10.1093/ref:odnb/17104.

57. Horne, *Theological Lectures*, p. iv.

58. Bloxam, *Register*, vol. VI, pp. 153–9; Darwall-Smith, 'The Monks of Magdalen', pp. 263–4. See Oxford University Archives, WPβ/7/1, for Wheeler's election. For Durell, see Nicholas Pocock, rev. Richard Sharp, 'Durell, David (1728–1775)', *Oxford Dictionary of National Biography* (2004), https://doi.org/10.1093/ref:odnb/8312. Neil W. Hitchin, 'The Politics of English Bible Translation in Georgian Britain', *Transactions of the Royal Historical Society*, 6th ser., 9 (1999), 67–92. For Tracy, see M. St John Parker, 'Tracy, John, seventh Viscount Tracy of Rathcoole, 1722–1793', *Oxford Dictionary of National Biography* (2004), https://doi.org/10.1093/ref:odnb/71345. Founder's Kin at All Souls gave first claim to fellowships to descendants of the college's founder, Archbishop Henry Chichele (1362?–1443).

59. See Foster, *Alumni Oxoniensis 1715–1886*, vol. IV, p. 1534, for Wheeler's progress.

60. Joyce M. Horn, *John Le Neve: Fasti Ecclesiae Anglicanae 1541–1857*, vol. 8: *Bristol, Gloucester, Oxford and Peterborough Dioceses* (London: University of London, Institute of Historical Research, 1996), p. 99. He resigned the Poetry Professorship in 1776.

61. Thanks to Judith Curthoys for this information.

62. Wellcome Institute, MS 4990. Whether Wheeler lectured in other years is unknown. It is improbable that he did so after being selected as Regius Professor in 1776.

63. Wheeler himself had minimal experience of running a parish. The living of Ewelme (where he was buried) came with the Regius Professorship. He also held the Magdalen rectory of Candlesby cum Scremby, Lincs., 1772–83. https://theclergydatabase.org.uk/jsp/persons/CreatePersonFrames.jsp?PersonID=38520

64. Horne, *Theological Lectures*, p. 269, Lecture xi.

65. Ibid., p. 272.
66. [James Hurdis], *A Word or two in vindication of the University of Oxford, and of Magdalen College in particular, from the posthumous aspersions of Mr Gibbon*, sl (1800?), p. 36.
67. *Gentleman's Magazine* 53 (1783), 629. According to William Newcome, Bishop of Waterford & Lismore in 1783 and formerly vice-principal of Hertford College, Wheeler possessed 'excellent parts and severe application, sound taste and judgement, exact and extensive learning. He was at once an elegant and deep Scholar, a Critic, and a Philosopher'. Newcome, a biblical scholar, was writing to Henry Bathurst, a colleague of Wheeler's as a canon of Christ Church (Horne, *Theological Lectures*, p. x).
68. Wheeler was insistent that he wanted to make emendations himself to any published version of his theological lectures. It was not until 1819 that one volume containing fifteen of the lectures was published, edited by his friend and contemporary Thomas Horne, formerly Fellow of Trinity College. There were plans to publish up to three volumes of Wheeler's works, but the response to the first volume was so poor, in terms of both reviews and sales, that Thomas Horne's son thought it inadvisable to publish any more. See Magdalen College Archives, MC: PR30/1/C3/8.
69. Robin Darwall-Smith and Peregrone Horden (eds), 'The Unloved Century. Georgian Oxford Reassessed', special issue of *History of Universities* 35/1 (2022).
70. See, among many titles, Mary Fissell and Roger Cooter, 'Exploring Natural Knowledge. Science and the Popular', in *The Cambridge History of Science*, vol. 4, ed. Roy Porter (Cambridge, UK: Cambridge University Press, 2003), pp. 129–58; essays in Alan Q. Morton (ed.), 'Science Lecturing in the Eighteenth Century', special issue of the *British Journal for the History of Science* 28 (1995); Larry Stewart, *The Rise of Public Science: Rhetoric, Technology, and Natural Philosophy in Newtonian Britain, 1660–1750* (Cambridge, UK: Cambridge University Press, 1992); Geoffrey V. Sutton, *Science for a Polite Society: Gender, Culture, and the Demonstration of Enlightenment* (Boulder, CO: Westview, 1995).

CHAPTER 4

Further reading

Maurice Daumas, *Scientific Instruments of the Seventeenth and Eighteenth Centuries and their Makers* (London: B. T. Batsford, 1972)

Larry Stewart, *The Rise of Public Science: Rhetoric, Technology, and Natural Philosophy in Newtonian Britain, 1660 Amended.1750* (Cambridge, UK: Cambridge University Press, 1992)

Jim Bennett and Sofia Talas (eds), *Cabinets of Experimental Philosophy in Eighteenth-Century, Europe* (Leiden: Brill, 2013)

Thomas L. Hankins, *Science and the Enlightenment* (Cambridge, UK: Cambridge University Press, 1985)

Michael Hoskin (ed.), *The Cambridge Illustrated History of Astronomy* (Cambridge, UK: Cambridge University Press, 1997)

Notes and references

1. Ruth Wallis, 'Cross-currents in Astronomy and Navigation: Thomas Hornsby, FRS (1733–1810)', *Annals of Science* 57(3) (2000), 219–40, https://doi.org/10.1080/00033790050074147.
2. Ruth Wallis, 'Hornsby, Thomas (1733–1810)', Oxford Dictionary of National Biography, retrieved 21 Sep. 2022, from https://www.oxforddnb.com/view/10.1093/ref:odnb/9780198614128.001.0001/odnb-9780198614128-e-13805.
3. Royal Society Archives JBO/25/48.
4. Now in the collection of the History of Science Museum, Oxford, inventory number 44831, https://hsm.ox.ac.uk/collections-online#/item/hsm-catalogue-9419, retrieved 31 Aug 2022.

5. Thomas Hornsby, 'An inquiry into the quantity and direction of the proper motion of Arcturus; with some remarks on the diminution of the obliquity of the ecliptic', *Philosophical Transactions* 63 (1773), pp. 93–125, at p. 94; 'An account of the observations of the transit of Venus and of the eclipse of the sun, made at Shirburn Castle and at Oxford', *Philosophical Transactions*, 59 (1769), 172–82, at p. 175. In observing the solar eclipse of 1 April 1764, he mentions also a 12-ft refractor by Bird, an 18-inch reflector and a 9-inch reflector with a micrometre by Dollond, but some at least of these were assembled for the occasion, Thomas Hornsby, 'Observations on the eclipse of the sun, April 1, 1764: in a letter to the Right Honourable James Earl of Morton, Pres. R. S.', *Philosophical Transactions*, 54 (1764), pp. 145–9, at pp. 145–6.

6. Royal Astronomical Society MSS Radcliffe A.1.12.

7. The advantage of the 96-part division was that, when the radius had been struck off the arc to give 64 parts (equivalent to 60°), division down to single degrees could be performed by successive bisection.

8. J. A. Bennett, 'Equipping the Ratcliffe Observatory: Thomas Hornsby and his instrument-makers', in *Making Instruments Count*, ed. R. G. W. Anderson, J. A. Bennett, and W. F. Ryan (Aldershot: Variorum, 1993), pp. 232–41.

9. Royal Astronomical Society, RAS MSS Ratcliffe A.7.

10. Ibid.

11. A. V. Simcock, *The Ashmolean Museum and Oxford Science 1683–1983* (Oxford: Museum of the History of Science, 1984), pp. 11–13.

12. *The Oxford Ten-Year Book, made up to the end of the Year 1860* (Oxford: John Henry and James Parker, 1863), pp. 58–9.

13. Wallis, 'Cross-currents', p. 221.

14. Jim Bennett, 'The Lost Cabinet of Experimental Philosophy of the University of Oxford', in *Cabinets of Experimental Philosophy in Eighteenth-Century Europe*, ed Jim Bennett and Sofia Talas (Leiden: Brill, 2013), pp. 69–78.

15. Bodleian Library, MS Radcliffe Trust d.9, f. 30.

16. An appraisal of Hornby's character and teaching style is at Wallis, 'Cross-currents', pp. 237–40.

17. Bodleian Library, MS Radcliffe Trust d.9, f. 32.

18. Hornsby, 'An inquiry', p. 105.

19. This means that he sat on the Board of Longitude, which was responsible for any awards in recognition of a method for finding longitude at sea, or advances towards such a solution.

20. Royal Astronomical Society, RAS MSS Radcliffe A.1.43.

21. Thomas Hornsby, 'A discourse on the parallax of the sun', *Philosophical Transactions*, 53 (1763), 467–95.

22. Ibid., p. 495.

23. Thomas Hornsby, 'On the transit of Venus in 1769', *Philosophical Transactions*, 55 (1765), 326–44.

24. Ibid., p. 334.

25. John Harris, *Navigantium atque itinerantium bibliotheca. Or, a Complete Collection of Voyages and Travels* (London: T. Woodward, 1744–8).

26. Perhaps the *L'Hemisphere Meridional pour voir plus distinctement les Terres Australes* (Amsterdam: Covens and Mortier, 1742).

27. Hornsby, 'On the transit', pp. 342–4.

28. Ibid., p. 344.

29. Hornsby, 'An account'.

30. Bradley E. Schaefer, 'The Transit of Venus and the Notorious Black Drop Effect', *Journal for the History of Astronomy*, 32 (2001), 325–36.

31. Hornsby, 'An account', p. 176.

32. Harry Woolf, *The Transits of Venus: A Study of Eighteenth-Century Science* (Princeton: Princeton University Press, 1959), pp. 143, 148–9.

33. Ibid., p. 177.
34. Thomas Hornsby, 'The quantity of the sun's parallax as deduced from the observations of the transit of Venus, on June 3, 1769', *Philosophical Transactions*, 61 (1771), 574–9.
35. This is very close to the modern value for the solar parallax of 8.794 seconds.
36. Hornsby, 'The quantity', pp. 578–9.
37. Royal Astronomical Society, RAS MSS Radcliffe A.1.59.
38. William Wales, 'Journal of a voyage, made by order of the Royal Society, to Churchill River, on the north-west coast of Hudson's Bay; of thirteen months residence in that country; and of the voyage back to England; In the years 1768 and 1769', *Philosophical Transactions*, 60 (1770), pp. 100–136.
39. Bodleian Library, MS Radcliffe Trust d.12, f. 132v.
40. Wales, 'Journal of a voyage', p. 115.
41. 'Take no one's word'.
42. Hornsby, 'An inquiry'. For comments by Hornsby on his paper, see a letter to an unnamed correspondent (internal evidence suggests the fifth Duke of Marlborough), Royal Astronomical Society, RAS MSS Radcliffe A.1.47.
43. Hornsby, 'An inquiry', p. 125.
44. Bennett, 'Equipping'.
45. One mural quadrant is at the History of Science Museum, Oxford (inventory no. 30919); the other is at the Science Museum, London. The History of Science Museum also holds parts of the equatorial sector (inventory no. 17236), the transit instrument (inventory no. 96686), and the zenith sector (inventory no. 84526).
46. The History of Science Museum, Oxford, has a thermometer (inventory no. 29413) and a barometer (inventory no.22842).
47. Bodleian Library, MSS. Radcliffe Trust. d. 9-23; f. 4; e. 11–12.
48. Robert T. Gunther, *Early Science in Oxford*, vol. 11 (Oxford: printed for the author, 1937), pp. 401–11.
49. Bodleian Library, MS Radcliffe Trust. d. 16.
50. Gunther, *Early Science*, vol. 11, p. 407.
51. Ibid., pp. 402–3.
52. Simcock, *The Ashmolean*, pp. 11–12; J. A. Bennett, S. A. Johnston, and A. V. Simcock, *Solomon's House in Oxford: New Finds from the First Museum* (Oxford: Museum of the History of Science, 2000), pp. 18–20.
53. Thomas Bugge, *Journal of a Voyage through Holland and England*, trans. M. Dybdahl and K. M. Pedersen, ed. K. M. Pedersen (University of Aarhus: History of Science Department, 1997), pp. 212–17.
54. Bennett, 'The lost cabinet', pp. 69–73.
55. 'A Catalogue of the Philosophical Apparatus belonging To The Rev^d: D^r: Hornsby of Oxford', Bodleian Library, Oxford, MS Top. Oxon. c. 236, ff. 3–13.
56. Bennett, 'The lost cabinet', pp. 71–3.
57. Bodleian Library, MS Radcliffe Trust d.18, f. 2^v; see also f. 22.
58. Bodleian Library, MS Radcliffe Trust d. 16, ff. 36–9, 62^v–3, 158–9, 160–7, 180–94, 196–202, and *passim*; ibid., d.18, ff. 101–1^v, 103–3^v, 105–5^v, 137–8.
59. MS Radcliffe Trust d. 9, f. 64^v.
60. Ibid., f. 73^v.
61. Ibid., f. 93.
62. Ibid., f. 105.
63. Ibid., f. 198.
64. Ibid., f. 202.
65. Simcock, *The Ashmolean*, between pp. 12 and 13.

66. MS Radcliffe Trust d.15, f. 10^v.

67. Ibid, f. 18.

68. Bodleian Library, MS Radcliffe Trust d.16, f. 209^v. Hornsby had a correspondence with William Herschel, J. A. Bennett, 'Catalogue of the Archives and Manuscripts of the Royal Astronomical Society, *Memoirs of the Royal Astronomical Society* 85 (1978), 1–90, see p. 79.

69. Royal Astronomical Society, RAS MSS Herschel W.1/13.H.23.

70. J. A. Bennett, '"On the Power of Penetrating into Space": The Telescopes of William Herschel', *Journal for the History of Astronomy* 7 (1976), 75–108, at pp. 75–6.

71. Royal Astronomical Society, RAS MSS Herschel W.1/13.H.25, 26, 27. Note also ibid., H1/1, p. 12.

72. Royal Astronomical Society, RAS MSS Herschel W.1/13.H.28.

73. Royal Astronomical Society, RAS MSS Herschel W.1/13.H.29.

74. Michael Hoskin, *Discoverers of the Universe: William and Caroline Herschel* (Princeton: Princeton University Press, 2011), pp. 50–1.

75. 'A Catalogue of the Philosophical Apparatus', f. 7^v.

76. Bodleian Library, MS Radcliffe Trust d.11, f. 24.

77. Ibid., f. 27.

78. Ibid., f. 28.

79. Simcock, *The Ashmolean*, between pp. 12 and 13.

80. Wallis, 'Cross-currents', p. 236.

81. Royal Astronomical Society, RAS MSS Radcliffe A.2, *passim*.

82. Royal Astronomical Society, RAS MSS Radcliffe A.2.138.

83. Royal Astronomical Society, RAS MSS Radcliffe A.2.48.

84. Royal Astronomical Society, RAS MSS Radcliffe A.1.175.

85. Wallis, 'Cross-currents', pp. 227, 230–6.

86. H. Knox-Shaw, J. Jackson, and W. H. Robinson (eds), *The Observations of the Reverend Thomas Hornsby . . . made with the Transit Instrument and Quadrant at the Radcliffe Observatory, Oxford in the years 1774 to 1798* (London: Oxford University Press, 1932); Wallis, 'Cross-currents', pp. 236–7. Hornsby's observing books are in the archives of the Royal Astronomical Society, RAS MSS Hornsby 1–3; Bennett, 'Catalogue', p. 39.

87. Arthur A Rambaut, 'Notes on the unpublished observations made at the Radcliffe Observatory, Oxford, between the years 1774 and 1838; with some results for the year 1774', *Monthly Notices of the Royal Astronomical Society* 60 (1900), 265–93.

88. Ibid., p. 268.

89. Knox-Shaw et al. (eds), *The Observations*, p. 7.

90. Ibid., pp. 7–8.

91. Ibid., p. 9. The equivalent in Celsius is −19.

92. George Robert Michael Ward, *Oxford University statutes*, 2 vols (London: William Pickering, 1845–51), vol 1, p. 22; a revised statute of 1839 required the lecturer 'to expound the authors of the best repute in physics', ibid., vol. 2, p. 236.

93. Robert Plot, *The natural history of Oxford-shire: being an essay toward the natural history of England* (Oxford, 1677); Robert Plot, *The natural history of Stafford-shire* (Oxford, 1686).

94. Gunther, *Early Science*, vol. 11, p. 410. He also issued a longer, printed syllabus, ibid., p. 411. The title chosen by Hornsby echoes that of Priestley's book, *Experiments and Observations on Different Kinds of Air* (London, 1774), and later volumes.

95. Later, as 'gases', named respectively as carbon dioxide, nitric oxide, hydrogen, and oxygen, the 'acid and alkaline' airs dealt with in his surviving lectures being 'vitriolic acid' (sulphur dioxide), 'nitrous acid' (nitrogen dioxide), 'marine acid' (hydrogen chloride), and 'alkaline' (ammonia). See also Bodleian Library, MS Radcliffe Trust d.12.

96. Bodleian Library, MS Radcliffe Trust d.12.

97. Ibid., f. 44.
98. Notes for possible Sedleian lectures are 'On earths' (Bodleian Library, MS Radcliffe Trust d.22), 'On physiology' (ibid., d.23); 'On heat and metals' (ibid., d.21); 'On chemistry and medicine' (ibid., f.4); 'On the nature of air' (ibid, d.13) (with more physical content than ibid., d.12).

CHAPTER 5

Further reading

Cooke appears in the *Oxford Dictionary of National Biography*, as well as in the register of Oxford alumni, *Alumni Oxonienses*. A further, very brief, published obituary of Cooke appeared in *The Gentleman's Magazine, and Historical Review*, July 1853, p. 94. His presence in the archives, however, is not extensive, and there is certainly no centralized repository of his papers. Nevertheless, small traces of him do survive in various archives, as indicated in the notes below.

Acknowledgements

I am very grateful to a number of Oxford college archivists for their help in accessing materials, both in person and as scans: Richard Allen (Magdalen), Verity Parkinson (Merton), Harriet Patrick (Corpus Christi), and Julian Reid (Merton and Corpus Christi). I am similarly grateful to Elizabeth Wells, archivist of Westminster School, for scanned material. I benefitted greatly from the assistance of Faye McLeod, Oxford's Keeper of the University Archives, and from that of the staff of the Shakespeare Birthplace Trust in Stratford-upon-Avon. Thanks must go to Richard Parkinson for discussions about Jane Austen, to Mark McCartney, Nigel Aston, and Christopher Stray for their useful comments on a draft of this chapter, and to Elizabeth Miller for facilitating a trip to Cubbington. The Leigh/Cooke/Austen family tree in Figure 5.1 was created using Family Echo (https://www.familyecho.com/).

Notes and references

1. Surrey History Centre, Woking; Surrey Church of England Parish Registers; Reference: BKG/1/2 (St Nicholas, Great Bookham). Samuel Cooke was also rector of Cotsford in Oxfordshire.
2. The two main branches of the Leigh family, the Leighs of Adlestrop in Gloucestershire and the Leighs of Stoneleigh Abbey near Kenilworth in Warwickshire, both descended from the Sir Thomas Leigh (*c*.1504–71), who had been Lord Mayor of London in the 1550s. In 1643, his great-grandson on the Warwickshire line, also called Thomas, was created 1st Lord Leigh of Stoneleigh, but this title died out five generations later, whereupon the Stoneleigh estates passed to the Gloucestershire branch of the family in the person of George Leigh Cooke's uncle, the Rev. Thomas Leigh (1734–1813), rector of Adlestrop and Broadwell (also in Gloucestershire). When he died without children, the Stoneleigh estate passed further to his nephew James Henry Leigh (1765–1823) of Adlestrop, who would eventually present George Leigh Cooke to the parish of Cubbington, which was part of the Stoneleigh estate. In 1839, the Leigh barony was revived and given to James Henry Leigh's son Chandos Leigh (1791–1850). See John Burke and John Bernard Burke, *A Genealogical and Heraldic History of the Extinct and Dormant Baronetcies of England, Ireland and Scotland*, 2nd ed. (London: Scott, Webster, and Geary, 1841), pp. 307–8. See also the valuable biographical notes at the end of Deirdre Le Faye (ed.), *The Letters of Jane Austen*, 3rd ed. (New York: Oxford University Press, 1995), and the family trees included in Deirdre Le Faye, *A Chronology of Jane Austen and her Family* (Cambridge, UK: Cambridge University Press, 2006).
3. Theophilus Leigh had migrated to Balliol from Corpus Christi College. It is said that he was elected Master of Balliol (perhaps through the influence of his uncle, James Brydges, 8th Baron Chandos, then Visitor of the college) as a temporary solution to the Balliol fellows' inability to choose one of their own. Unexpectedly, he then served as Master from 1726 until his death in 1785. See: John

Jones, *Balliol College: A History*, 2nd ed. (Oxford: Oxford University Press, 1997), ch. 13; Samuel Cooke appears in the list of fellows of Balliol on p. 328.

4. The two brothers appear to have maintained a close connection throughout their lives: for example, when George was to take up the parish of Cubbington, it was Theophilus who was dispatched to report on the state of the vicarage (Shakespeare Birthplace Trust, Stratford-upon-Avon: DR18/3/17/3/15, T. L. Cooke to J. H. Leigh, 26 December 1820). Indeed, when Theophilus died in 1846, the London *Morning Post* mistakenly reported the death as George's (*The Morning Post*, issue 22,733, Wednesday 14 October 1846, p. 3a); a correction and apology were printed the following day (*The Morning Post*, issue 22,734, Thursday 15 October 1846, p. 6e).

5. Mary is named in the will that her mother made in May 1820 (The National Archives, Kew: PROB-11-1718-460: https://discovery.nationalarchives.gov.uk/details/r/D165528), and was evidently unmarried at that time, but there are few clear records of her thereafter. The latest is in a letter written from St Leonards-on-Sea by her brother Theophilus in 1845, where he notes in passing that he is 'here at my Sisters' (British Library, London: Correspondence of Rev. Philip Bliss, Add MS 34575, ff. 650–1, T. L. Cooke to P. Bliss, 7 October [1845]).

6. At the time that Cooke entered Corpus Christi, the president was Dr John Cooke (1734–1823), and it has been asserted that he was George Leigh Cooke's uncle (Thomas Fowler, *The History of Corpus Christi College with lists of its members* (Oxford: Printed for the Oxford Historical Society at the Clarendon Press, 1893), p. 303), but I have been unable to clarify this relationship.

7. The highlights of Cooke's career are noted briefly in his entries in the *Oxford Dictionary of National Biography* and *Alumni Oxonienses*. His progression through Church positions is outlined in the online Clergy of the Church of England Database: https://theclergydatabase.org.uk/jsp/persons/CreatePersonFrames.jsp?PersonID=10114

8. Merton College Archives, Blackwell Collection: Fletcher and Hanwell day book 1798–1800, p. 546.

9. See Christopher D. Hollings, 'Two Books on the Elements of Algebra', in *Mathematical Book Histories. Printing, Provenance, and Practices of Reading*, ed. Philip Beeley and Ciarán Mac an Bhaird (Basel: Birkhäuser, 2023).

10. For much of the eighteenth century, the Savilian Professorship of Geometry had been occupied by incumbents whose interests were much more astronomical than mathematical, or who apparently lacked any mathematical credentials whatsoever (cf. the Sedleian Professors of that century). However, Cooke's arrival in Oxford in 1797 had coincided with the appointment to the Savilian Professorship of Abraham Robertson, who had previously deputized for his absentee predecessor and was considerably more active on the mathematical front than most of the other professors in the century before him (at least until he exchanged the geometry chair for that of astronomy in 1810). See Allan Chapman and Christopher Hollings, 'A Century of Astronomers: From Halley to Rigaud', in *Oxford's Savilian Professors of Geometry: The First 400 Years*, ed. Robin Wilson (Oxford: Oxford University Press, 2022), pp. 55–91.

11. Specifically Thomas Reid's *Elements of the Philosophy of the Human Mind* (1785) and *Essays on the Powers of the Human Mind* (1803), and Dugald Stewart's *Elements of the Philosophy of the Human Mind* (1792).

12. See the family trees mentioned in note 2 or else the slightly simpler one in Irene Collins, *Jane Austen and the Clergy* (London: Hambledon, 2002), p. 6.

13. Besides Great Bookham, these encounters often took place in London, Bath, or Brighton; after 1800, following a serious illness suffered by Samuel Cooke in the 1790s, he and his wife spent much of the year in the latter two places (Joyce Hemlow, Althea Douglas, Peter Hughes, and Curtis D. Cecil (eds), *The Journals and Letters of Fanny Burney*, 12 vols (Oxford: Clarendon Press, 1972–84), vol. 5, letter 469, n. 3). Jane Austen certainly visited Great Bookham in June 1814 (see letter 102 in Le Faye, *Letters of Jane Austen*) around the time of the publication of *Mansfield Park*, which was praised by the Cookes, as Jane reported to her sister Cassandra: 'they admire Mansfield Park

exceedingly. Mr Cooke says "it is the most sensible Novel he ever read"—and the manner in which I treat the Clergy, delights them very much' (letter 101). The prospect of an earlier visit (in 1799) had not been met with much enthusiasm, however: Cassandra Cooke had just published a Gothic novel entitled *Battleridge: An Historical Tale, Founded on Facts*, 2 vols (London: C. Cawthorn, 1799), which the extended family were expected to read; it has been suggested that aspects of the Gothic satire of *Northanger Abbey* may have derived from *Battleridge* (Claire Tomalin, *Jane Austen: A Life*, revised and updated ed. (London: Penguin, 2000), p. 151; Susan Allen Ford, '"Real Solemn History" and Cassandra Leigh Cooke's *Battleridge*', *Persuasions*, no. 41 (2019), 75–91). Glowing personal assessments of Samuel and Cassandra Cooke may be found in the letters of Fanny Burney, who lived in Great Bookham during the 1790s (Hemlow et al., *Journals and Letters of Fanny Burney*, vol. 3, letter 122; vol. 4, letters 263 and 309).

14. Letter 44 in Le Faye, *Letters of Jane Austen*.

15. Letter 71 in Le Faye, *Letters of Jane Austen*.

16. Letter 70 in Le Faye, *Letters of Jane Austen*.

17. As one of Jane Austen's biographers has put it: 'Svelte and harmless, he [Theophilus Leigh Cooke] typified another aspect of new Regency manners—not coarse wit or drunken laughter, which she could pardon, but a cold politeness, a stylish rudeness that did not conceal indifference and had the effect of an icy insult' (Park Honan, *Jane Austen: Her Life* (London: Phoenix Giant, 1987), p. 294). Contrasting views of Theophilus, George, and Mary Cooke had been conveyed by Fanny Burney over a decade earlier, in December 1797: 'The eldest son […] is a remarkably pleasing young man: the younger seems as sulky as the sister is haughty'; in expressing regret at leaving Bookham, Burney had noted: 'I do not absolutely include the fair young Lady in my sorrow!' (Hemlow et al., *Journals and Letters of Fanny Burney*, vol. 4, letter 263). By March 1799, however, Burney's view of Mary Cooke had improved, if only a little, when she described her as 'sensible, but stiff & cold, & by no means equally amiable in her disposition *now*, though I think improving, & opening into something better' (Hemlow et al., *Journals and Letters of Fanny Burney*, vol. 4, letter 309). By 1811 Jane Austen herself was becoming concerned about how Mary was to find a husband while looking after her ailing mother—but she took comfort in the fact that Bookham now had two curates, 'so that I think she must fall in love with one or the other' (letter 74 in Le Faye, *Letters of Jane Austen*).

18. Sir J. T. Coleridge, *A Memoir of the Rev. John Keble, M.A., late vicar of Hursely*, 2 vols (Oxford and London: James Parker & Co., 1869), p. 16.

19. J[ohn] A[ndrew] H[amilton], 'Coleridge, Sir John Taylor', *Dictionary of National Biography* (London; Oxford: Oxford University Press, 1885–1900), vol. 11. Cooke remained in contact with Coleridge for the remainder of his life (Coleridge, *Memoir of the Rev. John Keble*, p. 16), and also corresponded with Coleridge's son John Duke Coleridge (1820–94) (British Library: Add MS 86266, ff. 237–40). See note 62.

20. Arthur Penrhyn Stanley, *The Life of Thomas Arnold, D.D., abridged and newly edited, with notes, etc.* (London: Hutchinson & Co., [1903] [originally 1844]), p. 8–9; Coleridge, *Memoir of the Rev. John Keble*, p. 16. We note that all three of Cooke's sons attended Rugby School while Thomas Arnold was headmaster there: *Rugby School Register: From 1675 to 1867 inclusive* (Rugby: W. Billington; London: Whittaker and Co., 1867), pp. 98, 109, 118.

21. Thomas Charles-Edwards and Julian Reid, *Corpus Christi College, Oxford: A History* (Oxford: Oxford University Press, 2017), pp. 269, 273.

22. Indeed, the office and the duty still exist (http://www.corpusjcr.org/wp-content/uploads/2019/06/CCC-JCR-Standing-Orders-28.05.19.pdf), though a haiku is now deemed acceptable if the poet laureate has been too busy with their proper studies to fulfil the traditional role (http://www.corpusjcr.org/jcr-life/poet-laureate/).

23. Corpus Christi College Archives: E/1/5—Book of Songs and Odes, 20 Nov 1812–Jul 1852. Other early records of the JCR, mostly statutes, resolutions, and minutes, also survive as E/1/1–4. An

account of the foundation and traditions of the Corpus Christi JCR may be found in A.S., 'The Junior Common Room', *The Pelican Record* 2(3) (March 1892), 86–9.

24. Corpus Christi College Archives: E/1/5, f. 17$^\text{v}$.

25. Corpus Christi College Archives: E/1/5, f. 1$^\text{r}$.

26. Corpus Christi College Archives: E/1/3, f. 42$^\text{v}$.

27. This is noted in the ode for 1844, the year in which Cooke's son matriculated at Corpus Christi: 'Old Codger Cooke, who ne'er forsook the room whence sprung his fame, // An olive-branch, full stout and staunch, has sent to bear his name [. . .]' (Corpus Christi College Archives: E/1/5, f. 19$^\text{v}$).

28. 'Rev. G. L. Cooke, B.D.', *The Gentleman's Magazine, and Historical Review* (July 1853), 94. The same obituary is also the source of the detail about Cooke being 'the founder, and for many years secretary, of the original Literary Dining Club', on which little further information seems to be available. The only corroboration of this detail that I have so far found is a single letter of 1821 in the archives of Westminster School (GB 2014 WS-05-ELM-02-3-26) in which Cooke, as secretary of the dining club, welcomes a new member.

29. *Jackson's Oxford Journal*, issue 2987 (Saturday 28 July 1810), 3b. Wherever he used the title, Cooke consistently signed himself 'Sedleian Reader', but is elsewhere, equally consistently, referred to as 'Professor'.

30. The surviving university records of this period are in general much less detailed where decision-making is concerned than those of later decades: the minutes of university council meetings, for example, include only headline issues (often just one per meeting), and are very much *institutional* minutes, little concerned with individuals (royalty and university leaders excepted).

31. Edmund Venables, rev. M. C. Curthoys, 'Parsons, John (1761–1819)', *Oxford Dictionary of National Biography*, https://doi.org/10.1093/ref:odnb/21466; John Jones, 'Sound Religion and Useful Learning: The Rise of Balliol under John Parsons and Richard Jenkyns, 1798–1854', in *Balliol Studies*, ed. John M. Prest (London: Leopard's Head Press, 1982), pp. 89–124. See also M. C. Curthoys, 'The Examination System', in *The History of the University of Oxford*, vol. VI: *Nineteenth-Century Oxford, Part 1*, ed. M. G. Brock and M. C. Curthoys (Oxford: Clarendon Press, 1997), pp. 339–74.

32. *Alumni Oxonienses*.

33. Vivian H. H. Green, 'Routh, Martin Joseph (1755–1854)', *Oxford Dictionary of National Biography*, https://doi.org/10.1093/ref:odnb/24182.

34. Oxford University Archives: WP/β/7/2.

35. *Jackson's Oxford Journal*, issue 3034 (Saturday 22 June 1811), 3b.

36. Evidence of Cooke having delivered his lectures in 1840, for example, may be found in the wording of an Oxford news item in a Cambridge newspaper: 'The Sedleian Professor of Natural Philosophy began a course of lectures on Newton's *Principia*, on Tuesday last' (Cambridge Independent Press, vol. 31, issue 1447 (Saturday 23 May 1840), p. 4b).

37. Shakespeare Birthplace Trust, Stratford-upon-Avon: DR18/3/17/3/12, G. L. Cooke to J. H. Leigh, 13 January 1820. We note that this letter, and Cooke's other letters of this period, was written from New College Lane, where he was perhaps living in the house traditionally occupied by the Savilian Professor of Astronomy. Not then being used by the incumbent professor (Abraham Robertson, 1751–1826), this was being let out by the university (H. E. Bell, 'The Savilian Professors' Houses and Halley's Observatory at Oxford', *Notes and Records of the Royal Society* 16(2) (1961), 179–86). In this case, Cooke's neighbour would have been his close colleague in mathematical teaching, Stephen Peter Rigaud (1774–1839), the Savilian Professor of Geometry. See also note 74.

38. For example, J. B. Morrell, who commented: 'Though sinecuring was non-existent at Cambridge, George Leigh Cooke maintained that comfortable tradition at Oxford'. ('Science and the Universities', *History of Science* 15 (1977), 145–52 at p. 150)

39. *Report of Her Majesty's Commissioners appointed to inquire into the State, Discipline, Studies, and Revenues of the University and Colleges of Oxford: together with the Evidence, and an Appendix* (London: W. Clowes and Sons for Her Majesty's Stationery Office, 1852), Evidence, p. 189. This being said, Cooke was advertising a course of lectures on Newton's *Principia* as late as May 1851 (*Jackson's Oxford Journal*, issue 5115 (Saturday 10 May 1851), 3).

40. On Oxford examinations (particularly those in mathematics), see John Fauvel, 'Georgian Oxford', in *Oxford Figures: Eight Centuries of the Mathematical Sciences*, ed. John Fauvel, Raymond Flood, and Robin Wilson, 2nd ed. (Oxford: Oxford University Press, 2013), pp. 181–201.

41. British Library, London: Peel Papers, vol. CLXII. Correspondence with Dr. C. Lloyd, Bishop of Oxford, 1818–1826, Add MS 40342, ff. 358–9, G. L. Cooke to R. Peel, 15 June 1826.

42. Ibid.

43. Ibid.

44. British Library, London: Peel Papers, vol. CCVII. General correspondence, 20 May–7 July 1826, Add MS 40387, ff. 253–4, R. Peel to G. L. Cooke, 29 June 1826.

45. British Library, London: Peel Papers, vol. CLXII. Correspondence with Dr. C. Lloyd, Bishop of Oxford, 1818–1826, Add MS 40342, ff. 360–1, C. Lloyd to R. Peel, 2 July 1826.

46. Oxford University Archives: WP/γ/2/8 (Mathematical scholarships).

47. George Leigh Cooke, *The first three sections and part of the seventh section of Newton's Principia, with a preface recommending a geometrical course of mathematical reading, and an introduction on the atomic constitution of matter, and the laws of motion* (Oxford and London: John Henry Parker, 1850).

48. George Leigh Cooke, *A few remarks recommending a geometrical course of mathematical studies for the majority of Oxford students, comprising a statement of the atomic constitution and general Properties of matter* (Oxford: John Henry Parker, 1849), 'To the reader'.

49. Cooke, *First three sections*, p. 1.

50. Cooke, *First three sections*, p. 2.

51. Ibid.

52. Cooke, *First three sections*, p. 6.

53. Ibid.

54. Cooke, *First three sections*, p. 7.

55. J. M. Dubbey, 'The Introduction of the Differential Notation to Great Britain', *Annals of Science* 19(1) (1963), 37–48; Brigitte Stenhouse, 'Mary Somerville's Early Contributions to the Circulation of Differential Calculus', *Historia Mathematica* 51 (2020), 1–25.

56. [B. Powell], *A short treatise on the principles of the differential and integral calculus* (Oxford: University Press, 1830); Bartholomew Price, *A treatise on the differential calculus, and its application to geometry: founded chiefly on the method of infinitesimals* (London: George Bell, 1848).

57. W. Tuckwell, *Reminiscences of Oxford*, 2nd ed. (London: Smith, Elder & Co., 1907), p. 204. Cf. the parallel situation in Cambridge with regard to new notation in examinations: [Augustus De Morgan], 'Cambridge differential notation. On the notation of the differential calculus, adopted in some works lately published in Cambridge', *Quarterly Journal of Education* 8 (1834), 100–10.

58. A rare example of a non-political connection is his membership of Oxford's nineteenth-century scientific association, the Ashmolean Society, of which he was elected a member in 1847 (*Proceedings of the Ashmolean Society*, no. XXIV (1847), p. 153). It is in a drawing of a meeting of the Ashmolean Society in 1823 that we find the only surviving depiction of Cooke (Fauvel, 'Georgian Oxford', p. 193).

59. These events were captured in the Corpus Christi JCR ode for 1846 (written by Charles Blackstone): 'The world's gone mad with politics, & all are in a bustle // For Free-Trade or Protection, for STANLEY, PEEL, or RUSSELL; // But our old CODGER COOKE right well both parties did combine, Sir, // When he gave "Protection" to our Rights, & full "Free Trade" in wine, Sir [...]' (Corpus Christi College Archives: E/1/5, f. 23v).

60. British Library, London: Correspondence of Rev. Philip Bliss, Add MS 34581, ff. 94–5, G. L. Cooke to P. Bliss, 16 March [1847].

61. Oxford University returned two MPs to Parliament from 1603 until the abolition of the constituency in 1950.

62. *Jackson's Oxford Journal*, issue 4909 (Saturday 29 May 1847), 3b; British Library, London: Coleridge Family Papers, Add MS 86266, ff. 237–40, G. L. Cooke to J. D. Coleridge, 5 and 8 July 1847.

63. The voting preferences of the professors were summarized in *The Times* (12 March 1829, p. 3a). Opposition to Catholic emancipation is a frequently recurring theme in the records mentioned in note 30.

64. M. G. Brock, 'The Oxford of Peel and Gladstone', in *The History of the University of Oxford*, vol. VI: *Nineteenth-Century Oxford, Part 1*, ed. M. G. Brock and M. C. Curthoys (Oxford: Clarendon Press, 1997), pp. 7–71.

65. J. E. Austen Leigh, *A Memoir of Jane Austen, to which are added Lady Susan and fragments of two other unfinished tales by Miss Austen* (London: Richard Bentley & Son, 1882), p. 77.

66. London Metropolitan Archives; London Church of England Parish Registers; Reference Number: P82/GEO1/021 (St George, Bloomsbury, Camden); *Jackson's Oxford Journal*, issue 3262 (Saturday 28 October 1815), 3b.

67. The children were Helen Ann (1816–69), Samuel Hay (1817–77), George Theophilus (1819–94), Caroline Cassandra (1822–1907), William Hay (1824–46), and Julia (1828–1918). None of the daughters married, and apparently lived out their lives on the income, some of it deriving from their Hay grandparents, granted to them in their father's will (see also note 86). The two eldest sons followed their father into clerical careers (Samuel Hay as a Student of Christ Church and at parishes in Cheshire and Kent, and George Theophilus as a Fellow of Magdalen and vicar of Beckley—see note 84), whilst the youngest joined the army and died in India in his early 20s (F. Clark, *The East-India register and army list, for 1847* (London: Wm. H. Allen & Co., 1847), Bombay, p. 131).

68. See note 2. Cooke had also been rector of Porlock in Somerset for two months at the beginning of 1811, but appears to have given this up in favour of Wyck Rissington: see the Clergy of the Church of England Database (note 7). Note that both Wyck Rissington and Broadwell with Adlestrop are close enough to Oxford that Cooke might occasionally have visited them.

69. *The historical register of the University of Oxford, being a supplement to the Oxford University Calendar, with an alphabetical record of university honours and distinctions, completed to the end of Trinity Term, 1888* (Oxford: Clarendon Press, 1888), p. 43.

70. On the Austen family and Cubbington, see William Jarvis, 'A note on the Rev. James Austen', *The Jane Austen Society: Report for the Year 1981*, reproduced in *Collected Reports of the Jane Austen Society 1976–1985* (Alton, Hants: The Jane Austen Society, 1989), pp. 179–81; Robert Bearman, 'Henry Austen and the Cubbington living', *Persuasions*, no. 10 (1988), 22–6. On the parish of Cubbington more generally, see L. F. Salzman (ed.), *A History of the County of Warwick*, Volume 6: *Knightlow Hundred* (London: Institute of Historical Research, 1951), pp. 74–8 (British History Online: http://www.british-history.ac.uk/vch/warks/vol6/pp74-78, accessed 23 March 2022); see also Chris Pickford and Nikolaus Pevsner, *Warwickshire*, revised ed., The Buildings of England (New Haven; London: Yale University Press, 2016), pp. 308–9.

71. Shakespeare Birthplace Trust, Stratford-upon-Avon: DR18/3/17/3/8–9, G. L. Cooke to J. H. Leigh, 22 December [1819] and C. Cooke to J. H. Leigh, 28 December 1819. On the parishes of Broadwell and Adlestrop, see C. R. Elrington (ed.), *A History of the County of Gloucester*, Volume 6 (London: Institute of Historical Research, 1965), pp. 49–59, 8–16 (British History Online: https://www.british-history.ac.uk/vch/glos/vol6/pp49-59; https://www.british-history.ac.uk/vch/glos/vol6/pp8-16, both accessed 23 March 2022); see also David Verey and Alan Brooks, *Gloucestershire I: The Cotswolds*, The Buildings of England (New Haven; London: Yale University Press, 2002), pp. 131–133, 200–2. Some years earlier, Cassandra Cooke had similarly lobbied the Earl of

Liverpool on behalf of her elder son (British Library, London: Liverpool Papers, vol. XLIV, Add MS 38233, ff. 177–8, C. Cooke to 1st Earl of Liverpool, 29 August 1799; vol. XLVII, Add MS 38236, f. 113, C. Cooke to 1st Earl of Liverpool, 30 July 1802).

72. The extensive surviving correspondence concerning Cooke's taking on of these parishes may be found at the Shakespeare Birthplace Trust, Stratford-upon-Avon: DR18/3/17/3 Cubbington living. On the parish of Hunningham, see Salzman, *History of the County of Warwick*, pp. 117–20 (British History Online http://www.british-history.ac.uk/vch/warks/vol6/pp117-120, accessed 23 March 2022); see also Pickford and Pevsner, *Warwickshire*, p. 369. One complication upon which much ink was spent was the question of whether ecclesiastical law would compel Cooke to give up either Wyck Rissington or Broadwell upon accepting Cubbington; in the end, he gave up both. Indeed, questions had previously been raised as to whether Cooke should be allowed to hold Wyck Rissington and Broadwell simultaneously, and he was only able to do so thanks to a special dispensation from the Archbishop of Canterbury (Shakespeare Birthplace Trust, Stratford-upon-Avon: DR18/8/19 Adlestrop, Glos.: Rectory papers), whose granting provoked disapproval in some quarters (see, for example, letter 89 in Le Faye, *Letters of Jane Austen*).

73. Shakespeare Birthplace Trust, Stratford-upon-Avon: DR18/3/17/3/12, G. L. Cooke to J. H. Leigh, 13 January 1820.

74. This is the year in which the 'Rev. Cooke' who had paid poor rates in the parish of St Peter-in-the-East since November 1815 stopped doing so (Oxfordshire History Centre, Oxford: St Peter-in-the-East Parish Records, Poor Rate Book 1811–16 PAR213/5/F2/3, f. 191^v and Poor Rate Book 1820–7 PAR213/5/F2/5, ff. 124^v, 133^r). St Peter-in-the-East included New College Lane (see note 37). Although the poor rate records do not give first names, nor precise addresses, it seems likely that this is our Cooke, not only because he first appears in the records around the time of George Leigh Cooke's marriage, when he would have moved out of Corpus Christi, but also because his entries in the record always appear alongside those labelled 'Rigaud'.

75. Shakespeare Birthplace Trust, Stratford-upon-Avon: DR18/3/17/3/14, G. L. Cooke to J. H. Leigh, 22 May 1820.

76. G. F. Peppitt, *Cubbington: Reflections on village life, Domesday to 1970*, millennium ed. (Kenilworth: The Pleasaunce Press, 2000), p. 104.

77. Shakespeare Birthplace Trust, Stratford-upon-Avon: DR18/3/17/6 Cubbington School, 1846. A draft conveyance for the land (DR18/3/17/6a) features detailed annotations in Cooke's handwriting. Peppitt, *Cubbington*, p. 104 states that Cooke contributed £535 of the total £907 needed for the new buildings.

78. *Oxford Society for the Relief of Distressed Travellers and Others, Report*, Oxford: Munday and Slatter, 1820, p. 5; *Annual Report of the National Society for Promoting the Education of the Poor in the Principles of the Established Church*, London, 1812, p. 78. Cooke also donated funds to provide new school rooms in Coventry (*Coventry Herald*, issue 866 (Friday 19 November 1824), 3b), and to preserve Shakespeare's tomb and monument in Stratford-upon-Avon (*Leamington Spa Courier*, vol. 9, issue 402 (Saturday 16 April 1836), 1b).

79. *Leamington Spa Courier*, vol. 9, issue 390 (Saturday 23 January 1836), 1c; *Leamington Spa Courier*, vol. 13, issue 655 (Saturday 6 February 1841), 2d. Cooke was also president of the Warwick and Leamington Church Union (J. M. Neale, *The life and times of Patrick Torry, D.D., Bishop of Saint Andrew's, Dunkeld, and Dunblane, with an appendix on the Scottish liturgy* (London: Joseph Masters, 1856), p. 360), and chair of the committee overseeing Leamington Spa's Warneford Hospital (*Leamington Spa Courier*, vol. 19, issue 1061 (Saturday 7 October 1848), 2b).

80. *Leamington Spa Courier*, vol. 22, issue 1199 (Saturday 24 May 1851), 1c. Cooke appears as president of the society in his entry in the list of life members of the British Association for the Advancement of Science that is appended to the report of the Association's 1837 meeting (*List of members of the British Association for the Advancement of Science. To which are added a list of the officers, rules of*

the *Association, and contents of the reports already published* (London: Richard and John E. Taylor, 1838), p. 5).

81. *Leamington Spa Courier*, vol. 7, issue 291 (Saturday 1 May 1834), 2d; *Aris's Birmingham Gazette*, issue 5050 (Monday 3 September 1838), 2.

82. *Report of the first and second meetings of the British Association for the Advancement of Science; at York in 1831, and at Oxford in 1832: including its proceedings, recommendations, and transactions* (London: John Murray, 1833), p. 611; see also note 80.

83. On Beckley, see Mary D. Lobel (ed.), *A History of the County of Oxford*, Volume 5, *Bullingdon Hundred* (London: Institute of Historical Research, 1957), pp. 56–76 (British History Online: https://www.british-history.ac.uk/vch/oxon/vol5/pp56-76, accessed 29 March 2022); see also Alan Brooks and Jennifer Sherwood, *Oxfordshire: North and West*, The Buildings of England (New Haven; London: Yale University Press, 2017), pp. 133–6. On the Cookes in Beckley, see Peter M. Wheeler (ed.), *Beckley Reflects: The Village Greets the Millennium by Remembering Its Past* (Beckley: Grove Farmhouse, 1999), pp. 195–9.

84. George Theophilus Cooke's tenure as vicar of Beckley came to a rather dramatic end in 1894 with his accidental death by carbolic acid poisoning: 'Distressing death of the vicar of Beckley: accidental poisoning', *Jackson's Oxford Journal*, issue 7346 (Saturday 6 January 1894), 5g.

85. Cooke's grave, which he shares with his wife Anne (d. 1869), stands just to the west of Beckley church within a cluster of other Cooke graves, including those of his brother Theophilus, son George, and daughters Helen, Caroline, and Julia. The Cooke family is also commemorated in several inscriptions and windows within the church building.

86. To give just one example, Cooke's youngest daughter, Julia, appears in the 1901 census as 'living on [her] own means' with three servants in a house in North Oxford (The National Archives, Kew: Census Returns of England and Wales, 1901; Class: RG13; Piece: 1381; Folio: 35; Page: 7). Cooke's will (PROB 11/2170/220) is available online from The National Archives: https://discovery.nationalarchives.gov.uk/details/r/D46686.

CHAPTER 6

Further reading

Price appears in the *Oxford Dictionary of National Biography*, as well as in the published register of Oxford alumni, *Alumni Oxonienses*. He is the subject of several published obituaries, including short notes in *The Times* (no. 35713 (30 December 1898), 4b), *Jackson's Oxford Journal* (no. 7606 (31 December 1898), 7g), *Nature* (59(1523) (5 January 1899), 229–30), *The Athenæum* (no. 3715 (7 January 1899), 20), *The Oxford Magazine* (17(9) (25 January 1899), 149), and the *Monthly Notices of the Royal Astronomical Society* (59(5) (February 1899), 228–9), as well as slightly longer pieces in the *Proceedings of the London Mathematical Society* (30 (November 1898), 332–4, information supplied by Price's former student and Oxford's Waynflete Professor of Pure Mathematics, E. B. Elliott), and in the *Proceedings of the Royal Society of London* (75 (1905), 30–4, by Oxford's Professor of Experimental Philosophy, R. B. Clifton). A more recent short account is Robin Wilson, 'Twinkle, Twinkle, Little Bat!', *The Pembrokian*, no. 35 (July 2011), 16–17. As a prominent figure, both within Oxford and without, Price is represented in a number of archives, particularly in the form of correspondence relating to Oxford University Press. The most extensive primary materials concerning Price may be found in the Bartholomew Price Papers at Pembroke College, Oxford (PMB/S/14). General background on nineteenth-century Oxford mathematics may be found in the writings of Keith Hannabuss, most particularly 'Mathematics', in *The History of the University of Oxford*, vol. VII: *Nineteenth-Century Oxford, Part 2*, ed. M. G. Brock and M. C. Curthoys (Oxford: Clarendon Press, 2000), pp. 433–55.

Acknowledgements

I am very grateful to Amanda Ingram, archivist of Pembroke College, Oxford, for scanning materials for me during the pandemic, and subsequently for allowing me access to the Bartholomew Price Papers in person, as well as for the detail about Pembroke's loan to William Price (see note 46). I am similarly grateful to Richard Allen, archivist of Magdalen College, Oxford, for scans from the E. B. Elliott Papers, and to Adam Crothers of the library of St John's College, Cambridge, for scanned materials from the Isaac Todhunter Papers. Thanks must also go to the staff of the Gloucestershire Archives, to Tessa Shaw, formerly of The Queen's College Library, Oxford, to Michael Riordan and Amy Ebrey of The Queen's College Archives, and to Robin Wilson for details about Price. I am particularly grateful to Mark McCartney and Christopher Stray for their useful comments on an early draft of this chapter.

Notes and references

1. Gloucestershire Archives, Gloucester; Gloucestershire Church of England Parish Registers; Reference Number: Gdr/V1/302 (Church of St James, Coln St Dennis). On the parish of Coln St Dennis, see C. R. Elrington (ed.), *A History of the County of Gloucester*, vol. 8 (London: Institute of Historical Research, 1968), pp. 28–34 (British History Online: https://www.british-history.ac.uk/vch/glos/vol8/pp28-34, accessed 5 July 2022); see also David Verey and Alan Brooks, *Gloucestershire I: The Cotswolds*, 3rd ed., The Buildings of England (New Haven; London: Yale University Press, 2002), pp.300–1.
2. Gloucestershire Archives, Gloucester; Gloucestershire Church of England Parish Registers; Reference Number: Gdr/V1/342 (Church of St James, Coln St Dennis). Since William Price took up his first non-college ecclesiastical post in 1810, it seems reasonable to suppose that he married around this same time. However, his various links within Berkshire, Oxfordshire, and Gloucestershire, in conjunction with a lack of information about Mary Price's maiden name or place of birth, make it difficult to identify a marriage record with any certainty.
3. In *Alumni Oxonienses* and at https://theclergydatabase.org.uk/jsp/persons/DisplayPerson.jsp?PersonID=146136 (accessed 5 July 2022), as well as in the printed sermon from his funeral (Gloucestershire Archives, Gloucester: P97/MI/2).
4. The children of William and Mary Price were Ann (1812–75), William Henry (1813–89), Mary (1814–95), Henrietta (1815–76), Harriett (1817–1907), Bartholomew (1818–98), and Charlotte Sophia (1823–1903). Following Mary's death in 1828, William remarried, though a marriage record is again difficult to identify with certainty, as are any other details of his second wife, Harriett. Together, they had a further five children: John (March–April 1831), Harriett Maria (April–June 1832), George (1834–?), James (1836–?), and Adelaide (1839–1900).
5. The story is related in Elrington, *History of the County of Gloucester*, with reference to a plaque in the church. In 1957, the clock was replaced by an electric one donated by members of the Price family. A list of subscriptions and expenses for the restoration of the clock in 1839 appears at the end of the more general accounts book for the Church of St James the Great in Coln St Dennis (Gloucestershire Archives, Gloucester: P97/VE/2/6).
6. William Henry Price matriculated at Pembroke in 1831, aged 18 (*Alumni Oxonienses*).
7. Douglas Macleane, *Pembroke College*, University of Oxford college histories (London: F. E. Robinson, 1900), pp.158–9; see also the entry for Pembroke College in H. E. Salter and Mary D. Lobel (eds), *A History of the County of Oxford*, vol. 3: *The University of Oxford* (London: Institute for Historical Research, 1954), pp. 288–97 (British History Online: http://www.british-history.ac.uk/vch/oxon/vol3/pp288-297, accessed 5 July 2022).
8. *The historical register of the University of Oxford, being a supplement to the Oxford University Calendar, with an alphabetical record of university honours and distinctions, completed to the end of Trinity Term, 1888* (Oxford: Clarendon Press, 1888), p. 387.

9. The details of the scholarships are set out, for example, in *Historical register 1888*, pp. 119–20.

10. On the scholarships, see the comments in Keith Hannabuss, 'Mathematics' in *The History of the University of Oxford*, vol. VII: *Nineteenth-Century Oxford, Part 2*, ed. M. G. Brock and M. C. Curthoys (Oxford: Clarendon Press, 2000), pp. 433–55 at p. 446. Price's scholarship examination duties of the 1860s, for instance, are noted in passing by his sometime co-examiner Charles L. Dodgson: Edward Wakeling (ed.), *Lewis Carroll's Diaries: The Private Journals of Charles Lutwidge Dodgson (Lewis Carroll)*, vol. 4 (Luton: Lewis Carroll Society, 1997), p. 150; vol. 5 (1999), p. 51.

11. *Report of Her Majesty's Commissioners appointed to inquire into the State, Discipline, Studies, and Revenues of the University and Colleges of Oxford: together with the Evidence, and an Appendix* (London: W. Clowes and Sons for Her Majesty's Stationery Office, 1852), Evidence, pp. 59–67.

12. Pembroke College, Oxford: Bartholomew Price Papers, PMB/S/14/6, R. L. Cotton to B. Price, 14 April 1853.

13. 'University intelligence', *The Times*, issue 21405 (Monday 18 April 1853), 3f.

14. *Alumni Oxonienses*; *Historical register 1888*, p. 120.

15. Tony Simcock, 'Laboratories and Physics in Oxford Colleges, 1848–1947', in *Physics in Oxford 1839–1939: Laboratories, Learning, and College Life*, ed. Robert Fox and Graeme Gooday (Oxford: Oxford University Press, 2005), pp. 119–68 at p. 129.

16. Dale was elected a member in May 1839, and was on the society's committee by 1845 (*Proceedings of the Ashmolean Society*, no. XVI (1839), 9; *Proceedings of the Ashmolean Society*, no. XXII (1845), 107). Over the following years, he presented several papers to the society; see, for example, *Proceedings of the Ashmolean Society*, no. XXIII (1846), 123–5 (on elliptic polarization); *Proceedings of the Ashmolean Society*, no. XXV (1848), 199 (on an experiment involving 'a delicate galvanometer').

17. Cotton politely acknowledged receipt of the commission's inquiries with regard to Worcester College in a letter consisting of a single sentence, but did not otherwise respond (*Report of Her Majesty's Commissioners*, Evidence, p. 378). On Cotton, see Peter B. Nockles, 'Cotton, Richard Lynch (1794–1880)', *Oxford Dictionary of National Biography*, https://doi.org/10.1093/ref:odnb/6423.

18. Vivian H. H. Green, 'Routh, Martin Joseph (1755–1854)', *Oxford Dictionary of National Biography*, https://doi.org/10.1093/ref:odnb/24182.

19. 'Death of the Warden of All Souls', *Jackson's Oxford Journal*, no. 5470 (Saturday 27 February 1858), 5c.

20. Bartholomew Price, *A treatise on the differential calculus, and its application to geometry: founded chiefly on the method of infinitesimals* (London: George Bell, 1848); Bartholomew Price, *A treatise on the differential calculus, and its applications to algebra and geometry: founded chiefly on the method of infinitesimals*, vol. I. of *A treatise on infinitesimal calculus; containing differential and integral calculus, calculus of variations, applications to algebra and geometry, and analytical mechanics* (Oxford: Oxford University Press, 1852; 2nd ed., 1857).

21. For example: 'University intelligence', *The Morning Chronicle* (London), no. 26918 (Saturday 2 April 1853), 4e.

22. Pembroke College, Oxford: Bartholomew Price Papers, PMB/S/14/6, R. L. Cotton to B. Price, 14 April 1853.

23. Oxford University Archives: WP/γ/24/6, Hebdomadal Board Minutes, 1841–54, p. 316.

24. See, for example, 'University intelligence', *The Times*, no. 22002 (Thursday 15 March 1855), 5f.

25. 'University intelligence', *The Times*, no. 21434 (Saturday 21 May 1853), 6f.

26. 'University intelligence', *The Times*, no. 21645 (Monday 23 January 1854), 10d; 'University intelligence', *The Times*, no. 21729 (Monday 1 May 1854), 10e.

27. 'University intelligence', *The Times*, no. 21882 (Thursday 26 October 1854), 6f.

28. See, for example, 'University intelligence', *The Times*, no. 22185 (Monday 15 October 1855), 6e; 'University intelligence', *The Times*, no. 22588 (Tuesday 27 January 1857), 10f; *Oxford University*

Gazette, supplement (2) to no. 690 (Thursday 13 November 1890), 133; 'University intelligence', *The Times*, no. 26070 (Thursday 12 March 1868), 9f. The lecture courses advertised by various Oxford professors were parodied in a short piece in *Punch* in 1865, which included the following: 'The Sedleian Professor will give his usual entertainment of ventriloquism, and the beauties of the magnet, assisted by a Magic Lantern. Half-price at nine o'clock' (*Punch* (Saturday 6 May 1865), 185).

29. A typical annual report, consisting of a single short, descriptive paragraph may be found, for example, in *Oxford University Gazette*, XXI(711) (Tuesday 26 May 1891), 501.

30. There are two versions of the history: the original edition (Douglas Macleane, *A history of Pembroke College, Oxford, anciently Broadgates Hall, in which are incorporated short historical notices of the more eminent members of this house* (Oxford: Printed for the Oxford Historical Society at the Clarendon Press, 1897)) and a shorter version that was published as part of a series of Oxford college histories (Douglas Macleane, *Pembroke College*, University of Oxford College Histories (London: F. E. Robinson & Co., 1900)). For biographical notes on Macleane and a critique of his historical writings, see John Platt, 'The College History', *Pembroke College Record* (1998–2000), 54, 59–60.

31. Pembroke College, Oxford: Douglas Macleane Papers, PMB/T/5/5 (1/10), P. Hedgeland to D. Macleane, 17 July 1893. By the 1890s, Hedgeland was Vicar of Penzance and Prebendary of Exeter; see *Alumni Oxonienses*.

32. Pembroke College, Oxford: Douglas Macleane Papers, PMB/T/5/5 (1/24), 'Orger Recollections', f. iii. Orger went on to observe, however, that 'the material for this purpose which the College supplied was seldom promising'. By the 1890s, Orger was Vicar of Hougham-by-Dover and a former fellow and tutor of St Augustine's College, Canterbury; see *Alumni Oxonienses*.

33. John Gilmore, 'Mitchinson, John (1833–1918)', *Oxford Dictionary of National Biography*, https://doi.org/10.1093/ref:odnb/47175.

34. Pembroke College, Oxford: John Mitchinson Papers, PMB/S/15/85, 'Oxford Memories', p. 8. Internal evidence indicates that the memoirs were written in 1916. Mitchinson also recounts that in later years he reminded Price of this conversation ('ruefully adding, "and it all came true"'), although the latter had no memory of it.

35. Pembroke College, Oxford: Douglas Macleane Papers, PMB/T/5/5 (1/24), 'Orger Recollections', f. iii.

36. Pembroke College, Oxford: Douglas Macleane Papers, PMB/T/5/5 (1/22), 'Rev. Dr. Bartrum Recollections', pp. 5–6. At the beginning of the 1890s, Bartrum was the recently retired headmaster of Berkhamsted School in Hertfordshire; see *Alumni Oxonienses*.

37. Pembroke College, Oxford: Douglas Macleane Papers, PMB/T/5/5 (1/21), G. Gainsford to D. Macleane, u.d.

38. This happened on at least three occasions: 1864 (*Alumni Oxonienses*), 1873 (see note 45), and 1891 (Pembroke College, Oxford: Bartholomew Price Papers, PMB/S/14/115, Memorandum by Evan Evans, Master of Pembroke College, nominating Bartholomew Price as Vicegerent of the College, 5 November 1891).

39. See, for example, *Report of Her Majesty's Commissioners*, Report, p. 95.

40. For a thorough study of this gradual relaxation of restrictions and its repercussions, see: B. Duckenfield, *Changes in the Celibacy Rule at the Colleges of Oxford and Cambridge Universities*, PhD thesis, Kingston University, 2008.

41. Ibid., p. 168.

42. Macleane, *History of Pembroke College*, pp. 456–7.

43. General Register Office, London: England and Wales Civil Registration Indices, Q3 1857, vol. 5b, p. 79 (St Thomas, Devon).

44. Bartholomew and Amy Eliza Price raised seven children: Amy Maud (1858–1923), William Arthur (1860–1954), Mary Eliza Mabel (1862–1939), Alice Margaret (1864–?), Rose Emelyn (1866–94), Elinor Rokeby (1868–1956), and Bartholomew George (1870–1947).

45. Pembroke College, Oxford: John Mitchinson Papers, PMB/5/15/18, E. Moore to J. Mitchinson, 9 November 1873.

46. For a survey of the five Mastership elections that Pembroke went through in the nineteenth century, see John Platt, '19th Century Mastership Elections', *Pembroke College Record* (1998–2000), 50–4. It is possible that Price had had the Mastership in his sights for some time, for there is a hint that his father had missed out on the post in 1809, and that this had remained a sore point within the Price family. This suggestion is found in a letter written by Louis Henry Hall (1842–1905) to Douglas Macleane in 1900 (Pembroke College, Oxford: Douglas Macleane Papers, PMB/T/5/29 (3 of 5), L. H. Hall to D. Macleane, 12 January 1900). L. H. Hall was the grandson of the man who did win the 1809 Mastership contest: George William Hall (1770–1843). He related to Macleane the family lore that although G. W. Hall 'was always most friendly with William Price, the family always considered he ought to have been elected in 1809 Master of Pembroke'. William Price's departure for Coln St Dennis in 1810 is at least consistent with a disappointed candidate leaving Oxford (Pembroke leant him £500 to rebuild the parsonage), but otherwise there is no evidence for this claim, nor that Bartholomew Price held the view attributed here to the family. The careers of members of the Hall and Price families were somewhat intertwined throughout the nineteenth century: L. H. Hall's uncle, Edward Duncan Hall (1826–74), a sometime fellow of Pembroke, was rector of Coln St Dennis between the incumbencies of William Price and William Henry Price (see note 53). This is no coincidence though: from at least as early as 1810, Coln St Dennis had been under the patronage of Pembroke (Elrington, *History of the county of Gloucester*). This would have given Bartholomew Price even more reason to remain in contact with the parish of his birth. Indeed, in 1848, he loaned the church £100 to carry out extensive restorations (Gloucestershire Archives, Gloucester: P97/VE/2/6), and he also preached a sermon there on at least one occasion (this was printed as *A sermon preached at the parish church of Coln Saint Dennis, Gloucestershire, on Sunday, January 1st, 1854* (London: A. and G. A. Spottiswoode, 1854)). Price's will, of 18 May 1892, makes reference to property in Coln St Dennis (https://probatesearch.service.gov.uk/).

47. In a letter to Mitchinson at the beginning of 1864, Jeune reflected upon the qualities of the candidates (Pembroke College, Oxford: John Mitchinson Papers, F. Jeune to J. Mitchinson, 1 January 1864). Mitchinson later wrote that had Thomas Frederick Henney (1810–60), seemingly the third member of a triumvirate of senior fellows which also included Evans and Price, still been alive, 'there would have been an unanimous election' (Pembroke College, Oxford: John Mitchinson Papers, PMB/S/15/85, 'Oxford Memories', p. 8). On Henney and Evans, see *Alumni Oxonienses*; neither has an entry in the *Oxford Dictionary of National Biography*.

48. Although he did not stand, Mitchinson's own candidacy for the Mastership apparently had some support among the student body. John William Horsley (1845–1921), who was an undergraduate at Pembroke at the time of the election, penned the following verse (in reference to Mitchinson's short stature): 'Remember when you choose a Head // How small the lump that leavens: // We won't have Evans at any price, // And as for Price, Oh 'eavens!' (Platt, 'Mastership elections', p. 52). Horsley later recalled this verse in a letter to Macleane, to which a punning Latin adaptation was added (by Macleane?) in red ink: 'Pricium pro magistro odi, // Evanus est petitor, O di!' (Pembroke College, Oxford: Douglas Macleane Papers, PMB/T/5/5 (1/14), J. W. Horsley to D. Macleane, 30 August n.y.). Price himself thought that Mitchinson's candidacy ought not to be 'silently ignored' (Pembroke College, Oxford: John Mitchinson Papers, PMB/S/15/11, B. Price to J. Mitchinson, 25 February 1864). Early in his memoirs, Mitchinson promised to say something about Price's eventual successful election to the Mastership, noting that it 'shall be duly chronicled, if my existence here

is so far prolonged' (Pembroke College, Oxford: John Mitchinson Papers, PMB/S/15/85, 'Oxford Memories', p. 9). Sadly, the memoirs end without covering that material.

49. Pembroke College, Oxford: John Mitchinson Papers, PMB/S/15/85, 'Oxford Memories', p. 7. This is a view that tallies with those supplied to Macleane in the 1890s by former Pembroke students: their comments about Evans were rather more lukewarm than those concerning Price. For example, the poet Richard Watson Dixon (1833–1900, matric. 1852) remarked that Evans was 'able enough, but very slow' (Pembroke College, Oxford: Douglas Macleane Papers, PMB/T/5/5 (1/15), R. W. Dixon to D. Macleane, 12 September 1893). Philip Hedgeland (see note 31) was also critical of Evans.

50. Pembroke College, Oxford: Douglas Macleane Papers, PMB/T/5/5 (1/22), 'Rev. Dr. Bartrum Recollections', p. 7.

51. Pembroke College, Oxford: John Mitchinson Papers, PMB/S/15/85, 'Oxford Memories', p. 8.

52. Ibid. College oral history has done Evans a disservice by casting him as the Master who did indeed vote for himself, although this was not the case (Platt, 'Mastership elections', p. 50).

53. Platt, 'Mastership elections', p. 51; J. P. D. Dunbabin, 'Finance and Property', in *The History of the University of Oxford*, vol. VI: *Nineteenth-Century Oxford, part 1*, ed. M. G. Brock and M. C. Curthoys (Oxford: Clarendon Press, 1997), pp. 375–440 at pp. 419–20. William Henry Price appears to have left Oxford around this time. From 1861, he was rector of Somerton in Oxfordshire, but in 1874 he exchanged this for the rectorship of his father's old parish of Coln St Dennis, which he held until the end of his life (*Alumni Oxonienses*; plaque in church).

54. Pembroke College, Oxford: John Mitchinson Papers, PMB/S/15/11, B. Price to J. Mitchinson, 25 February 1864.

55. Even allowing for the fact that Price was not the most legible of scribes at the best of times.

56. Pembroke College, Oxford: John Mitchinson Papers, PMB/S/15/11, B. Price to J. Mitchinson, 25 February 1864.

57. See, for example, Joshua Bennett, 'A History of "Rationalism" in Victorian Britain', *Modern Intellectual History* 15(1) (2018), 63–91.

58. Pembroke College, Oxford: John Mitchinson Papers, PMB/S/15/11, B. Price to J. Mitchinson, 25 February 1864.

59. In his letter to Mitchinson in January 1864 (see note 47), Jeune had stressed the need for the next Master of Pembroke to be a good church, university, and college man, particularly when 'faith and unbelief are [. . .] in battle'. Platt ('Mastership Elections', p. 51) speculates that this may be an oblique reference to Price's views. On Powell, see Pietro Corsi, *Science and Religion: Baden Powell and the Anglican Debate, 1800–1860* (Cambridge, UK: Cambridge University Press, 1988).

60. One of these may have been the following: Bartholomew Price, *The presence of God in the Creation: a sermon preached before the University of Oxford, at St. Mary's Church, on the afternoon of Sunday, Oct. 14, 1855* (London: A. and G. A. Spottiswoode, 1855).

61. Pembroke College, Oxford: John Mitchinson Papers, PMB/S/15/11, B. Price to J. Mitchinson, 25 February 1864.

62. Macleane, *Pembroke College*, p. 246; Platt, 'Mastership Elections', p. 52.

63. Pembroke College, Oxford: John Mitchinson Papers, PMB/5/15/18, E. Moore to J. Mitchinson, 9 November 1873.

64. See, for example, the comments of E. Moore to J. Mitchinson of July 1879, quoted in Platt, 'Mastership Elections', p. 52.

65. Ibid.

66. Pembroke College, Oxford: John Mitchinson Papers, PMB/S/15/24, A. T. Barton to J. Mitchinson, 16 December n.y. [*c*.1879].

67. Ibid.

68. Pembroke College, Oxford: Douglas Macleane Papers, PMB/T/5/29 (3 of 5), L. H. Hall to D. Macleane, 12 January 1900.

69. 'The late Master of Pembroke', *The Oxford Magazine*, 10(6) (Wednesday 25 November 1891), 96. Price officiated at the funeral: *Jackson's Oxford Journal*, no. 7237 (Saturday 5 December 1891), 8b. The rumours reported by Barton may nevertheless have had some basis: if the mooted higher office was no sinecure, then it is possible that Evans declined it. According to Barton, it was only under pressure that Evans had accepted the Vice Chancellorship of the university in 1878; newspaper reports of that appointment do seem to hint at some initial uncertainty in the arrangement: 'It is now definitely settled that . . .' ('The universities', *Pall Mall Gazette*, no. 4202 (Friday 9 August 1878), 7).

70. Macleane, *Pembroke College*, p. 256.

71. Platt, 'Mastership Elections', p. 53. On Grose, see J. S. Cotton, rev. M. C. Curthoys, 'Grose, Thomas Hodge (1845–1906)', *Oxford Dictionary of National Biography*, https://doi.org/10.1093/ref:odnb/33589.

72. Another Mastership contest earlier in the century (the one that had ultimately seen Jeune appointed) had also had to be referred to the Visitor: Platt, 'Mastership Elections', p. 50.

73. Pembroke College, Oxford: John Mitchinson Papers, PMB/S/15/38, M. Creighton to J. Mitchinson, 3 January 1892 [misdated 1891]. Creighton's major work was a multivolume history of the papacy during the Reformation: M. Creighton, *A history of the papacy during the period of the Reformation*, 5 vols (London: Longmans, Green, 1882–94). Some interest may also have stemmed from the fact that he was himself in correspondence with the Prime Minister at that time over ecclesiastical appointments, and would be named Bishop of Peterborough that year.

74. Pembroke College, Oxford: John Mitchinson Papers, PMB/S/15/38, M. Creighton to J. Mitchinson, 3 January 1892 [misdated 1891].

75. Ibid. In taking up the Mastership, Price does seem to have been conscious of his age. In replying to a letter of congratulations from G. G. Stokes, Price commented upon the fact that he was taking up the office 'at a time of life when we should rather seek rest and relaxation from work'. He noted, however, that 'there are occasions on which inclination has to give way to duty, and calls are made upon one when we should prefer to be left alone' (Cambridge University Library: Stokes Papers Add MS 7656, P675, B. Price to G. G. Stokes, 5 February 1892). Similar sentiments appear in a letter of the following day from Price to the mediaeval historian Thomas Frederick Tout (University of Manchester Library: Papers of Thomas Frederick Tout, TFT/1/970/2, B. Price to T. F. Tout, 6 February 1892).

76. Given Price's involvement with such high-level matters as the Cleveland Commission in the 1870s, it is likely that he and Salisbury were acquainted personally.

77. Letters of congratulation survive from the publisher and future Provost of Worcester College C. H. Daniel, and from the Master of Balliol Benjamin Jowett (Pembroke College, Oxford: Bartholomew Price Papers, PMB/S/14/116, C. H. Daniel to B. Price, *c*.15 January 1892; PMB/S/14/117, B. Jowett to B. Price, 16 January 1892).

78. Pembroke College, Oxford: Douglas Macleane Papers, PMB/T/5/5 (1/22), 'Rev. Dr. Bartrum Recollections', p. 7.

79. 'In Memoriam: Bartholomew Price, D.D.', *The Oxford Magazine*, 17(9) (25 January 1899), 149.

80. Macleane, *History of Pembroke College*, p. 346.

81. W. R. Ward, *Victorian Oxford* (London: Frank Cass & Co., 1965), p. 216.

82. *Oxford Magazine*, 17(9) (25 January 1899), 149.

83. Ibid.

84. Pembroke College, Oxford: John Mitchinson Papers, PMB/S/15/85, 'Oxford Memories', p. 9. Price certainly did dabble in the business world: he was a shareholder in the Great Western Railway (Society of Genealogists, London: Great Western Railway Shareholder Index), and also in Silver's rubber factory in East London (John A. Tully, *Silvertown: The Lost Story of a Strike That Shook London and Helped Launch the Modern Labor Movement* (New York: Monthly Review Press, 2014),

pp. 182–3). It is interesting to note the slight reflection here of a generational difference between Price and his Sedleian predecessor: whereas Price has railway shares, Cooke had shares in the canal and at times voiced opposition to the building of the railways (*Journals of the House of Commons*, vol. 101, part II: 5 June 1846 to 12 January 1847, pp. 810–11).

85. *Special report from the Select Committee on the Oxford and Cambridge Universities Education Bill: together with the Proceedings of the Committee, Minutes of Evidence, and Appendix* (London, 1867), Evidence, qs. 2116, 2133–5. The question of university extension also fell within the committee's purview, and Price spoke in favour of this, finding that the standard of and interest in mathematics among grammar school boys were generally better than those of boys from public schools (q. 2200).

86. *A statement concerning the business and finance of the University [of Oxford], and proposals for the administration of the same (a paper prepared at the request of the consolidated committee and now under consideration)* (Oxford, 1867; Bodleian Library, Oxford: G. A. Oxon 8° 705 (1)). Some years later, Price was a co-author, along with R. G. Livingstone and G. Wood, of another financial report, this time assessing the costs faced by undergraduates at Pembroke (*Report of the committee upon the expenses of the undergraduates* (Oxford, 1886; Bodleian Library, Oxford: G. A. Oxon 8° 1929). One of the suggestions of this report was that individual students' college accounts might be made public, as 'a salutary check upon extravagance' (p. 8).

87. *Statement concerning the business and finance of the University*, p. 10.

88. Much of Price's official correspondence concerning OUP was destroyed (Pembroke College, Oxford: Bartholomew Price Papers, PMB/S/14/1 (1/7), J. D. Fleeman to P. H. Sutcliffe, u.d. [*c.* April 1975]), but what does survive has been drawn upon in detail by histories of the press, in particular the multivolume *History of Oxford University Press*, ed. Ian Anders Gadd, Simon Eliot, William Roger Louis, and Keith Robbins, 4 vols (Oxford: Oxford University Press, 2013–17), which we rely upon heavily here.

89. The university press was also known as the 'Clarendon Press' because its printing works were in the Clarendon Building in central Oxford, but nowadays that label is reserved for a particular imprint of the press. By the mid-nineteenth century, the press consisted of two parts: the Learned Press, which produced books in small print runs, often using hand-presses, and the Bible Press, which operated steam-powered presses to produce much larger print runs; see Robert Banham, 'The Workforce', in *The History of Oxford University Press*, vol. II: *1780 to 1896*, ed. Simon Eliot (Oxford: Oxford University Press, 2013), pp. 175–225. On the Bible Press specifically, see Amy Flanders and Stephen Colclough, 'The Bible Press', in *History of Oxford University Press*, vol. II, pp. 357–401.

90. Mark Curthoys, 'The Press and the University', in *History of Oxford University Press*, vol. II, pp. 27–75, at p. 52.

91. Simon Eliot, 'The Evolution of a Printer and Publisher', in *History of Oxford University Press*, vol. II, pp. 77–113, at p. 99.

92. As early as 1872, Price's salary was raised to £1,250, following a decision by the Delegates that he should take on 'a complete superintendence of all the businesses' (Curthoys, 'The Press and the University', p. 60).

93. This caused occasional frustrations, such as when Price was unable to get any decisions from the Delegates during the Long Vacation (John Feather, 'Authors and Publishers', in *History of Oxford University Press*, vol. II, pp. 321–54 at p. 336). Even when away from Oxford himself, Price tried to keep up with press business, such as when he was holidaying in north Wales in August 1880 (Simon Eliot, 'Machines, Materials and Money', in *History of Oxford University Press*, vol. II, pp. 115–73 at pp. 119–20). Price worked particularly closely with one of the press's commercial partners, the London publisher Alexander Macmillan (1818–96), frequently travelling to London for meetings (Eliot, 'Machines, Materials and Money', pp. 119–20; Mary Hammond, 'The London Connection', in *History of Oxford University Press*, vol. II, pp. 277–319).

94. Price adopted a somewhat sarcastic tone in two letters of 1875, for example, one to the geologist Nevil Story Maskelyne (1823–1911) and another to the classicist Ingram Bywater (1840–1914), enquiring as to when they might submit 'any portion' of their respective texts (Feather, 'Authors and Publishers', pp. 335–6). Maskelyne's text, a treatise on crystallography, was not published until 1895; Bywater's projected history of philosophy never appeared.

95. Curthoys, 'The Press and the University', p. 61. This attitude is particularly in evidence in Price's involvement with the launch of the *Oxford English Dictionary*: although he was in favour of the *Dictionary*, he insisted that it had to be profitable. For an overview of Price's role in the establishment of the *Dictionary*, see Elizabeth Knowles, 'Dictionaries and other Works of Reference', in *History of Oxford University Press*, vol. II, pp. 601–30; for a more detailed treatment, see Peter Gilliver, *The Making of the Oxford English Dictionary* (Oxford: Oxford University Press, 2016). Price features quite prominently in the catalogue of the correspondence of James Murray (1837–1915), the first editor of the *OED* (Bodleian Library, Oxford: MSS Murray).

96. Curthoys, 'The Press and the University', p. 75.

97. On the press's educational output, see Christopher Stray, 'Educational Publishing', in *History of Oxford University Press*, vol. II, pp. 473–511 and Robert Fraser, 'Educational Books', in *History of Oxford University Press*, vol. III: *1896 to 1970*, ed. Wm. Roger Louis (Oxford: Oxford University Press, 2014), pp. 443–69.

98. Jonathan R. Topham, 'Science, Mathematics, and Medicine', in *History of Oxford University Press*, vol. II, pp. 513–57 at p. 533.

99. Ibid. The very poor income that Thomson and Tait's book brought them was a point of contention between Price and the authors (ibid., p. 539).

100. Ibid., pp. 541–2.

101. Eliot, 'The evolution of a Printer and Publisher', p. 109; Hammond, 'The London Connection', p. 317.

102. Eliot, 'The Evolution of a Printer and Publisher', p. 110; William Whyte, 'Oxford University Press, 1896–1945', in *The History of Oxford University Press*, vol. III, pp. 59–96 at p. 65.

103. Ibid., p. 66.

104. Peter H. Sutcliffe, *The Oxford University Press: An Informal History* (Oxford: Clarendon Press, 1978), p. 65.

105. Ibid., p. 66. As Hammond ('The London Connection', p. 316) observes, 'Gell has been much maligned in the history of the Press'.

106. Price, *Differential calculus* (1848), p. iv.

107. Ibid., p. iii.

108. His use of a London publisher rather than the university press may have been part of this, although it is worth noting that in the first half of the nineteenth century, prior to Price's tenure as Secretary, the press published few textbooks (Stray, 'Educational Publishing', pp. 473–4).

109. Price, *Differential calculus* (1848), p. iii.

110. [B. Powell], *A short treatise on the principles of the differential and integral calculus* (Oxford: University Press, 1830).

111. On the introduction of differential notation in Britain, see J. M. Dubbey, 'The Introduction of the Differential Notation to Great Britain', *Annals of Science* 19(1) (1963), 37–48; Brigitte Stenhouse, 'Mary Somerville's Early Contributions to the Circulation of Differential Calculus', *Historia Mathematica* 51 (2020), 1–25.

112. Price's old-fashioned approach was later noted by his student and future colleague E. B. Elliott in reminiscences about his undergraduate days (Magdalen College, Oxford: Papers of Edwin Bailey Elliott, F34/A1/4, Notes for an untitled lecture to the Oxford Mathematical Society on 'mathematics and mathematicians in the Oxford of various times', p. 9).

113. Augustus De Morgan, *The differential and integral calculus*, Library of Useful Knowledge (London: Baldwin and Cradock for the Society for the Diffusion of Useful Knowledge, 1836). On De Morgan's use of limits, see Joan L. Richards, 'Augustus De Morgan, the History of Mathematics, and the Foundations of Algebra', *Isis* 78(1) (1987), 6–30.

114. Price, *Differential calculus* (1848), p. iv. On Cauchy and analysis, see Judith V. Grabiner, *The Origins of Cauchy's Rigorous Calculus* (Cambridge, MA: MIT Press, 1981).

115. Price, *Differential calculus* (1848), p. iv.

116. One version of Price's differential calculus text was praised by John Herschel, who saw it as indicating the growth of interest in higher mathematics in Oxford, but it is not clear which version—the relevant letter (Pembroke College, Oxford: Bartholomew Price Papers, PMB/S/14/53, 2 of 6) appears to be dated 'April 25, 1823', which of course makes no sense. All other surviving correspondence between Price and Herschel (at Pembroke and the Royal Society) dates from the 1860s. Herschel would later hail the second edition of Price's volume III as 'no doubt [...] becoming the standard work for the English student for many years to come' (Pembroke College, Oxford: Bartholomew Price Papers, PMB/S/14/53, 6 of 6, J. F. W. Herschel to B. Price, 24 December 1868).

117. Price, *Differential calculus* (1852), p. vi.

118. Ibid., p. ix.

119. See, for example, Joan L. Richards, 'The Art and the Science of British Algebra: A Study in the Perception of Mathematical Truth', *Historia Mathematica* 7 (1980) 343–65.

120. Price, *Differential calculus* (1852), p. vi.

121. It was replaced by a chapter on the calculus of operations, based on the work of Duncan Gregory (1813–44), whom Price acknowledged in his new Preface. On Gregory's work in this area, see Patricia R. Allaire and Robert E. Bradley, 'Symbolical Algebra as a Foundation for Calculus: D. F. Gregory's Contribution', *Historia Mathematica* 29(4) (2002), 395–426.

122. Bartholomew Price, *A treatise on infinitesimal calculus; containing differential and integral calculus, calculus of variations, applications to algebra and geometry, and analytical mechanics*, vol. II: *integral calculus and calculus of variations* (Oxford: University Press, 1854); 2nd ed. [integral calculus, calculus of variations, and differential calculus], 1865.

123. Price, *Infinitesimal calculus*, vol. II, p. iii. Price may have had the text by Baden Powell in mind (see note 110). This begins its treatment of integral calculus with the bald statement (p. 81): 'The Integral Calculus is the inverse of the Differential.' Powell acknowledged that not all integrands arise as the result of differentiation, but confined his attention to those that might be adapted into such a form—by writing them as infinite series, for example.

124. Price, *Infinitesimal calculus*, vol. II, p. iv.

125. Price's definition of integral calculus (p. 4) is more inclusive but a little less transparent than that of Powell (note 123): 'The *Integral Calculus* is the aggregate of the rules by which Integrals are determined, and the code of laws subject to which Differentials and Integrals in their mutual relations may be applied to questions of Geometry and Physics.'

126. Compare Price, *Infinitesimal calculus*, vol. II, p. 4 with Augustin-Louis Cauchy, *Résumé des leçons données a l'École royale polytechnique, sur le calcul infinitésimal* (Paris: de l'Imprimerie royale, 1823), p. 81 and with De Morgan, *Differential and integral calculus*, p. 99. Even the language of continuous functions was still reasonably new at this time, at least in a British university setting: the term had been used by De Morgan, but not by Powell. See Jesper Lützen, 'The Foundation of Analysis in the 19th century', in *A History of Analysis*, ed. Hans Niels Jahnke, History of Mathematics, vol. 24 (Providence, RI: American Mathematical Society; London: London Mathematical Society, 2003), pp. 155–95. Price's engagement with wider sources from abroad is once again in evidence: in commenting on the ongoing progress of mathematics, Price directed his readers towards the journals of Joseph Liouville (1809–82) and August Leopold Crelle (1780–1855)—respectively: *Journal de mathématiques pures et appliquées*, published in Paris, and *Journal für die reine und*

angewandte Mathematik, published in Berlin. Around this time, Price was also reading the Parisian *Comptes rendus de l'Académie des Sciences* (Cambridge University Library: Stokes Papers Add MS 7656, P653, B. Price to G. G. Stokes, 24 May 1856).

127. 'A Treastise on Infinitesimal Calculus', *The Athenæum*, no. 1419 (6 January 1855), 15. The review appeared anonymously, but Millard has subsequently been identified as the author: https://athenaeum.city.ac.uk/reviews/contributors/contributorfiles/MILLARD,J..html (accessed 1 August 2022).

128. Bartholomew Price, *A treatise on infinitesimal calculus; containing differential and integral calculus, calculus of variations, applications to algebra and geometry, and analytical mechanics*, vol. III: *statics, and dynamics of material particles* (Oxford: University Press, 1856; 2nd ed., 1868).

129. Ibid., p. v.

130. Ibid., pp. iv–v.

131. 'A Treastise on Infinitesimal Calculus', *The Athenæum*, no. 1486 (19 April 1856), 489–90. Famously a critic of the Cambridge examination system, De Morgan did not waste the opportunity to disparage the books used in Cambridge, all of them, he claimed, '*Examination Guides*, cut up into bits, ready dressed for the examiner'. He went on: 'Logic and mathematics are making daily advances on the Isis, while on the Cam all the attention is given to training for the problem papers.'

132. Bartholomew Price, *A treatise on infinitesimal calculus; containing differential and integral calculus, calculus of variations, applications to algebra and geometry, and analytical mechanics*, vol. IV: *the dynamics of material systems* (Oxford: University Press, 1861; 2nd ed., 1889). Volumes III and IV of Price's *Treatise on infinitesimal calculus* were also packaged as volumes I and II of a *Treatise on analytical mechanics*. The possibility of a fifth volume on 'hydromechanics, light, sound, and mechanics of molecules of a vibrating system' was raised by Price in a letter to John Herschel in October 1865, but the plans never came to anything (Royal Society, London: Herschel Papers, HS/14.57, B. Price to J. F. W. Herschel, 28 October 1865). This may have been in response to an earlier suggestion made by Herschel, who upon receiving a copy of volume IV had commented on the scope for a further volume on the propagation of waves (Royal Society, London: Herschel Papers, HS/23.358, J. F. W. Herschel to B. Price, 21 December 1861, copy), although many years earlier, Price had expressed his hope to G. G. Stokes that someone would compile a treatise containing 'all that has been done in the theory of Undulations' (Cambridge University Library: Stokes Papers Add MS 7656, P647, B. Price to G. G. Stokes, 23 March 1849).

133. *Nature*, 42(1094) (16 October 1890), 585–7 at p. 587. The mathematics students of Oxford may not in fact have read Price's treatises, however: E. B. Elliott later admitted that had not read either of Price's volumes III or IV, and suspected that the same was true of most students (Magdalen College, Oxford: Papers of Edwin Bailey Elliott, F34/A1/4, Notes for an untitled lecture to the Oxford Mathematical Society on 'mathematics and mathematicians in the Oxford of various times', p. 10).

134. Karen Hunger Parshall, 'James Joseph Sylvester', in *Oxford's Savilian Professors of Geometry: The First 400 Years*, ed. Robin Wilson (Oxford: Oxford University Press, 2022), 120–43.

135. Robin Wilson, 'G. H. Hardy and E. C. Titchmarsh', in Wilson (ed.), *Oxford's Savilian Professors of Geometry*, 144–81.

136. Andrew Warwick, *Masters of Theory: Cambridge and the Rise of Mathematical Physics* (Chicago: University of Chicago Press, 2003); Alex D. D. Craik, *Mr Hopkins' Men: Cambridge Reform and British Mathematics in the 19th Century* (London: Springer, 2007).

137. Pembroke College, Oxford: Douglas Macleane Papers, PMB/T/5/5 (1/21), G. Gainsford to D. Macleane, u.d. Gainsford, who matriculated at Pembroke in 1848 and had college rooms adjoining Price's, recalled how his musical ambitions had been thwarted by Price's coaching when he was asked (politely) not to play the piano on the frequent occasions when Price was seeing private pupils. See also PMB/T/5/5 (1/24), 'Orger Recollections', f. iii, and PMB/T/5/5 (1/22), 'Rev. Dr. Bartrum Recollections', pp. 6–7.

138. Pembroke College, Oxford: John Mitchinson Papers, PMB/S/15/8, J. Mitchinson to F. Jeune, 23 May 1863.

139. See, for example, the comments of Price's colleague, R. B. Clifton, Oxford's late-nineteenth-century Professor of Experimental Philosophy in his obituary of Price: *Proceedings of the Royal Society of London* 75 (1905), 30–4 at p. 30.

140. For an overview of the relationship between Price and Dodgson, see Edward Wakeling, 'Lewis Carroll and the Bat', *Antiquarian Book Monthly Review*, IX(7), issue 99 (July 1982), 252–9. See also the references to Price in Alexander L. Taylor, *The White Knight: A Study of C. L. Dodgson (Lewis Carroll)*, PhD thesis, University of Glasgow, 1952.

141. Edward Wakeling (ed.), *Lewis Carroll's Diaries*, 10 vols (Lewis Carroll Society, 1993–2007).

142. An account of the reading party can be found in a letter that Dodgson wrote to his sister from Whitby: see Morton N. Cohen (ed.), *The Letters of Lewis Carroll*, vol. 1: *ca. 1836–1885* (London: Macmillan, 1979), pp. 26–9. Although surviving references to them are extremely rare, it seems that Price may have organized such 'reading parties' on a regular basis: Dodgson confided to his diary in December 1856 that during the following Long Vacation, he hoped 'to take lodging wherever Price has his reading party and so get occasional help from him' with ongoing mathematical reading (Wakeling, *Lewis Carroll's Diaries*, vol. 2, p. 129).

143. Thomas Fowler, by then President of Corpus Christi College, quoted in 'Our Lewis Carroll Memorial', *The St. James's Gazette*, 36(5514) (Friday 11 March 1898), 7.

144. Dodgson's diary entry of Friday 16 March 1855, for example, reads: 'Walked with Price in the morning—a long and pleasant walk—discussed my new position etc.' (Wakeling, *Lewis Carroll's Diaries*, vol. 1, p. 91). Another instance in which Dodgson consulted Price is interesting because of the way it reflects on mathematical standards in Oxford: in May 1857, Dodgson was asked for his advice on whether a mathematically gifted young man should go to Oxford or Cambridge. Dodgson suggested Cambridge 'if his powers are really *high* in Mathematics' (ibid., vol. 3, p. 58). However, he also consulted Price who had the opposite opinion: 'He strongly advises Oxford, as he says he [the student] would be sure to succeed here, and might very possibly be outstripped in Cambridge' (ibid., vol. 3, p. 60). Dodgson also noted in his diary that '[a] higher consideration is the more liberal education which Oxford gives' (ibid.), but it is not clear whether this is his thought or Price's.

145. Two volumes of Lewis Carroll's games and puzzles that were published in the 1990s draw upon letters and solutions exchanged by Dodgson and Price that were then held by the Price family: Edward Wakeling, *Lewis Carroll's Games and Puzzles* (New York: Dover, 1992); Edward Wakeling, *Rediscovered Lewis Carroll Puzzles* (New York: Dover, 1995). A particularly contentious problem was the monkey and weight puzzle (Edward Wakeling, 'Recreational Mathematics', in *The Mathematical World of Charles L. Dodgson (Lewis Carroll)*, ed. Robin Wilson and Amirouche Moktefi (Oxford: Oxford University Press, 2019), pp. 141–76 at p. 153). Dodgson's stutter is acknowledged in many sources, and has been the subject of scholarly discussion (for example: J. De Keyser, 'The Stuttering of Lewis Carroll', in *Neurolinguistic Approaches to Stuttering: Proceedings of the International Symposium on Stuttering (Brussels, 1972)*, ed. Yvan Lebrun and Richard Hoops, The Hague; (Paris: Mouton, 1973), 32–6). Price's impediment is noted in Mitchinson's memoirs as having been triggered in particular by words beginning with an aspirate: Mitchinson recalled that when required to recite the *Te Deum* in the college chapel, Price would deliberately omit the line 'Holy, Holy, Holy' (Pembroke College, Oxford: John Mitchinson Papers, PMB/S/15/85, 'Oxford Memories', p. 13). Price's stutter is also alluded to in the obituary written by E. B. Elliott (*Proceedings of the London Mathematical Society* 30 (1898), 333), who also provided the following physical description of Price: 'somewhat above middle height, of spare frame and homely features, with an expression watchful of all but unfriendly to none.'

146. The line 'dined with the Prices' and variants thereof appears frequently in Dodgson's diaries. Photographs of several of Price's children are included in Edward Wakeling, *The Photographs of Lewis Carroll: A Catalogue Raisonné* (Austin: University of Texas Press, 2015).

147. Dodgson was elected a member of the Ashmolean Society in February 1859, probably having been proposed by Price, and went on to deliver several lectures to the society (see, for example, Taylor, *The White Knight*, pp. 50–1; Robin Wilson, *Lewis Carroll in Numberland: His Fantastical Mathematical Logical Life* (London: Penguin, 2009), pp. 34–5). In May 1866, Price communicated Dodgson's paper 'Condensation of determinants, being a new and brief method for computing their arithmetical values' to the Royal Society (it was published in volume 15 of the society's *Proceedings* in 1867; on the paper, see Francine Abeles, 'Determinants and Linear Systems: Charles L. Dodgson's view', *British Journal for the History of Science* 19(3) (1986), 331–5). In sending Dodgson's paper to G. G. Stokes, Secretary of the Royal Society, Price commented that it was a paper by 'an old pupil and friend [...] who is a very clever fellow' (University Library, Cambridge: Stokes Collection, Add MS 7656, P657, B. Price to G. G. Stokes, 14 May 1866).

148. Wakeling, 'Lewis Carroll and the Bat', p. 252.

149. In the main text, we will quietly sidestep the distinction in Oxford terminology between an 'examiner' and a 'moderator'. The term 'examiner' is reserved for examiners of the final honour school examinations, whereas a 'moderator' examines the 'moderations' exams that take place earlier in a candidate's degree, as a hurdle to overcome on the path towards finals. At different times, Price served both as an examiner and as a moderator (*Historical register 1900*, p. 814).

150. M. C. Curthoys, 'The Examination System', in *History of the University of Oxford*, vol. VI, pp. 339–74.

151. Curthoys ('Examination system', p. 352) estimates that prior to 1850, only 10–15% of candidates sat for honours in mathematics.

152. *The Times*, issue 22236 (Thursday 13 December 1855), 12d; *A letter from the Sedleian Professor of Natural Philosophy, to a candidate disappointed in the late mathematical examination* (Oxford: Baxter, 1855). The letter was also reproduced in part in *Jackson's Oxford Journal*, no. 5355 (Saturday 15 December 1855), 5a.

153. The class list does not appear with Price's letter in *The Times*, but it is printed next to the extracts from the letter in *Jackson's Oxford Journal*. The disappointed candidate, whom Price addressed simply as 'My dear H–', must have been the student who appears in the list as James P. Hicks of Lincoln College, one of the third-class candidates. This in turn is probably a misprint for John P. Hicks (1833–95), who had matriculated at Pembroke in 1851 before migrating to a scholarship at Lincoln College the following year (*Alumni Oxonienses*). He went on to a career as a barrister (Joseph Foster, *Men-at-the-Bar: A biographical hand-list of the members of the various Inns of Court, including Her Majesty's judges, etc.*, 2nd ed. (London, Aylesbury: Hazell, Watson, and Viney, 1885)).

154. William Spottiswoode, *Elementary theorems relating to determinants* (London: Longman, Brown, Green, and Longman, 1851); William Spottiswoode, *Meditationes analyticae*, 4 parts (London, 1847).

155. A. J. Crilly, 'Spottiswoode, William (1825–1883)', *Oxford Dictionary of National Biography*, https://doi.org/10.1093/ref:odnb/26171.

156. *Alumni Oxonienses*; W. J. Lewis, *Notes on the history of the parish of North Wraxhall, co. Wilts: with a life of the late rector Francis Harrison, M.A. at one time Fellow, Dean and Tutor of Oriel College, Oxford* (London: Society for Promoting Christian Knowledge, 1913).

157. *Alumni Oxonienses*.

158. *The Times*, no. 22271 (Wednesday 23 January 1856), 11a; also reproduced in *Jackson's Oxford Journal*, no. 5361 (Saturday 26 January 1856), 4e-f.

159. *Correspondence on the subject of the late second public examination in the mathematical schools* (Oxford: J. H. and J. Parker, 1856). Price's letter to Ashpitel of 1 January 1856 was eventually also published in *Jackson's Oxford Journal*, no. 5362 (Saturday 2nd February 1856), 4f.

160. *Correspondence on the subject of the late second public examination*, p. 7.

161. Ibid., p. 8.

162. Ibid.

163. Ibid., p. 10.

164. Ibid., pp. 9–10.

165. Ibid., p. 10.

166. Ibid.

167. Ibid., p. 21.

168. A date-stamp on a British Museum copy of the pamphlet, available on Google Books, gives the date of accession as 20 March 1856.

169. *The Times*, no. 22273 (Friday 25 January 1856), 10c; also reproduced alongside the printing of Ashpitel's letter in *Jackson's Oxford Journal* (see note 159).

170. At this time, for example, competitive examinations were just starting to be used for entry into the civil service (Richard Willis, *Testing times: A History of Vocational, Civil Service and Secondary Examinations in England since 1850* (Leiden; Boston: Brill, 2013)).

171. *A reply to the second letter of the Sedleian professor on the subject of the recent examination in the final mathematical schools* (Oxford: J. Vincent, 1856).

172. *A rejoinder to the reply of the Rev. F. Ashpitel, M.A. Brasenose College: one of the examiners, on the subject of the recent examination in the final mathematical schools* (Oxford: John Henry and James Parker, 1856).

173. Ibid., p. 5.

174. Ibid., p. 8.

175. Ibid., p. 3.

176. Ibid., pp. 17–8.

177. A. V. Simcock, 'Walker, Robert (1801–1865)', *Oxford Dictionary of National Biography*, https://doi.org/10.1093/ref:odnb/38098. Some slight mystery surrounds Harrison's departure as an examiner. The term of office of an examiner was usually two years, so it is not clear why Harrison's had expired. A biographer of Harrison later noted that he never spoke of what had happened, but that there was a general feeling within the university that he had been treated harshly (Lewis, *Notes on the history of the parish of North Wraxhall*, pp. 3–4). Unfortunately, these few vague words are all that we know of the matter. Although Harrison remained active within the university, and has received credit for his role in promoting the study of mathematics (ibid., p. 4), it was not until 1867 that he again took on examining duties for finals (though he did serve as a moderator (see note 149) on several occasions before then: *Historical register 1900*, p. 710).

178. He did, however, serve as a moderator (see note 149) in the mid-1860s (*Historical register 1900*, p. 588).

179. General Register Office, London: England and Wales Civil Registration Indices, Q3 1857, vol. 5b, p. 79 (St Thomas, Devon). Several letters from Spottiswoode to Price survive in the Price papers at Pembroke (PMB/S/14); Spottiswoode was one of the few correspondents to address Price as 'Dear Bat'.

180. The society existed until 1901, when it merged with the Ashmolean Natural History Society; see Elizabeth Megan Price, *Town and Gown: Amateurs and Academics: The Discovery of British Prehistory, Oxford 1850–1900: A Pastime Professionalised*, DPhil thesis, University of Oxford, 2007. Thanks to the energetic influence of William Buckland (1784–1856), geology was a dominant subject within the activities of the society (N. A. Rupke, 'Oxford's Scientific Awakening and the Role of Geology', in *History of the University of Oxford*, vol. VI, pp. 543–62).

181. *Proceedings of the Ashmolean Society*, no. XXI (1844), 35, 73. Price was treasurer at least as early as 1863 (Bodleian Library, Oxford: MS Dep. c. 657, MSS. of papers read at, and proceedings of, meetings of the Ashmolean Society, ff. 159–60).

182. *Proceedings of the Ashmolean Society*, no. XXIII (1846), 143–5.

183. B. Price, 'On the principle of virtual velocities', in William Spottiswoode, *Meditationes analyticae, part II* (London, 1847). Material on virtual velocities can also be found in the third volume of Price's *Treatise on infinitesimal calculus*.

184. *Proceedings of the Ashmolean Society*, no. XXIII (1846), 145; *Proceedings of the Ashmolean Society*, no. XXIV (1847), 152–3.

185. *Proceedings of the Ashmolean Society*, no. XXIV (1847), 177–9.

186. Bodleian Library, Oxford: MS Dep. c. 657, MSS. of papers read at, and proceedings of, meetings of the Ashmolean Society, f. 183. Price's new-found interest in probabilities also appears in his correspondence around this time: in October 1865, for example, he asked John Herschel whether he had any interest in the subject (Royal Society, London: Herschel Papers, HS/14.57, B. Price to J. F. W. Herschel, 28 October 1865). One wonders whether this interest could originally have been sparked by a mathematical query that Price had received in the mid-1850s: following recent comments by John Henry Newman on 'the carelessness of our baptisms', the unidentified correspondent had asked whether Price could calculate the probabilities of priests and bishops being among the unbaptized (Pembroke College, Oxford: Bartholomew Price Papers, PMB/S/14/7, c.1855).

187. *Proceedings of the Ashmolean Society*, no. XXVI (1849), 227–8. The paper was subsequently printed as *An essay on the relation of the several parts of a mathematical science to the fundamental idea therein contained; the substance of which was read before the Ashmolean Society on the evening of May 14, 1849* (Oxford: Ashmolean Society, 1849).

188. Price's paper probably prompted a paper that Powell presented to the society later that year, which was subsequently printed as *On necessary and contingent truth: considered in regard to some primary principles of mathematical and mechanical science* (Oxford: Ashmolean Society, 1849).

189. *Proceedings of the Ashmolean Society*, no. XXVI (1849), 228.

190. Price appears as Vice President in 1856, and became a member of council in 1859. His committee memberships (as recorded in the association's annual reports) cover topics including terrestrial magnetism (1862), tidal observations (1870), the improvement of geometrical teaching (1873), and the tabulation of mathematical functions (1889).

191. *Report of the seventeenth meeting of the British Association for the Advancement of Science; held at Oxford in June 1847* (London: John Murray, 1848), notices and abstracts, p. 5.

192. *Report of the thirty-first meeting of the British Association for the Advancement of Science; held at Manchester in September 1861* (London: John Murray, 1862), notices and abstracts, pp. 6–9.

193. *Report of the thirty-fifth meeting of the British Association for the Advancement of Science; held at Birmingham in September 1865* (London: John Murray, 1866), notices and abstracts, p. 7. Only the titles of these two lectures survive, but the one on Taylor's Theorem is likely to be what sparked J. J. Sylvester's 1865 communication 'on Professor Price's modification of Arbogast's method' (ibid., p. 9).

194. *Report of the thirtieth meeting of the British Association for the Advancement of Science; held at Oxford in June and July 1860* (London: John Murray, 1861), notices and abstracts, pp. 1–3. A similar lecture on a similar theme would later praise Price for the brevity of his address: H. H. Turner, 'The characteristics of the observational sciences', *Science, N.S.*, 34(872) (15 September 1911), 321–38 at p. 322.

195. Price alluded in this context to 'a complaint made by an eminent philosopher on the decay of mathematical knowledge in Great Britain, and especially in that of physico-mathematical knowledge' (*Report of the thirtieth meeting of the British Association for the Advancement of Science*, p. 3). This

is almost certainly a reference to Charles Babbage's *Reflections on the decline of science in England, and on some of its causes* (London: B. Fellowes, 1830).

196. Price usually confined himself to political issues concerning the universities. A rare example of him stepping into wider matters is his signing of a letter condemning the treatment of Jews in Russia ('Oxford University and the persecution of the Jews in Russia', *The Times*, no. 30434 (Saturday 18 February 1882), 4a–b).

197. Examples of Price's high-profile contacts include the creator of the penny post, Sir Rowland Hill (1795–1879), who arranged for Price to be shown the Stamp Office machinery (Pembroke College, Oxford: Bartholomew Price Papers, PMB/S/14/25, Sir R. Hill to E. Hill, 21 July 1886, letter of introduction for B. Price), and the sometime Governor of the Bank of England, Hucks Gibbs, 1st Baron Aldenham (1819–1907) (Pembroke College, Oxford: Bartholomew Price Papers, PMB/S/14/150, Lord Aldenham to B. Price, 4 June 1897). Price's membership of the Athenaeum Club in London suggests that he knew how to move in the 'right' circles (Pembroke College, Oxford: Bartholomew Price Papers, PMB/S/14/64, Sir J. Hooker to B. Price, 19 April 1875).

198. Pembroke College, Oxford: Bartholomew Price Papers, PMB/S/14/6, W. E. Gladstone to B. Price, 25 February 1853.

199. Pembroke College, Oxford: Bartholomew Price Papers, PMB/S/14/48, W. E. Gladstone to B. Price, 20 December 1871. Price accepted the Commissionership on the understanding that the purpose of the Commission was *enquiry* only (British Library, London: Gladstone Papers, vol. CCCXLVII, Add MS 44432, ff. 300–1, B. Price to W. E. Gladstone, 22 December 1871). On the Cleveland Commission, see Christopher Harvie, 'From the Cleveland Commission to the Statutes of 1882', in *History of the University of Oxford*, vol. VII, pp. 67–96.

200. Letters from Price, often on Clarendon Press notepaper, are scattered throughout the Gladstone Papers in the British Library.

201. *Royal Commission on Scientific Instruction and the Advancement of Science*, vol. I: *first, supplementary, and second reports, with minutes of evidence and appendices* (London: Eyre and Spottiswoode, 1872), pp. 211–20. Price was also proposed by Gladstone's Opposition as a member of a further University Commission of 1877, but his name was rejected by Disraeli's Conservative government ('The Universities Bill', *Nature*, 16(392) (3 May 1877), 1–2). When Gladstone's Liberal Party returned to power in 1880, Price offered Gladstone his 'hearty congratulations', noting that '[m]any of us here felt under a weight for the last few years and now feel that we can breathe freely' (British Library, London: Gladstone Papers, vol. CCCLXXVIII, Add MS 44463, f. 203, B. Price to W. E. Gladstone, 26 April 1880).

202. Royal Society, London: Certificates of election and candidature for Fellowship of the Royal Society, EC/1852/10.

203. *Proceedings of the Royal Society of London* 75 (1905), 30–4 at p. 34.

204. Royal Society, London: Referees' reports on scientific papers submitted to the Royal Society for publication, vols. 3–7. Price was particularly kind to a penurious Northumbrian schoolteacher, William Shanks, whose computational papers he communicated to the Royal Society (they appeared in volumes 15 and 16 of the society's *Proceedings*), and for whom he sought a small grant from the society (Royal Society, London: Miscellaneous Correspondence received by the Royal Society on official business, MC/8/58).

205. Pembroke College, Oxford: Bartholomew Price Papers, PMB/S/14/22, Sir E. Sabine PRS to B. Price, 20 June 1865. News of the Visitorship was received gladly by the Astronomer Royal, Sir George Biddle Airy (Pembroke College, Oxford: Bartholomew Price Papers, PMB/S/14/23, Sir G. B. Airy to B. Price, 30 August 1865).

206. *Monthly Notices of the Royal Astronomical Society* 59(5) (February 1899), 228–9.

207. See note 88.

208. Pembroke College, Oxford: Bartholomew Price Papers, PMB/S/14/4, Dr Brennecke to B. Price, c.1851. For another enquiry addressed to Price, see note 186.

209. For example, in 1861 Sylvester asked for Price's help in solving a problem in statics (Pembroke College, Oxford: Bartholomew Price Papers, PMB/S/14/14, J. J. Sylvester to B. Price, 11 March 1861), and a few years later the engineer William Rankine (1820–72) wrote to Price regarding hydrodynamics (ibid., PMB/S/14/28, W. J. M. Rankine to B. Price, 8 August 1867). Charles Dodgson's diaries make reference to a variety of mathematical problems on which he consulted Price, and the prolific textbook writer Isaac Todhunter (1820–84) asked for Price's advice on the calculus of variations (ibid., PMB/S/14/19, I. Todhunter to B. Price, 6 October 1864), a subject on whose recent history, with reference to Price's treatise, he had recently written (I. Todhunter, *A history of the progress of the calculus of variations during the nineteenth century* (Cambridge: Macmillan, 1861)). He also provided Price with comments on volume IV of his *Infinitesimal calculus* which were subsequently incorporated into the new edition (St John's College, Cambridge: Papers of Isaac Todhunter, Notebook A9, ff. 118–23). Also regarding mathematical correspondence in the opposite direction, in a letter of January 1854, Price sought the help of G. G. Stokes with the solution of a differential equation (Cambridge University Library: Stokes Papers Add MS 7656, P650, B. Price to G. G. Stokes, 23 January 1854).

210. *Proceedings of the London Mathematical Society* 30 (1898), 333; *Nature* 26(663) (13 July 1882), 263); *Nature* 4(85) (15 June 1871), 129. On the early activities of the Oxford Mathematical Society (founded in 1888 by J. J. Sylvester), see the assorted papers in a scrapbook concerning university societies that is preserved in the Bodleian Library (G. A. Oxon b. 147), and the minute book kept by the society's first secretary E. B. Elliott (Magdalen College, Oxford: E. B. Elliott Papers, F34/A1/1).

211. Royal Society, London: Herschel Papers; Cambridge University Library: Stokes Papers Add MS 7656; Kelvin Papers Add MS 7342. Herschel, Stokes, and Thomson are also represented in the Price Papers at Pembroke (PMB/S/14/37, PMB/S/14/40, PMB/S/14/53, PMB/S/14/95, PMB/S/14/96, PMB/S/14/106, PMB/S/14/139). The connection between Price and Stokes is particularly striking: they were rough contemporaries, both applied mathematicians who were appointed to major chairs at around the same time (in Stokes's case, Cambridge's Lucasian Professorship of Mathematics, in 1849), both eventually rose to the Masterships of their respective Oxbridge colleges (coincidently, Pembroke in each case), and both were effective academic administrators, though Stokes was more active in research than Price. They carried out a warm correspondence over decades, visited each other at home and in college, and also frequently encountered one another in official settings, particularly at the Royal Society, of which Stokes was a long-standing Secretary. There were those in the British scientific establishment who viewed Stokes's heavy administrative workload for the Royal Society as a waste of his time and talents (Stokes was as incapable as Price was of treating his various roles as sinecures), and yet Price had a different view of academic administration. Upon Stokes's election as President of the Royal Society in November 1885, Price heartily congratulated his friend: 'no one can be so conversant with the business as you are' (Cambridge University Library: Stokes Papers Add MS 7656, RS1885, B. Price to G. G. Stokes, 17 November 1885). On Stokes, see Mark McCartney, Andrew Whitaker and A. S. Wood (eds), *George Gabriel Stokes: Life, Science and Faith* (Oxford: Oxford University Press, 2019).

212. Pembroke College, Oxford: PMB/S/14/9, PMB/S/14/16, PMB/S/14/24, PMB/S/14/36. Another correspondent worth mentioning in passing is the philosopher Francis William Newman, whom we encountered in Chapter 5 as a mathematically talented examinee of George Leigh Cooke. In the 1860s, Newman sought Price's advice on mathematical reading (PMB/S/14/32–5).

213. An example of Price's networking is in his offering to find college rooms for William Thomson for the 1847 British Association meeting even though they had never been in contact before: 'I hope you will excuse my addressing you this without an introduction' (Cambridge University Library: Kelvin Papers Add MS 7342, P136, B. Price to W. Thomson, 14 June 1847).

214. Pembroke College, Oxford: PMB/S/14/21, PMB/S/14/24. As well as supporting Boole's fellowship of the Royal Society, Price had also proposed him for an honorary DCL in Oxford in 1859 (Des MacHale, *George Boole: His Life and Work* (Dublin: Boole Press, 1985), p. 218).

215. Price's books also seem to have reached at least a small continental European readership, as a single surviving letter from a Swedish reader attests (Pembroke College, Oxford: Bartholomew Price Papers, PMB/S/14/46, A. D. Wackerbarth to B. Price, 31 May 1871).

216. Pembroke College, Oxford: Bartholomew Price Papers, PMB/S/14/109, C. Hermite to B. Price, 13 May 1890; PMB/S/14/112, C. Hermite to B. Price, 2 June 1891; Marta Menghini, 'The Role of Projective Geometry in Italian Education and Institutions at the End of the 19th Century', *International Journal for the History of Mathematics Education* 1(1) (2006), 35–55 at p. 49.

217. Pembroke College, Oxford: Bartholomew Price Papers, PMB/S/14/114, J. J. Sylvester to B. Price, 6 June 1891.

218. He was consulted, for example, on the organization of the university's physiological laboratory (Bodleian Library, Oxford: Papers collected by Hugh Sinclair, MS Eng. c. 8050, B. Price to J. B. Saunderson, 26 April 1893), whose earlier establishment he had supported ('The Physiological Laboratory and Oxford medical teaching', *Nature*, 31(801) (5 March 1885), 414).

219. Price appears to have delegated all his preaching at Gloucester to a deputy, 'mistrusting the carrying power of his thin high pitched voice, and the slight impediment in his utterance' (Pembroke College, Oxford: John Mitchinson Papers, PMB/S/15/85, 'Oxford Memories', p. 9; see also note 145). The Dean and Chapter benefitted instead from Price's business acumen in balancing the Cathedral accounts; Price lived in Gloucester during the Long Vacations (*Monthly Notices of the Royal Astronomical Society*, 59(5) (February 1899), 229). In 1903, a clock made by the Arts and Crafts designer Henry Wilson and dedicated to Price was installed in the North Transept of the Cathedral.

220. J. A. Stewart, rev. C. A. Creffield, 'Fowler, Thomas (1832–1904)', *Oxford Dictionary of National Biography*, https://doi.org/10.1093/ref:odnb/33228. Fowler was a former pupil of Price's, and had known him at least since the Whitby reading party that Charles Dodgson had also attended (see note 143); one author has referred to Fowler as 'Price's protégé' (Topham, 'Science, mathematics, and medicine', pp. 539–40).

221. Bodleian Library, Oxford: G. A. Oxon c. 284, 'Pembroke College Scrapbook', f. 2b.

222. Changes in college statutes driven by the reforms of the 1850s had led to the creation of the new category of 'honorary fellow' at Queen's, and Price was one of the first people to be elected to such a position, in June 1868 (The Queen's College, Oxford: College Register M 1862–1873, 11 June 1868; Price's letter of acceptance is recorded on 22 October 1868).

223. A ticket and a menu for the meal are preserved in the scrapbook cited in note 221.

224. Besides Fowler, the college heads were J. E. Sewell (New College) and H. F. Pelham (Trinity); two other signatories were future Provosts of Oriel (C. L. Shadwell and L. R. Phelps). The professors were W. Ince (Regius Professor of Divinity), W. Odling (Waynflete Professor of Chemistry), R. B. Clifton (Professor of Experimental Philosophy), and W. Esson (Savilian Professor of Geometry). Also on the list is T. H. Grose, Fellow of Queen's, whose name had been proposed in 1891 as an external candidate for the Mastership of Pembroke.

225. 'The approaching retirement of Professor Price at Oxford', *The Manchester Guardian* (Monday 27 June 1898), 5g.

226. Ibid.

227. General Register Office, London: England and Wales Civil Registration Indices, Q4 1898, vol. 3a, p. 481 (Oxford).

228. 'Funeral of the Master of Pembroke', *Jackson's Oxford Journal*, no. 7607 (Saturday 7 January 1899), 3e. Copies of the order of service for the funeral are preserved in the scrapbook cited in note 221, and also in the Pembroke College Archives (PMB/S/14/154).

229. Cambridge University Library: Stokes Papers Add MS 7656, P644, A. E. Price to G. G. Stokes, 18 January [1899].

230. On this point more generally, see Arthur Engel, 'Emerging Concepts of the Academic Profession at Oxford 1800–1854', in *The University in Society*, vol. I: *Oxford and Cambridge from the 14th to*

the *Early 19th Century*, ed. Lawrence Stone (Princeton, NJ: Princeton University Press, 1974), pp. 305–51.

231. Although we largely sidestep the point here, Price also had a part to play in the establishment of physics teaching in Oxford (Fox and Gooday, *Physics in Oxford 1839–1939*). One obituary described him as 'the best and surest friend of Natural Science' (*Oxford Magazine*, 17(9) (25 January 1899), 149). In some correspondence of April 1871, he was asked, along with H. J. S. Smith and W. Esson, to clarify the dividing line between physics and mixed mathematics, so that the teaching duties of a new Reader in Physics might properly be defined (Pembroke College, Oxford: Bartholomew Price Papers, PMB/S/14/44–5).

CHAPTER 7

Further reading

Margaret E Rayner, 'The 20th century' in *Oxford Figures*, ed. John Fauvel, Raymond Flood, and Robin Wilson (Oxford: Oxford University Press, 2013), pp. 303–323.

John Heard, *From Servant to Queen. A Journey through Victorian Mathematics* (Cambridge: Cambridge University Press, 2019).

June Barrow-Green, 'Cambridge Mathematicians' Responses to the First World War' in *The War of Guns and Mathematics*, ed. David Aubin and Catherine Goldstein (Providence: American Mathematical Society, 2014), pp. 59–124.

E.A. Milne, 'Augustus Edward Hough Love (1863–1940)', *Biographical Memoirs of Fellows of the Royal Society* 3 (9) (1941), pp. 466–482.

Notes and references

1. E. A. Milne, 'Augustus Edward Hough Love, 1863–1940', *Journal of the London Mathematical Society* 16 (1941), 69–80 at p. 70.
2. Beach's teaching was of the classical style with an emphasis on the quadrivium. He nurtured such a dislike of science that he resigned in 1889 when the first science master was appointed at the school. See S. Jones, 'School Was Forced to Find a New Home', *Black Country Bugle* (28 November 2018), p. 12. Students passing the Cambridge Mathematical Tripos were divided into three classes according to merit: Wranglers (first class), Senior Optimes (second class), Junior Optimes (third class).
3. Milne, 'Augustus Edward Hough Love', p. 70.
4. For a discussion of the development of the Mathematical Tripos and the role of coaching, see Andrew Warwick, *Masters of Theory: Cambridge and the Rise of Mathematical Physics* (Chicago: University of Chicago Press, 2003).
5. Between 1882 and 1902 Webb coached 100 students to top 10 places. For a detailed description of Webb's coaching style, see Warwick, *Masters of Theory*, pp. 247–52.
6. A. R. Forsyth, 'Old Tripos Days at Cambridge', *The Mathematical Gazette* 19 (1935), 162–79 at p. 175; Anon, 'Obituary. Robert Rumsey Webb', *Monthly Notices of the Royal Astronomical Society* 97 (1937), 283.
7. Arthur Berry made his career at Cambridge, pursuing an interest in economics as well as in mathematics. The third wrangler in 1885 was George Richmond who, like Berry, was a scholar at King's College, and who also made his career in Cambridge.
8. Since 1885 the Smith's Prizes have been judged on the basis of an essay rather than by examination. Although Love's winning essay of 1887 no longer appears to exist, its title, 'The small free vibrations of a thin elastic shell, and on the free and fixed vibrations of an elastic spherical shell containing a given mass of liquid', is closely related to titles of papers published by Love soon afterwards: A. E. H. Love, 'The free and forced vibrations of an elastic spherical shell containing a given mass of liquid', *Proceedings of the London Mathematical Society* 19 (1889), 170–207; A. E. H. Love, 'The small free

vibrations and deformation of a thin elastic shell', *Philosophical Transactions of the Royal Society of London. Series A* 179 (1889), 491–546. The latter paper generated some controversy between Love and Lord Rayleigh which is discussed in detail in C. R. Calladine, 'The Theory of Thin Shell Structures 1888-1988', *Proceedings of the Institution of Mechanical Engineers* 202 A3 (1988), 141–9. For a history of the Smith's Prizes, see J. E. Barrow-Green, '"A Corrective to the Spirit of too Exclusively Pure Mathematics": Robert Smith and his Prizes at Cambridge University', *Annals of Science* 56 (1999), 271–316.

9. G. H. Hardy, *A Mathematician's Apology* (Cambridge: Cambridge University Press, 1940), p. 147.

10. Godfrey's year, 1895, was recognized as being a particularly brilliant one, with Thomas Bromwich topping the list and John Hilton Grace and Edmund Taylor Whittaker being bracketed equal second.

11. J. Larmor, 'Augustus Edward Hough Love', *The Eagle* 52 (1941), 62–3 at p. 62.

12. C[harles] H[enry] T[hompson], 'Professor A. E. H. Love', *The Queen's College Record* II, 8 (1941), 11–12 at p. 11.

13. That is, eros (romantic love) or agape (God's divine love).

14. See *The Australian*, 3 December 1898, p. 1266.

15. Most of Love's lectures were given in the Electrical Laboratory, which was in the Townsend Building of the Clarendon Laboratory; see Milne, 'Augustus Edward Hough Love', p. 72.

16. Love taught the following courses at Oxford:

    1898–9 Gravitational attraction and potential theory

    1900–38 Advanced applied mathematics—analytical dynamics—analytical statics—attractions and electrostatics—differential and integral calculus—differential equations—dynamics—electricity and magnetism—Fourier series—geometrical optics—harmonic analysis—hydrodynamics—hydrostatics—introduction to mathematical physics—mechanics of deformable bodies—potential theory—relativity (1927)—spherical harmonics—tensor calculus (1931)—waves and sound—problem classes

    1939 Gravitational attraction and potential theory—electricity and magnetism—dynamics—hydrodynamics. Informal problem classes: electricity and magnetism; electrodynamics; hydrodynamics.

17. See Milne, 'Augustus Edward Hough Love', p. 72.

18. See Calladine, 'The theory of thin shell structures', p. 142.

19. See Thompson, 'Professor A. E. H. Love', p. 12.

20. A. E. H. Love, *Theoretical Mechanics. An Introductory Treatise on the Principles of Dynamics with Applications and Numerous Examples* (Cambridge: University Press, 1897).

21. G. T. Walker, 'Theoretical Mechanics', *The Mathematical Gazette* 1(13) (1898), 173–4 at p. 173. Walker, the senior wrangler in 1889, was a lecturer at Trinity College Cambridge.

22. S. Brodetsky, 'Statics, Dynamics, and Hydrodynamics', *Nature* 110 (1921), 243–4 at p. 244. The price of the book in 1921 is equivalent to approximately £55 in 2023.

23. A. E. H. Love, *Elements of the Differential and Integral Calculus* (Cambridge: University Press, 1909), p. v.

24. Anon, 'Elements of the Differential and Integral Calculus by A. E. H. Love', *The Mathematical Gazette* 5 (1910), 316–17 at p. 317.

25. See Milne, 'Augustus Edward Hough Love', pp. 78–80.

26. A. E. H. Love, 'On the collapse of boiler flues', *Proceedings of the London Mathematical Society* 24 (1893), 208–19.

27. A. E. H. Love, 'On recent English researches in vortex motion', *Mathematische Annalen* 30 (1887), 326–44.

28. A. E. H. Love, 'The free and forced vibrations of an elastic spherical shell containing a given mass of liquid', *Proceedings of the London Mathematical Society* 19 (1889), 170–207.

29. Between February and September 1887, Love exchanged several letters with Klein about the content of the article (Cod. Ms. F. Klein 10: 871–80, Manuscript Division, Niedersächsische Staats- und Universitätsbibliothek Göttingen). His paper heralded the wider dissemination on the Continent of British work on hydrodynamics. See R. Tobies and D. E. Rowe, *Korrespondenz Felix Klein—Adolf Mayer*, Teubner Archiv zur Mathematik 14 (Leipzig: Teubner, 1990), p. 161. Love retained an association with Klein throughout his career, being one of the sponsors of the 1912 portrait of Klein by the leading German impressionist Max Liebermann. See R. Tobies, *Felix Klein. Visions for Mathematics, Applications, and Education*, trans. V. A. Pakis (Basel: Birkhäuser, 2021), p. 616.

30. Letter from Love to Klein, 6 July 1887. Cod. Ms. F. Klein 10: 876, Manuscript Division, Niedersächsische Staats- und Universitätsbibliothek Göttingen.

31. A. E. H. Love, 'Hydrodynamik: Physikalische Grundlegung' (IV. 15), 'Hydrodynamik: Theoretische Ausführungen' (IV. 16), in *Encyklopädie der mathematischen Wissenschaften* (Leipzig, 1901).

32. See, for example, L. N. G. Filon, 'Mathematics of Elasticity', *Nature* 105 (1920), 511–2.

33. For a detailed discussion of the contents of *Elasticity*, see Milne, 'Augustus Edward Hough Love', pp. 73–6.

34. It was, for example, recommended reading for mathematics undergraduates at Edinburgh University.

35. A. E. H. Love, *A Treatise on the Mathematical Theory of Elasticity*, 2 vols (Cambridge: University Press, 1892–3), vol. 1, p. ix.

36. Greenhill was Professor of Mathematics at the Royal Military Academy, Woolwich.

37. A. G. Greenhill, 'Mathematical Elasticity', *Nature* 47 (1893), 529–30 at p. 529.

38. G. B. Mathews, 'A Standard Treatise on Elasticity', *Nature* 74 (1906), 74–5 at p. 75.

39. Lamb's *Hydrodynamics* of 1895 was first published as *A Treatise on the Motions of Fluids* in 1878.

40. E. B. Wilson, 'A Treatise on the Mathematical Theory of Elasticity', *Bulletin of the American Mathematical Society* 34 (1928), 242–3.

41. See Tobies, *Felix Klein*, p. 302.

42. Letter from Love to Klein, 4 February 1901. Cod. Ms. F. Klein 10: 881, Manuscript Division, Niedersächsische Staats- und Universitätsbibliothek Göttingen.

43. It was the proposal of a translation with the possibility of including corrections that prompted the production of the second edition. See Love's letter to Klein of 4 February 1901 (note 42).

44. A. E. H. Love, *Lehrbuch der Elastizität*, tr. A. Timpe (Leipzig; Berlin: B. G. Teubner, 1907), p. V.

45. It was said that *Elasticity* had been translated into 'several languages': see Milne, 'Augustus Edward Hough Love', p. 76. To date, the only translation other than German to come to light is Russian: *Математическая теория упругости*, Москва (1935).

46. Letter from Francis Simon to Sybrens Ruurds de Groot, 15 February 1953. Papers of Sir Francis (Franz) Eugen Simon, Royal Society Archives, FS/7/2/179. I am grateful to Christopher Hollings for drawing this letter to my attention.

47. A. E. H. Love, 'Elasticity', in *Encyclopaedia Britannica*, 11th ed. (Cambridge: Cambridge University Press, 1911). Love also wrote articles on 'Functions of real variables', 'Infinitesimal calculus', and 'Calculus of variations' for the same edition.

48. A. E. H. Love, 'Address to the Mathematical and Physical Section of the British Association for the Advancement of Science', *Report of the British Association for the Advancement of Science, Leicester 1907* (1908), 427–3; *Nature* 76 (1907), 327–32.

49. See A. E. H. Love, 'The Yielding of the Earth to Disturbing Forces', *Proceedings of the Royal Society of London. Series A* 82 (1909), 73–88, and A. E. H. Love, 'Discussion on "Earth Tides"', *Report of the British Association for the Advancement of Science, Winnipeg 1909* (1910), 408–9.

50. The Adams Prize is named after John Couch Adams. It was founded in 1848 to commemorate Adams' discovery of the planet Neptune two years earlier. A subject was set every two years and it

was open to graduates of Cambridge. Today the prize is awarded annually to UK-based researchers under the age of 40.

51. A. E. H. Love, *Some Problems of Geodynamics being an Essay to which the Adams Prize in the University of Cambridge was Adjudged in 1911* (Cambridge: Cambridge University Press, 1911).

52. Isostasy is the principle that the Earth's crust is floating on its mantle, rather like an iceberg floating on water.

53. E. B. Wilson, 'Some Problems of Geodynamics, being an Essay to which the Adams' Prize in the University of Cambridge was adjudged in 1911', *Bulletin of the American Mathematical Society* 20 (1914), 432–4 at p. 433.

54. Anon, 'Science. Our Library Table. Some Problems of Geodynamics by A. E. H. Love', *The Athenaeum* 4397 (1912), 133.

55. For Love's introduction of 'Love waves', see Love, *Some Problems of Geodynamics*, pp. 176–81; and for a discussion, see Milne, 'Augustus Edward Hough Love', pp. 76–7.

56. Lord Rayleigh, 'On waves propagated along the plane surface of an elastic solid', *Proceedings of the London Mathematical Society* 1 (1885), 4–11.

57. Sir William Bragg, 'Address of the President', *Proceedings of the Royal Society of London. Series B.* 124 (1937), 395–6 at p. 396.

58. A. E. H. Love, 'Biharmonic analysis, especially in a rectangle, and its applications to the theory of elasticity', *Journal of the London Mathematical Society* 3 (1928), 144–56; *Proceedings of the London Mathematical Society* (2) 29 (1929), 189–242. The biharmonic equation is a fourth-order partial differential equation related to the Laplace equation.

59. H. W. Richmond (ed.), *Textbook of Anti-Aircraft Gunnery*, vol. 1 (London: HMSO, 1924), pp. 233–4.

60. See J. E. Barrow-Green, 'Cambridge Mathematicians' Responses to the First World War', in *The War of Guns and Mathematics*, ed. D. Aubin and C. Goldstein (Providence, RI: American Mathematical Society, 2014), pp. 59–124 at pp. 89–97.

61. A. E. H. Love and F. B. Pidduck, 'Lagrange's ballistic problem', *Philosophical Transactions of the Royal Society of London. Series A* 222 (1922), 167–226. Pidduck, an Oxford graduate in physics, was a research fellow at Queen's College. He had spent the war at Woolwich Arsenal working on ballistics.

62. R. V. Jones, 'Lindemann beyond the Laboratory', *Notes and Records of the Royal Society of London* 41 (1987), 191–210 at pp. 193–4.

63. See Milne, 'Augustus Edward Hough Love', p. 72.

64. See note 47.

65. A. E. H. Love, 'Boussinesq's problem for a rigid cone', *Quarterly Journal of Mathematics* 10 (1939), 161–75.

66. See A. N. Sneddon, 'Boussinesq's problem for a rigid cone', *Mathematical Proceedings of the Cambridge Philosophical Society* 44 (1948), 492–507 at pp. 492–3.

67. R. V. Southwell, Referee's Report on the paper by A. E. H. Love 'The Stress Produced in a Semi-Infinite Solid by Pressure on Part of the Boundary', [June 1929]. Royal Society Archives, RR/39/76. A. E. H. Love, 'The Stress Produced in a Semi-Infinite Solid by Pressure on Part of the Boundary', *Philosophical Transactions of the Royal Society of London. Series A* 228 (1929), 377–420.

68. Letter from Love to Larmor, 24 September 1902. GB 275 (Misc)/LO2/18, St John's College Library, Cambridge. Larmor had been senior wrangler in 1880, and the two had become friends after Larmor returned to St John's in 1885, the year in which Love was second wrangler. At the date of the letter, Larmor was a fellow and mathematical lecturer at St John's College and Secretary of the Royal Society. He would be elected to the Lucasian Chair in Cambridge the following year.

69. The meeting in Belfast was the annual meeting of the British Association for the Advancement of Science. Forsyth was Chair of the Teaching of Elementary Mathematics Committee of which Love and Larmor were both members.

70. See Milne, 'Augustus Edward Hough Love', p. 72.

71. *Proceedings of the Fifth International Congress of Mathematicians*, 2 vols (Cambridge: Cambridge University Press, 1913), vol. 1, p. 43. Unfortunately, no archives exist for the congress, so it is impossible to ascertain the extent of the tasks of the Secretaries. The two volume *Proceedings* can be found online at https://www.mathunion.org/icm/proceedings.

72. See A. E. H. Love, 'The application of the method of W. Ritz to the theory of the tides', in *Proceedings of the Fifth International Congress of Mathematicians*, vol. 2, pp. 202–8.

73. C. Runge, 'The mathematical training of the physicist in the university', in *Proceedings of the Fifth International Congress of Mathematicians*, vol. 2, pp. 598–607 at p. 605.

74. Anon, '*The* Fifth International Congress of Mathematicians', *Nature* 90 (1912), 4–6 at p. 6.

75. Letter from Love to Larmor, 30 May 1909. GB 275 (Misc)/LO2/44, St John's College Library, Cambridge. The paper under review was by Henry Ronald Hassé, then an assistant lecturer in Liverpool, and it had the title 'The equations of electrodynamics and the null influence of the earth's motion on optical and electrical phenomena'. See *Proceedings of the London Mathematical Society* 7 (1909), ix. The reviewers mentioned were Charles Niven, Hector Macdonald, Horace Lamb, Harry Bateman, and Ebenezer Cunningham.

76. A. E. H. Love, 'Mathematical research', *Proceedings of the London Mathematical Society* (2) 14 (1915), 178–88.

77. See ibid., 187–8.

78. Letter from Love to Larmor, 11 November 1903. GB 275 (Misc)/LO2/22, St John's College Library, Cambridge. The Cambridge Philosophical Society had proposed the project, but it never got off the ground. See. R. A. Sampson, 'On editing Newton', *Monthly Notices of the Royal Astronomical Society* 84 (1924), 378–83.

79. A. E. H. Love, Referee's Report on the paper by Karl Pearson, 'On the Kinetic Accumulation of Stress, illustrated by the Theory of Impulsive Torsion', 12 July 1900. Royal Society Archives, RR/15/97.

80. K. Pearson, 'On the kinetic accumulation of stress, illustrated by the theory of impulsive torsion', *Proceedings of the Royal Society of London* 67 (1901), 222–4.

81. A. E. H. Love, Referee's Report on the paper by G. H. Bryan and W. E. Williams 'The longitudinal stability of aerial gliders', 10 July 1903. Royal Society Archives, RR/16/27.

82. G. H. Bryan and W. E. Williams, 'The longitudinal stability of aerial gliders', *Proceedings of the Royal Society of London* 73 (1904), 100–16.

83. For a discussion of the Bryan–Williams paper, see T. J. M. Boyd, 'One Hundred Years of G. H. Bryan's *Stability in Aviation*', *Journal of Aeronautical History* Paper No. 4 (2011), 97–115 at pp. 104–6.

84. See Bragg, 'Address of the President', pp. 395–6. The Sylvester medal is named for James Joseph Sylvester and was first awarded in 1901 to Henri Poincaré.

85. A. E. H. Love, 'On the oscillations of a rotating liquid spheroid and the genesis of the Moon', *Report of the British Association for the Advancement of Science, Bath 1888* (1889), 562–3.

86. The correspondence consists of 67 letters from Love to Larmor, dating from 1899 to 1930. Much of it relates to Royal Society or London Mathematical Society business, as well as their own mathematical work.

87. Letter from Love to Larmor, 28 April 1899. GB 275 (Misc)/LO2/1, St John's College Library, Cambridge.

88. Letter from Love to Larmor, 4 October 1907. GB 275 (Misc)/LO2/35, St John's College Library, Cambridge.

89. See Milne, 'Augustus Edward Hough Love', p. 72.

90. See Calladine, 'The theory of thin shell structures', p. 142.

91. The election to the fellowship at Queen's College only became possible in 1927 due to a change in the statutes.
92. E. A. Milne, 'Augustus Edward Hough Love', *The Eagle* (St John's College, Cambridge) 52 (1941), 60–4 at p. 60.
93. The information about the sketch comes from Calladine, 'The theory of thin shell structures', p. 142. For information about Thornhill, see J. Dunning-Davies, 'Charles Kenneth Thornhill (1917–2007)', *Progress in Physics* 4 (2007), 115–16.
94. See, for example, Milne, 'Augustus Edward Hough Love' (*The Eagle*).
95. See Milne, 'Augustus Edward Hough Love' (*Journal of the London Mathematical Society*), p.78.

CHAPTER 8

Further reading

No formal biography of Chapman exists. His obituaries are:

T. G. Cowling, 'Sydney Chapman, 1888-1970', in *Biographical Memoirs of Fellows of the Royal Society*, vol. 17 (London: The Royal Society, 1971), pp. 53–89

T. G. Cowling and V. C. A. Ferraro, 'Obituary. Sydney Chapman', *QJRAS*, 13 (1972), 464–78 as well as Cowling's historical memoir

T. G. Cowling, 'Astronomer by Accident', *ARA&A*, 23 (1985), 1–18

The reader should assume these are the source of factual material in the text. In a series of short papers, Akasofu has provided important perspectives:

S. I. Akasofu, 'Chapman and Alfvén: A Rigorous Mathematical Physicist versus an Inspirational Experimental Physicist', *EOS* 84 (2003), 269–74

S. I. Akasofu, 'The Scientific Legacy of Sydney Chapman', *EOS*, 92 (2011), 281

S. I. Akasofu, 'Space Physics in the Earliest Days, as I Experienced It', *Perspectives of Earth and Space Scientists* (2019), AGU

S. I. Akasofu, 'A Biographical Sketch Based on the Book "Chapman Eighty, from his Friends"', *Perspectives of Earth and Space Scientists*, doi.org/10.1029/2020CN000135 (2020).

Southwood and de Moortel et al. give a concise summary of the Chapman/Birkeland/Alfvén disputes:

D. J. Southwood, 'From the Carrington Storm to the Dungey Magnetosphere', in *Magnetospheric Plasma Physics: The Impact of Jim Dungey's Research*, Astrophysics and Space Science Proceedings 41, ed. Stanley W. H. Cowley, David Southwood, and Simon Mitton (London: Springer, 2015), pp. 253–71.

D. J. Southwood, 'Kristian Birkeland: The Great Norwegian Scientist That Nobody Knows', The Birkeland Lecture, in *The Norwegian Academy of Sciences Yearbook* (Norwegian Academy of Sciences, 2017), pp. 255–70.

I. de Moortel, Isobel Falconer, and Robert Stack, 'Alfvén on Heating by Waves', *A&G*, 61(2) (2020), 2.34–9.

The AGU has published a number of retrospective articles, especially in C. S. Gillmor and J. R. Spreiter, *Discovery of the Magnetosphere* (AGU Publications, 1997) and articles in *Journal of Geophysical Research* 1 October 1994 and 1 May 1996, written by those participating in space science in the 1950s and 1960s.

Acknowledgements

I am grateful to Chris Hollings, Mark McCartney, Eric Priest, and David Southwood for reading and commenting on a preliminary version of this chapter, to David Southwood for tracking down the written version of Chapman's 1967 Birkeland symposium talk, Chris Hollings for obtaining the Oxford University records, Bernie Roberts for providing information about Peter Kendall, and the RAS librarian Sian Prosser who provided access to RAS Council minutes for 1940-4 at very short notice.

Notes and references

1. See Further reading
2. Cowling's obituary does not comment on how Chapman became Dr Chapman but the first use of D.Sc was on a paper published in the autumn of 1913, the year that Chapman won the prestigious Cambridge mathematics Smith's essay prize. J. Barrow-Green, "'A Corrective to the Spirit of too Exclusively Pure Mathematics": Robert Smith (1689–1768) and his Prizes at Cambridge University', *Annales of Science*, 56 (1999), 271–316 discusses the importance of this prize in Cambridge mathematical circles and one may assume that this had some impact on the award of his D.Sc. The PhD as known today did not exist in Cambridge in 1913.
3. His obituaries provides his complete publication list of over 400 papers covering a wide range of topics.
4. A plasma is an ionized gas where electromagnetic forces are dominant. Plasmas can have low ionization (e.g. the solar photosphere and the ionosphere) or be fully ionized (e.g. the solar corona, solar wind, and magnetosphere). A plasma can be described by the distribution functions of its components through the seven-dimensional (three space, three velocity space, and time) Boltzmann equation which is intractable without assumptions. Taking the first three moments of the Boltzmann equation leads to the equations of magnetohydrodynamics (MHD), which reflect the conservation of mass, momentum, and energy with closure through an equation of state: they are the equations of hydrodynamics extended to include the Lorentz force. The MHD equations are useful for describing large-scale properties but are analytically tractable only in highly symmetric situations, and in general their solution relies on computational modelling. Note that the formal framework of MHD was only established in the late 1930s and 1940s by, amongst others, T. G. Cowling and H. Alfvén.
5. Titles in parentheses such as (Sir) denote the award was made after the time being discussed.
6. Chapman had more success with geomagnetic observations at RGO than another distinguished scientist. Tom Gold writes that in 1956 after constructing a solar cosmic ray observatory, Gold was told by the Astronomer Royal ((Sir) Richard Woolley) to 'dismantle the building forthwith since such observations were not part of astronomy' (see note 12). This led to Gold's move to the USA.
7. The Chapman and Bartels volumes have a good description of this work. See note 9.
8. S. Chapman and V. C. A. Ferraro, 'A New Theory of Magnetic Storms', *Nature*, 126 (1930) 139–40.
9. S. Chapman and J. Bartels, *Geomagnetism, Vol 1&2*, (Oxford: Oxford University Press, 1940).
10. Space-based measurements arrived in 1958.
11. See Akasofu, 'Chapman and Alfvén', p. 274.
12. T. Gold, 'Early Times in the Understanding of the Earth's Magnetosphere', in *Discovery of the Magnetosphere*, ed. Gillmor and Spreiter (1997), pp. 77–82, at p. 78. The final sentence uses Chapman's words.
13. Akasofu, 'A Biographical Sketch'.
14. S. Chapman, 'Obituary Notices: Julius Bartels', *QJRAS* 6 (1965), 235–45. Bartels was at the Carnegie Institute of Terrestrial Magnetism at the time and in 1940 made his way home to Germany the long way: via Japan and the Trans-Siberian Railway.
15. Until 2014, it was customary for the president of the RAS to nominate their successor.
16. R. J. Tayler, *History of the Royal Astronomical Society*, vol. 2: *1920–1980* (Oxford: Blackwell, 1987).
17. The minutes of the RAS Council show that Chapman attended all council meetings while president, of order 7 per year. Despite the war, the council conducted its business as normal, with additional extensive discussions of firewatching and related matters. This includes the porter at the RAS apartments requesting a tin hat in April 1941, when the London Blitz had been going on for 7 months and was nearing its climax! Much of the RAS activity moved to Oxford at this time.

18. Though it should be noted that many involved in the war never talked about what they did. In his youth the author knew a number of such people.

19. For example, S. Chapman, 'The Kinetic Theory of a Gas Constituted of Spherically Symmetrical Molecules', *Phil. Trans. Roy. Soc. A* 211 (1912), 433–83; S. Chapman, 'On the Law of Distribution of Molecular Velocities, and on the Theory of Viscosity and Thermal Conduction, in a Non-uniform Simple Monatomic Gas', *Phil Trans. Roy. Soc A* 216 (1916), 279–348.

20. $K_n = \lambda_{mfp}/L$, where λ_{mfp} is a typical mean free path and L is the characteristic scale of the temperature, density, or velocity gradient.

21. Enskog's work was published as a Swedish dissertation in 1917 and subsequently in German journals. He was a school teacher for 13 years, before becoming a professor in Stockholm in 1930 (S. Chapman, 'Prof. David Enskog', *Nature*, 161 (1948), 193–4).

22. T. G. (Tom) Cowling (1906–90) collaborated extensively with Chapman in the 1930s on their classic textbook (see note 25). He studied for a PhD with E. A. Milne in Oxford and subsequently was taken on by Chapman at Imperial. He held subsequent posts at University College Swansea, University College Dundee (subsequently Queens College, Dundee, and part of the University of St Andrews until 1967), Manchester, and Bangor before finally settling as Professor of Applied Mathematics in Leeds. He was one of the founders of MHD and made major contributions to ionospheric conductivity, sunspots, stellar structure, and planetary dynamos. He is the author's 'academic grandfather'. His career is summarized in T. G. Cowling, 'Astronomer by Accident', *ARA&A*, 23 (1985), 1–18 and L. Mestel, 'T. G. Cowling, 17 June 1906–16 June 1990', in *Biographical Memoirs of Fellows of the Royal Society*, vol. 37 (London: The Royal Society, 1991), pp. 103–25.

23. T. G. Cowling, 'On the Radial Limitation of the Sun's Magnetic Field', *MNRAS*, 90 (1929), 140–54. Chapman continued to have a 'blind spot' about the existence of a large-scale magnetic field filling the heliosphere well into the 1950s.

24. Quoted from Cowling 'Astronomer by Accident', p. 7.

25. S. Chapman and T. G. Cowling, *The Mathematical Theory of Non-uniform Gases* (Cambridge: Cambridge University Press, 1970).

26. The book is not for the timid. One approach for those interested in just the results and the comparison of theory and experiment would be to look at Chapters 1–6 and then 12–14 in the 1970 edition. Although plasma transport is discussed in the final chapter, S. I. Braginskii, 'Transport Processes in a Plasma', *Reviews of Plasma Physics* 1 (1965), 205 is more satisfactory.

27. S. Chapman, 'Some Phenomena of the Upper Atmosphere', *Proc. Roy Soc A* 132 (1931), 353–74.

28. The Kew magnetic field measurements were reconstructed in 1938 by J. Bartels, 'Solar Eruptions and Their Ionospheric Effects—Classical Observation and Its New Interpretation', *Terr. Mag.* 42 (1937), 235–9, to which the reader is referred.

29. See H. S. Hudson, 'Carrington Events', *ARA&A* 59 (1971), 445–77.

30. The undisturbed magnetic field magnitude at Kew is of order 47,500 nT or 0.475 Gauss, so these perturbations are of order 1%.

31. The storm index, D_{st} (units nanotesla), is commonly used to measure the strength of magnetic storms. It is derived from multiple near-Equatorial measuring stations.

32. A terrella is a small, magnetized model of the Earth. It originated with William Gilbert in 1600 and was used extensively by Birkeland.

33. For historical background, see Southwood, 'Kristian Birkeland'.

34. S. Chapman, 'An Outline of a Theory of Magnetic Storms', *Proc. Roy. Soc. A* 95 (1918), 61–83.

35. F. A. Lindemann, 'A Note on the Theory of Magnetic Storms', *Phil Mag.* 38 (1919), 669–84. Frederick Lindemann (1886–1957) was created Baron (subsequently Viscount) Cherwell in 1941 and 1956, respectively. To avoid confusion, I refer to him as Lindemann.

36. Vincent Ferraro (1907–74) was the main collaborator with Chapman on his theory of magnetic storms. He began working at Imperial College in 1927 and eventually contributed enormously to

the theory: an excellent source for Ferraro's contribution is T. G. Cowling, 'Vincent Ferraro as a Pioneer of Hydromagnetics', *QJRAS* 16 (1975), 136–44. Moving onto what was then called hydro-magnetics (now MHD), he obtained what is now called Ferraro's law of isorotation in 1935. He moved to Kings College London, then to University College of the Southwest (now University of Exeter), and finally to Queen Mary College in 1952, and died in early 1974 after some years of illness.

37. S. Chapman and V. C. A. Ferraro, 'A New Theory of Magnetic Storms', *Nature* 126 (1930), 129–30.

38. S. Chapman and V. C. A. Ferraro, 'A New Theory of Magnetic Storms', *Terr. Mag.* 36 (1931a), 77–97; (1931b), 36, 171–86; (1932a), 37, 147–56; (1932b), 37, 421–9; (1933), 38, 79–96.

39. The thickness of the boundary is determined by the need for quasi-neutrality with modern models and simulations suggesting a thickness of a few 10s of kilometres. See, for example, P. J. Cargill and T. E. Eastman, 'The Structure of Tangential Discontinuities: Results of Hybrid Simulations', *JGR* 96 (1991), 13,763–79.

40. Julius Bartels (1899–1964) studied at Gottingen then taught meteorology at the Forestry High School in Eberswalde. From 1930 he was associated with the Carnegie Institute of Terrestrial Magnetism in Washington, DC, and in 1936 became director of Potsdam Geophysical Institute. He met Chapman in the 1920s and collaborated on the 1940 two-volume *Geomagnetism*. In 1945 he and his wife fled to Gottingen where after some inquiries about his relationship with the Nazis, he was appointed director of the Geophysical Institute and subsequently the Max Planck Institute for Aeronomy in Lindau. He made many important contributions to geomagnetic measurements. (Adapted from S. Chapman, 'Julius Bartels', *QJRAS* 6 (1965), 235–45.)

41. The two volumes are one book. The index, references, and many tables are all in volume 2. They were reprinted in 1951 and 1962 and can be found second hand.

42. Hannes Alfvén (1908–95) was the Nobel Physics laureate in 1970 for his fundamental work on plasma physics. He began his career in Uppsala, subsequently moving to Stockholm and the Royal Institute of Technology in 1940 (see R. S. Pease and S. Lundquist, 'Hannes Olof Gösta Alfvén, 30 May 1908–2 April 1995', in *Biographical Memoirs of Fellows of the Royal Society*, vol. 44 (London: The Royal Society, 1998), pp. 2–19, and C. G. Falthammar, 'The Scientific Legacy of Hannes Alfvén', *EOS* 93 (2012), 201). Along with Cowling, he was a pioneer of MHD, with his work on the 'frozen flux theorem' and the Alfvén wave (see de Moortel et al., 'Alfvén on Heating by Waves'). He subsequently spent two decades at University of California San Diego and proposed 'new paradigms' for plasma physics (Hannes Alfvén, *Cosmic Plasma* (Dordrecht: Reidel, 1981); S. Brush, 'Alfvén's Programme in Solar System Physics', *IEEE Trans Plasma Science* 20 (1992), 577).

43. See D. P. Stern, 'Forum', *EOS* 45 (2003), 488 and de Moortel et al., 'Alfvén on Heating by Waves').

44. A. J. Dessler and J. Wilcox, 'A Theory of Magnetic Storms and of the Aurorae', *EOS* 51 (1970), 180.

45. T. G. Cowling, 'On Alfvén's Theory of Magnetic Storms and of the Aurora', *Terrestrial Magnetism* 40 (1942), 209–14.

46. Quotation in Cowling, 'Astronomer by Accident', 11. Why Cowling agreed to write this article is unclear. Chapman was very productive during the early war years and could have done this himself. I assume Cowling was unaware of the history of Alfvén's submission to, and rejection by, *Terrestrial Magnetism*.

47. An entertaining episode is described in Southwood, 'From the Carrington Storm to the Dungey Magnetosphere'.

48. H. Alfvén, 'The Theory of Magnetic Storms and Auroras', *Nature* 167 (1951) 984.

49. D. F. Martyn, 'The Theory of Magnetic Storms and Aurora', *Nature* 167 (1951), 92–4.

50. For example, H. Alfvén, 'On the Theory of Magnetic Storms and Aurorae', *Tellus* 10 (1957), 104. This paper contains a generous acknowledgement of helpful discussions with Cowling. But any channel of communication with Chapman seems to have been abandoned.

51. See E. H. Vestine and S. Chapman, 'The Electric Current System of Geomagnetic Disturbances', *Terr. Mag*, 43 (1938), 351–82 and the volumes due to Chapman and Bartels.

52. A. J. Zmuda, J. H. Martin, and F. T. Heuring, 'Transverse Magnetic Disturbances at 1100 Kilometers in the Auroral Region', *JGR* 71 (1966), 5033–45.

53. Southwood, 'From the Carrington Storm to the Dungey Magnetosphere', and Southwood, 'Kristian Birkeland'.

54. Akasofu, 'Chapman and Alfvén'.

55. Cowling, 'Astronomer by Accident', 11 (emphasis added).

56. Alfvén, *Cosmic Plasma*, Reidel, 1981.

57. T. G. Cowling, A review of 'Cosmic Plasma', *GAFD*, 21 (1982), 324. The author also read this book in the early 1980s and was unimpressed. Re-reading it in 2022, his reaction is even more negative.

58. For example Southwood, 'From the Carrington Storm to the Dungey Magnetosphere', and Southwood, 'Kristian Birkeland'.

59. N. Fukushima, 'Some Topics and Historical Episodes in Geomagnetism and Aeronomy', *JGR* 99 (1994), 19,113–42. N. Fukushima, 'Unreasonable Discrimination of Birkeland's Current System in the History of Magnetic Storm Studies', *Proc. NIPR Symp. Upper Atmos. Phys.* 4 (1991) 108–15. The former is a proper discussion of Birkeland; the latter is not.

60. S. Borowitz, 'The Norwegian and the Englishman', *Phys. Perspectives* 10 (2008), 287–94. Although he studied the ionosphere in his early career, Borowitz had been an administrator for many decades. It is not clear what prompted his article.

61. This section makes use of Oxford University and Queen's College documents.

62. There is a memo written by Douglas Veale at the registry in 1933 following a meeting with F. A. Lindemann who argues for the possible appointment of a 'foreigner' at the next Sedleian vacancy. Lindemann named this person, but Veale did not give the name in his memo. Lindemann would have had in mind someone escaping from Nazi Germany and was also perhaps trying to nudge Love out. One can speculate that this may have been Erwin Schrödinger, whom Lindemann had met in Germany; Schrödinger was in Oxford in 1933 at the time his shared Nobel Prize (with Paul Dirac) was announced and held briefly a fellowship at Magdalen College. Nothing came of this and after a peripatetic few years Schrödinger ended up in Dublin, where the recently appointed Taoiseach Eamonn de Valera had founded the Dublin Institute for Advanced Study (DIAS). (Confidential note by Veale, 6 July 1933.)

63. One wonders why Milne had not put Chapman forward sooner. See earlier text.

64. The hypothesis was nullified by, amongst other things, measurements from a coalmine in Leigh, Lancashire, not far from Chapman's birthplace.

65. S. Chapman, 'Some Thoughts on Nomenclature', *Nature* 157 (1946), 405. S. Chapman, 'Upper Atmosphere Nomenclature', *JGR* 55 (1950), 395–9.

66. A situation familiar to present-day academics.

67. F. L. Korsmo, 'The Genesis of the International Geophysical Year', *Phys. Today* 60 (2007), 38.

68. Correlli Barnett, *The Lost Victory: British Dreams, British Realities 1945–1950* (London: Pan Books, 1995), and Correlli Barnett, *The Verdict of Peace: Britain between Her Yesterday and the Future* (New York: Macmillan, 2001).

69. HAO began in 1940 as an outgrowth of Harvard College Observatory with a site at 11,000 feet elevation at a molybdenum mine in Climax Colorado (see T. J. Bogdan, 'Donald Menzel and the Beginnings of HAO', *J. Hist. Astron.* 33 (2002), 157). It subsequently moved to Boulder, Colorado and became part of the National Center for Atmospheric Research (NCAR). The author spent 2½ enjoyable years as a young postdoc at HAO/NCAR in 1982–4.

70. See https://www.alaska.edu/uajourney/notable-people/fairbanks/sydney-chapman/

71. After wartime navy service including a major role in the development of the proximity fuse, Van Allen was at the Applied Physics Laboratory until 1950 when he moved to the University of Iowa where he remained. He was one of the great scientific figures in the second half of the twentieth century.

72. S. Chapman, 'The International Geophysical Year, 1957-1958', *Nature* 172 (1953), 327–9.

73. The oral interview and transcript can be found at https://kb.osu.edu/handle/1811/33918?show=full. I have quoted this at length because it conveys the sense of what Chapman accomplished with IGY.

74. Korsmo, 'The Genesis of the International Geophysical Year'.

75. From the 1998 interview (see note 73).

76. See Gillmor and Spreiter (eds), *Discovery of the Magnetosphere* (1997).

77. See, for example, J. A. Van Allen, 'Energetic Particles in the Earth's External Magnetic Field', in *Discovery of the Magnetosphere*, ed. Gillmor and Spreiter (1997), pp. 235–51 at p. 235.

78. Stephen E. Ambrose, *Eisenhower*, Vol. 2: *The President* (New York: Simon and Schuster, 1984).

79. S. Chapman, 'Notes on the Solar Corona and the Terrestrial Ionosphere', *Smithsonian Contribution to Astrophysics* 2 (1957), 1. The detection of highly ionized iron in the Sun's corona in the late 1930s had led to the realization that it was hot, of order 1 MK. Akasofu, 'The Scientific Legacy of Sydney Chapman', notes that this paper was rejected by the first journal to which it was submitted, hence the unusual place of publication.

80. E. N. Parker, 'Dynamics of the Interplanetary Gas and Magnetic Field', *ApJ* 128 (1958), 664–77.

81. E. N. Parker, 'Adventures with the Geomagnetic Field', in *Discovery of the Magnetosphere*, ed. Gillmor and Spreiter (1997), pp. 143–56.

82. T. Gold, 'Motions in the Magnetosphere of the Earth', *JGR* 64 (1959), 1219–24.

83. J. W. Dungey, 'Interplanetary Magnetic Fields and the Auroral Zone', *PRL* 6, (1961), 47–8.

84. J. W. Dungey, 'Memories, Maxims and Motives', *JGR*, 99 (1994), 19,189–97. S. W. H. Cowley, 'Dungey's Reconnection Model of the Earth's Magnetosphere: The First 40 Years', in *Magnetospheric Plasma Physics: The Impact of Jim Dungey's Research*, Astrophysics and Space Science Proceedings 41, ed. Cowley, Southwood, and Mitton (2015), pp. 1–32.

85. J. W. Dungey, 'Conditions for the Occurrence of Electrical Discharges in Astrophysical Systems', *Phil Mag* 44 (1953), 725–38. P. J. Cargill, 'Magnetic Reconnection in the Solar Corona: Historic Perspective and Modern Thinking', in *Magnetospheric Plasma Physics: The Impact of Jim Dungey's Research*, Astrophysics and Space Science Proceedings 41, ed. Cowley, Southwood, and Mitton (2015), pp. 221–51.

86. See Cowley, 'Dungey's Reconnection Model', p. 18 for the description of a particularly egregious effort of some scientists to denigrate his work.

87. R. H. Levy, H. E. Petschek, and G. L. Siscoe, 'Aerodynamic Aspects of the Magnetospheric Flow', *AIAA Journal* 2 (1964), 2065.

88. S. Chapman, 'The Extended Solar Corona', *IAU Symp.* 16 (1963), 235

89. S. I. Akasofu and S. Chapman, *Solar Terrestrial Physics* (Dordrecht: Reidel, 1972).

90. A. J. Dessler and E. N. Parker, 'Hydromagnetic Theory of Geomagnetic Storms', *JGR* 64 (1959), 2239–52.

91. S. Chapman and P. C. Kendall, 'Liquid Instability and Energy Transformation near a Magnetic Neutral Line: A Soluble Non-linear Hydromagnetic Problem', *Proc. Roy. Soc A* 271 (1963), 435–48.

92. Whether this was because he had no close collaborators after 1940 with the seniority to stand up to him and argue is unclear. However, in writing this chapter I have had some correspondence that suggests that there was rather too much unquestioning admiration from some.

CHAPTER 9

Further reading

The standard biographical introductions to Temple and Green are their respective Royal Society memoirs. Both contain a thorough engagement with their scientific contributions.

C. W. Kilmister, 'George Fredrick James Temple', *Biographical Memoirs of Fellows of the Royal Society of London* 40 (1994), 383–400. [*Kilmister*]

P. Chadwick, 'Albert Edward Green', *Biographical Memoirs of Fellows of the Royal Society of London* 47 (2001), 255–78. [*Chadwick*]

Acknowledgements

Thanks are due to the archivists of Queen's College, Oxford, Jesus College Cambridge, the University of Durham, and the University of California, Berkeley. I am grateful to the Abbott and monks of Quarr Abbey who welcomed an Irish Presbyterian into their midst for a few days in September 2021, and to Father Brian Kelly, Procurator of Quarr, who provided access to archive material relating to George Temple. Les Ruskell of the Farnborough Air Sciences Trust (FAST) library was extremely helpful in tracking down reports related to Temple's war work. Thanks to Alison Butler of Wesley Memorial Church, Oxford, who gathered together memories of Albert Green from members of the congregation. Finally, Peter Davies, Alan Day, John Kingman, Robin Knops, Ken Linsay, Sylvia Neumann, John and Hilary Ockendon, John Ringrose, Michael Sewell, and Dominic Welsh all generously shared, via phone call, email, or Skype there recollections of Temple and/or Green.

Notes and references

1. George Temple, 'Personal Record of Professor G Temple, FRS', p. 1, Quarr Abbey Archive [*Personal Record*].
2. George's paternal grandfather, John, is listed in the 1861 census as a 44-year-old shepherd living with his wife Ann, 44, and four children, the eldest of which is George's then 16-year-old father. 1861 Census of England, RG 9; Piece: 908; Folio: 121; Page: 43; GSU roll: 542,719. George's maternal grandfather, George, is listed in the 1861 census as a 53-year-old tailor and 'clerk of parish' living with his wife Elizabeth, 43, and three children, the youngest of which is George's then 4-year-old mother Fanny. 1861 Census of England, RG 9; Piece: 908; Folio: 117; Page: 35; GSU roll: 542,719.
3. 8 shillings and 9 pence would be worth approximately £35 in 2021.
4. The couple were married on 26 December 1896, with James listed as 'widowed'. England, Select Marriages, 1538–1973, item 9, p. 173, FHL file 1,278,891. James' first wife, Jane Ann, had died in in March 1896. England & Wales, Civil Registration Death Index, 1837–1915, vol. 1a, p. 117. They had married on the 19 May 1870. At that stage, when James was 25, he was listed as a signalman on the wedding certificate.
5. *Personal Record*, p. 4.
6. Ibid.
7. *Middlesex County Times* (Saturday 1 November 1913), 7. In an article entitled 'Opening of Ealing County School' it is noted that 'When the school commenced work in September, there were 111 boys on the roll'.
8. The scholarship granted George a free place at Ealing, and was announced in the *Ealing Gazette and West Middlesex Observer* on Saturday 19 July 1913, p. 6.
9. *Middlesex County Times* (Saturday 28 June 1913), 6.
10. *Middlesex County Times* (Saturday 1 August 1914), 7.
11. *Middlesex County Times* (Saturday 31 July 1915), 2 and (Saturday 5 August 1916), 7.
12. *Middlesex County Times* (Saturday 24 July 1915), 6. The article goes on to state that the total value of the grant was about £66, or £12 and 10 shillings per year.

13. Distinctions were in English, mathematics, advanced mathematics, heat, light and sound, chemistry, and French, with a special credit in oral French. *Middlesex County Times* (Saturday 28 July 1917), 7.

14. *The Kensington Post* (Friday 3 February 1933), 8, summarizes Goodall's career teaching at a number of schools before leaving Ealing County School in 1928 as Second Master to become Headmaster of Falmouth Grammar, and then to Sir George Monoux Grammar. He is listed as a B.Sc. and B.A. of London and B.Sc. of Reading, and as a 44-year-old bachelor. His popularity is noted in the *Middlesex County Times* (Saturday 8 August 1931), 8.

15. *Personal Record*, 7.

16. G. Temple, 'A generalisation of professor Whitehead's theory of relativity', *Proceedings of the Physical Society of London* **36** (1923), 192.

17. *Kilmister* states that the three papers concerned were 'A theory of relativity in which the dynamical manifold can be can formally presented upon the metrical manifold'. *Proc. Lon. Math. Soc.* **25** (1925) 414–16, 'On mass and energy', *Proc. Phys. Soc.* **37** (1925) 269–278, and Static and isotropic gravitational fields *Proc. Phys. Soc.* **37** (1926) 337–349. If Temple's recollection to Peter Neumann is correct this would imply that the work contained in the three papers was done over a relatively short period.

18. P. Neumann, 'Dom George Temple', *The Queen's College Record*, **6**(8) (1992), 12–16, at p. 13.

19. *Personal Record*, p. 10.

20. S. T. Shovelton, 'Prof. A.E. Jolliffe', *Nature*, 153 (1944), 488.

21. *Personal Record*, p. 10.

22. Ibid. Note Temple is not necessarily being sexist here as at this point Bedford was a women-only college.

23. L. G. Button, 'Harold Simpson (formerly Hilton)', *Bull. London Math. Soc.*, **8** (1976), 91–8, at p. 94.

24. G. Temple, 'Prof. J.G. Semple', *The Times* (Tuesday 5 November 1985), 16.

25. J. A. Tyrrell, 'John Greenlees Semple', *Bull. London Math. Soc.* **19** (1987), 378–86, at p. 378.

26. The fact that Temple cycled around the compound in a deerstalker is stated in a recorded interview with Father Gregory of Quarr Abbey on 11 September 2021.

27. *Personal Record*, p. 11.

28. Farnborough Air Sciences Museum Archives list 26 reports where Temple is either sole (17) or joint (9) author.

29. *Personal Record*, p. 49.

30. R. V. Jones, *Most Secret War: British Scientific Intelligence 1939–1945* (London: Hamish Hamilton, 1978), pp. 387–8.

31. *Kilmister*, p. 392. In assigning the work to Temple, Kilmister is also following R. V. Jones, as Temple, *Personal Record*, p. 11, which Kilmister almost certainly used, quotes an 1968 lecture by Jones where he makes the same point as he went on to make a decade later in *Most Secret War*.

32. The report which gives the 'back of an envelope' insight is the three-page RAE Technical Note 78, *Losses in Bomber Raids as affected by De-icing Equipment*, with the authorship listed only as The Staff of SME Dept. The report was probably authored in September 1942. The SME (Structures and Mechanical Engineering) Department was a large one, and it is reasonable to assume that only a small subgroup would have been involved in the writing of such a short, and relatively straightforward, document. The fact that Temple is on the distribution list for the report suggests he was not one of the authors. The more general, 19-page, report by Temple is RAE Technical Note SME 161, *Insurance Rates for Aircraft Accident Risks*, dated July 1943. Both reports are held by Farnborough Air Sciences Museum Archives under SME/TN/01 RAE SME TN 86 and SME/TN/02 RAE SME TN 161, respectively.

33. *Personal Record*, p. 14.

34. *Kilmister*, p. 393.

35. Neumann, 'Dom George Temple', p. 14.

36. J. F. C. Kingman, 'An Address Delivered on 9 May 1992 at the Memorial Service for George Temple in the Chapel of the Queen's College, Oxford', *The Queen's College Record*, **6**(8) (1992), 16–18, at p. 16.

37. *Personal Record*, p. 20.

38. Kingman, 'Address', p. 18.

39. S. L. Altmann and E. J. Bowen, 'Charles Alfred Coulson', *Biographical Memoirs of Fellows of the Royal Society of London* 20 (1974), 74–134, at p. 88.

40. *Personal Record*, p. 5.

41. Quarr Abbey Archive, 'Dom George Temple: Diary for 1944 used as a notebook' gives examples for Temple's writing. Comment on knowledge of Hebrew is contained in a recorded interview with Father Gregory of Quarr Abbey on 11 September 2021.

42. G. Temple, *The Classic and Romantic in Natural Philosophy: An Inaugural Lecture Delivered before the University of Oxford on 2 March 1954* (Oxford: Oxford University Press, 1954), p. 5.

43. Ibid., p. 8.

44. Ibid., pp. 8–9.

45. Ibid., p. 14.

46. G. Temple, 'The Growth of Mathematics: Presidential Address to the Mathematical Association', *The Mathematical Gazette* 41(337) (1957), 161–8, at p. 163.

47. G. Temple, 'Style and Subject in the Literature of Mathematics', in *Literature and Science: Proceedings of the Sixth Triennial Congress, Oxford, 1954* (Oxford: Blackwell, 1955), p. 16.

48. *Personal Record*, p. 34. The paper Temple is referring to is 'The Fundamental Paradox of the Quantum Theory', *Nature* 135 (1935), 957.

49. I use the word 'claiming' here as Temple's name does not appear as a member of the Inklings in standard studies of the group: Colin Duriez, *The Oxford Inklings* (Lion Hudson, 2015) or Humphrey Carpenter, *The Inklings* (London: Allen & Unwin, 1978). Irrespective of this Temple clearly knew Tolkien well enough for Tolkien in 1973 to give Temple a signed copy of the Allan & Ulwin *1974 J.R.R. Tolkien Calendar*. When Temple moved to Quarr he passed it onto the son of a mathematics colleague at Oxford, Dominic Welsh (<URL>https://www.tolkiencalendars.com/theCalendar.html</URL>, last accessed 7 September 2022).

50. *Personal Record*, p. 16.

51. Harry Lee Poe, *The Completion of C.S. Lewis* (Wheation, IL: Crossway, 2022), pp. 222–3.

52. Ibid., p. 202.

53. George Temple, 'St Philip's Begbroke, Thursday 25 July 1985', Draft of a tribute presumably to be read at Havard's funeral, Quarr Abbey Archive.

54. *Personal Record*, p. 15.

55. Interview with John and Hilary Ockendon, 13 August 2020. Hilary Ockendon described Temple as 'Very hands off as a supervisor . . . he was charming and absolutely delightful, but . . . if you weren't a very good student or if you were . . . a young and didn't know what you were doing student and you got assigned to Temple, most of them didn't survive. Not because he was nasty to them, but just because he didn't lead them.'

56. As part of his final farewell to Queen's College in 1982, he donated £2000 for the founding of the Temple Prize in Mathematics, which is still awarded to undergraduates of the college. The Queen's College, Oxford Governing Body Minute Book, 14.3.1981–12.10.1983, p. 160, 13 October 1982.

57. Private email from Father Brian Kelly of Quarr Abbey, 14 March 2022.

58. Interview with Father Gregory of Quarr Abbey on 11 September 2021.

59. G. Temple, 'Sets, Numbers and Taxa', *Expositiones Mathematicae* 2 (1984), 349–74, and 'Fundamental Mathematical Theories', *Philosophical Transactions A: Mathematical, Physical and Engineering Sciences* 354(1714) (1996), 1941–67.

60. Note entitled 'The Logistic Equation' for Father James Mitchell O.S.B., Quarr Abbey Archive. The material can be dated as some of it is written on the verso of pages advertising 1991 Royal Society of London events.
61. Temple's one publication in the area of theology is 'Conversation Piece at Cana', *Dominican Stud.* 7 (1954), 104–13. His one published foray into the boundaries between science and theology counterblasting the article showing a lack of understanding of basic mathematics is 'The Alleged Absurdity of Algebra', *New Blackfriars* 10(116) (1929), 1444–5.
62. G. Temple, 'Mathematics and Theology: A Sermon Preached in the Chapel of the College of St. John Baptist, Oxford, at Evensong on Sunday, 2 May 1971' and 'Mathematics and Theology' (undated), Quarr Abbey Archive,
63. Albert E Green's grandfather William J Green is listed in the 1881 census as a 'Labourer Linoleum Works' and in the 1901 census as simply a 'labourer'.
64. *Chadwick*, p. 257.
65. 1939 England and Wales Register, Enumeration District: AKCK, Metropolitan Borough of Hampstead, District 8/1, no.134.
66. Arthur Gray and Frederick Brittain, *A History of Jesus College Cambridge* (London: Heinemann, 1960), pp.197–8.
67. 'Obituary of Mr Arthur Gray', *The Times* (15 April 1940), 9.
68. Albert E Green, 1974 Timoshenko Medal Acceptance Speech, <URL>https://imechanica.org/node/179</URL>, (last accessed 9 September 2022).
69. D. R. Taunt, 'L. A. Pars', *Bulletin of the London Mathematical Society* 18 (1986), 505–6.
70. Green, 1974 Timoshenko Medal Acceptance Speech.
71. *The Times* (17 June 1932), 9. In Part I of the Tripos in 1932, 53 candidates were placed in the first class, 40 in the second class, and 28 in the third class, with a further 8 being allowed to have mathematics as a 'principal subject' for part of their BA ordinary degree.
72. Maurice V. Wilkes, *Memoirs of a Computer Pioneer* (Cambridge, MA: MIT Press, 1985), p. 18.
73. *The Times* (15 June 1934), 11.
74. George Batchelor, *The Life and Legacy of G.I.Taylor* (Cambridge, UK: Cambridge University Press (1996), pp. 219–20.
75. A. E. Green, 'The Equilibrium and Elastic Stability of a Thin Twisted Strip', *Proc. Roy. Soc. Lond. A* 154 (1936), 430–55.
76. Testimonial for A. E. Green by L. A. Pars, Durham University Records: Central Administration and Officers UND/CC1/V7.
77. Edgar Jones, *University College Durham: A Social History* (Merthyr Tydfil: Cambrian Printers, 1996), 255.
78. *Newcastle Evening Chronicle* (Tuesday 17 September 1940), 5. While Green is simply listed as 'Albert E Green. Field House Lane, Durham, registered without conditions', an unfortunate Mr Kenneth Goom, teacher at the Friend's School, Great Acton was berated by the judge for the Northern Area Conscientious Objectors' Tribunal with the words 'Would you like to teach geography with a map of the world marked German?'
79. Albert E Green, 1974 Timoshenko Medal Acceptance Speech https://imechanica.org/node/179 [Last accessed 21/4/22]
80. Letter from Green to Rivlin, 4 June 1958. Rivlin–Green correspondence, University of California, Berkeley, Special Collections BANC MSS 2011/248 (Carton 4, Folder 10).
81. Private email from R. Knops to the author, 20 July 2020.
82. The correspondence between Rivlin and Green, 155 letters written between 1956 and 1974, is held by the University of California, Berkeley, Special Collections BANC MSS 2011/248 (Carton 4, Folder 10).
83. *Chadwick*, p. 261.

84. Letter from Rivlin, 6 July 1961. Rivlin–Green correspondence, University of California, Berkeley, Special Collections BANC MSS 2011/248 (Carton 4, Folder 10).

85. *Chadwick*, p. 260.

86. W. B. Fisher, 'Honorary Degrees: Speeches of Presentation', *University of Durham Gazette* 16(2) (1969), 24.

87. John Ockendon stated that Naghdi visited every summer. Recorded interview with John and Hilary Ockendon, 13 August 2020. Green was also a regular visitor to Naghdi in Berkeley (*Chadwick*, p. 262).

88. *Chadwick*, p. 262.

89. Of the 83 papers published only 15 contained a third author.

90. *Chadwick*, p. 262, notes the three students as K. A. Lindsay, D. Nicol, and M. Troth. Lindsey remembers Green as 'an excellent supervisor' whom he met with every two weeks. Private email to the author from Ken Lindsey, 16 August 2020.

91. As a student Hilary Ockendon went to a 3rd-year undergraduate course on viscous flow and continuum models. She recalls 'good notes . . . which I used for years afterwards' from a lecturer who, if not inspiring in his delivery, was well organized. Recorded interview with John and Hilary Ockendon, 13 August 2020.

92. Private email from John Kingman, 5 July 2020.

93. Recorded interview with John and Hilary Ockendon, 13 August 2020.

94. Alan Day also worked in continuum mechanics and although Green and Day got on well together they did not collaborate. Day states that Green found it impossible to arrange for Naghdi to dine at Queen's, though it is not clear why this should have been problematic. Private email to the author from Alan Day, 8 August 2020.

95. Recorded interview with John and Hilary Ockendon, 13 August 2020.

96. Woods assessment of the disagreement between him and Green can be found in Leslie C. Woods, *Against the Tide: An Autobiographical Account of a Professional Outsider* (Bristol: IOP Publishing, 2000), pp. 259–63.

97. Recorded interview with John and Hilary Ockendon, 13 August 2020.

98. L. C. Woods, 'The Bogus Axioms of Continuum Mechanics', *Bulletin of the IMA* 17 (1981), 98–102, at p. 98.

99. A. E. Green, 'A Note on "Axioms of Continuum Mechanics"', *Bulletin of the IMA* 18 (1982), 7–9.

100. L. C. Woods, 'More on the Bogus Axioms of Continuum Mechanics', *Bulletin of the IMA* 18 (1982), 64–6, at p. 64.

101. Although the initial article by Woods was not explicitly targeted at Green, the view of John and Hilary Ockendon is that they were clearly 'broadsides against Albert' and that 'Les really liked to wind Albert up'. Recorded interview with John and Hilary Ockendon, 13 August 2020.

102. Green, 1974 Timoshenko Medal Acceptance Speech.

103. Hilary Ockendon felt that Leslie Woods 'was always slightly poking fun' at Green and that Green did 'get wound up by Les'. Recorded interview with John and Hilary Ockendon, 13 August 2020.

104. Green, 1974 Timoshenko Medal Acceptance Speech.

CHAPTER 10

Further reading

A comprehensive account of Brooke Benjamin's scientific work is to be found in the Royal Society *Biographical Memoir* (J. C. R. Hunt, 'Thomas Brooke Benjamin 15 April 1929–16 August 1995', in *Biographical Memoirs of Fellows of the Royal Society of London* 49 (London: Royal Society, 2003), pp. 39–67), the *Oxford Dictionary of National Biography* (M. S. Longuet-Higgins, 'Benjamin, (Thomas) Brooke

(1929–1995)', https://doi.org/10.1093/ref:odnb/60105), and an appreciation by Jerry Bona ('An Appreciation of T. Brooke Benjamin (1929–1995)', in *Mathematical Problems in the Theory of Water Waves: A Workshop on the Problems in the Theory of Nonlinear Hydrodynamic Waves, May 15–19, 1995, Luminy, France,* ed. F. Dias, J.-M. Ghidaglia, and J.-C. Saut, American Mathematical Society Contemporary Mathematics 200 (Providence, RI: American Mathematical Society, 1996), pp. xiii–xvi). His papers, including photographs, teaching and seminar notes, drafts of papers, unpublished manuscripts and calculations, committee papers, honours and awards, music, and poetry, are held by the Bodleian Library in Oxford (NCUACS 63/1/97), and his portrait, the black and white photograph taken in 1957 by Antony Barrington Brown that appears as our Figure 10.8, is in the archive of the National Portrait Gallery (NPG x104760).

Acknowledgements

In preparing this appreciation, the authors are indebted to Raymond Flood, Emeritus Fellow of Kellogg College, Oxford, and Jerry Bona, Professor at the University of Illinois at Chicago, for their interest and insightful advice. We are also grateful to Tamsin Reilly and the Mathematics support staff at the University of Bath Library for their invaluable help with sources. We are especially appreciative of the support of Brooke's widow, Natalia Benjamin (née Court).

Notes and references

1. https://www.livemozart.com/about-us
2. Sir Geoffrey (G. I.) Taylor OM, FRS was one of the great classical physicists of the twentieth century. The equipment in question was possibly a Perspex box which held liquid paraffin to help compensate for the optical distortion of the cylindrical surface of the apparatus. It was rectangular at the front and semi-circular at the back.
3. T. B. Benjamin and A. T. Ellis, 'The Collapse of Cavitation Bubbles and the Pressures Thereby Produced Against Solid Boundaries', *Philosophical Transactions of the Royal Society A* 260 (1966), 221–40.
4. J. K. Harvey, 'Some Observations of the Vortex Breakdown Phenomenon', *Journal of Fluid Mechanics* 14 (1962), 585–92.
5. J. Scott Russell, 'Report on Waves', in *Report of the Fourteenth Meeting of the British Association for the Advancement of Science, York, September 1844* (London: John Murray, 1845), pp. 311–90; O. Darrigol, 'The Spirited Horse, the Engineer, and the Mathematician: Water Waves in Nineteenth-Century Hydrodynamics', *Archive for History of Exact Sciences* 58 (2003), 21–95.
6. Sir George Gabriel Stokes, Cambridge Lucasian Professor of Mathematics 1849–1903; Sir George Biddell Airy, mathematician and Astronomer Royal from 1835 to 1881.
7. T. B. Benjamin, 'Instability of Periodic Wave Trains in Nonlinear Dispersive Systems', *Proceedings of the Royal Society A* 299 (1967), 59–75.
8. T. B. Benjamin and J. E. Feir, 'The Disintegration of Wave Trains on Deep Water. Part 1. Theory', *Journal of Fluid Mechanics* 27 (1967), 417–30.
9. A. I. Nekrasov, 'On Steady Waves', *Izvestiya Ivanovo-Voznesenskogo Politekhnicheskogo Instituta, part I*, 3 (1921), 52–65; ibid., part II, 6 (1922), 155–71 (both in Russian); T. Levi-Civita, 'Détermination Rigoureuse Desondes Permanentes d'ampleur Finie', *Mathematische Annalen* 93 (1925), 264–314.
10. A. Craik, 'George Gabriel Stokes on Water Wave Theory', *Annual Review of Fluid Mechanics* 37 (2005), 23–42 at p. 27. Craik's PhD was supervised by Brooke.
11. Russell, 'Report on Waves', pp. 332–3.
12. Yu. P. Krasovskii, 'On the Theory of Steady Waves of Finite Amplitude', *Zhurnal Vychislitelnoi Matematiki i Matematicheskoi Fiziki* 1 (1961), 836–55 (in Russian); English translation: *U.S.S.R. Computational Mathematics and Mathematical Physics* 1 (1962), 996–1018.

13. J. Leray and J. Schauder, 'Topologie et Équations Fonctionnelles', *Annales scientifiques de l'École Normale Supérieure. Troisième Série* 51 (1934), 45–78. A question raised by Leray in 1934 is one of the Clay Mathematics Institute Millennium Prize challenges for which there is a $1m prize for a successful solution: https://www.claymath.org/millennium-problems/ (J. Leray, 'Sur le Mouvement d'un Liquide Visqueux Emplissant l'Espace', *Acta Mathematica* 63(1) (1934), 193–248).

14. M. A. Krasnosel'skii, *Topological Methods in the Theory of Nonlinear Integral Equations* (Moscow: GITTL, 1956) (in Russian; English translation by A. H. Armstrong (Oxford: Pergamon Press, 1963)); M. A. Krasnosel'skii, *Positive Solutions of Operator Equations* (Moscow: Fizmatgiz, 1962) (in Russian; English translation by Richard E. Flaherty (Gronigen: P. Nordhoff, 1964)); O. A. Ladyzhenskaya, *Mathematical Problems in the Dynamics of a Viscous Incompressible Fluid* (Moscow: Fizmatgiz, 1961) (in Russian; English translation, 2nd ed., by Richard A. Silverman and John Chu, *The Mathematical Theory of Viscous Incompressible Flow* (New York; London; Paris: Gordon and Breach, 1969)); M. M. Vainberg, *Variational Methods for the Study of Nonlinear Operators* (Moscow: GITTL, 1956) (in Russian; English translation by Amiel Feinstein (San Fransisco: Holden-Day, 1964)).

15. Later Sir Michael Atiyah OM and President of the Royal Society, Fields Medal 1966, Copley Medal 1988, Abel Prize 2004.

16. T. B. Benjamin, 'The Alliance of Practical and Analytical Insights into the Nonlinear Problems of Fluid Mechanics', in *Applications of Methods of Functional Analysis to Problems in Mechanics: Joint Symposium IUTAM/IMU Held in Marseille, September 1–6, 1975*, Springer Lecture Notes in Mathematics 503 (Berlin: Springer-Verlag, 1976), pp. 8–29.

17. https://www.asap.unimelb.edu.au/bsparcs/aasmemoirs/mahony.htm

18. D. J. Korteweg and G. De Vries, 'On the Change of Form of Long Waves Advancing in a Rectangular Canal, and on a New Type of Long Stationary Waves', *Philosophical Magazine* 5(39) (1895), 422–43.

19. T. B. Benjamin, J. L. Bona, and J. J. Mahony, 'Model Equations for Long Waves in Nonlinear Dispersive Systems', *Philosophical Transactions of the Royal Society A* 272 (1972), 47–78.

20. J. L. Bona and R. Smith, 'The Initial-Value Problem for the Korteweg–de Vries Equation', *Philosophical Transactions of the Royal Society A* 278 (1975), 555–601.

21. T. B. Benjamin, 'The Stability of Solitary Waves', *Proceedings of the Royal Society A* 328 (1972), 153–83; J. L. Bona, 'On the Stability Theory of Solitary Waves', *Proceedings of the Royal Society A* 344 (1975), 363–74.

22. C. J. Amick and J. F. Toland, 'On Solitary Water Waves of Finite Amplitude', *Archive for Rational Mechanics and Analysis* 76(1) (1981), 9–95; C. J. Amick and J. F. Toland, 'On Periodic Water Waves and Their Convergence to Solitary Waves in the Long-Wave Limit', *Philosophical Transactions of the Royal Society A* 303 (1981), 633–69.

23. J. L. Bona, W. G. Pritchard, and L. R. Scott, 'A Comparison of Solutions of Two Model Equations for Long Waves', in *Fluid Dynamics in Astrophysics and Geophysics, Chicago, Ill., 1981*, Lectures in Applied Mathematics 20 (Providence, RI: American Mathematical Society, 1983), pp. 235–67.

24. T. B. Benjamin and J. C. Scott, 'Gravity-Capillary Waves with Edge Constraints', *Journal of Fluid Mechanics* 92(2) (1979), 241–67.

25. T. B. Benjamin and J. C. Scott, 'Waves in Narrow Channels: Faster Capillary Waves', *Nature* 276 (1978), 803–5.

26. T. B. Benjamin, 'Bifurcation Phenomena in Steady Flows of a Viscous Fluid. I. Theory', *Proceedings of the Royal Society A* 359 (1978), 1–26; 'II. Experiments', *Proceedings of the Royal Society A* 359 (1978), 27–43.

27. T. B. Benjamin and S. Bowman, 'Discontinuous Solutions of One-Dimensional Hamiltonian Systems', *Proceedings of the Royal Society A* 413 (1987), 263–95; T. B. Benjamin and P. J. Olver, 'Hamiltonian Structures Symmetries and Conservation Laws for Water Waves', *Journal of Fluid*

Mechanics 125 (1988), 137–85; T. B. Benjamin and T. J. Bridges, 'Reappraisal of the Kelvin–Helmholtz Problem. Part 1. Hamiltonian Structure', *Journal of Fluid Mechanics* 333 (1997), 301–25.

28. T. B. Benjamin, 'A New Type of Solitary Wave', *Journal of Fluid Mechanics* 245 (1992), 401–11.

29. T. B. Benjamin and T. Mullin, 'Anomalous Modes in the Taylor Experiment', *Proceedings of the Royal Society A* 377 (1981), 221–49.

30. T. B. Benjamin and A. D. Cocker, 'Liquid Drops Suspended by Soap Films. I. General Formulation and the Case of Axial Symmetry', *Proceedings of the Royal Society A* 394 (1984), 19–32; T. B. Benjamin and T. Mullin, 'Buckling Instabilities in Thin Layers of Viscous Fluid Subjected to Shearing', *Journal of Fluid Mechanics* 195 (1988), 523–40.

31. T. B. Benjamin and A. T. Ellis, 'Self Propulsion of Asymmetrically Vibrating Bubbles', *Journal of Fluid Mechanics* 212 (1990), 65–80.

32. T. B. Benjamin, 'High Profile Solution to Boffins Mystery', *Times Higher Education Supplement* 950 (18 January 1991), 17; T. B. Benjamin, 'The Right Sums', *Times Higher Education Supplement* 976 (19 July 1991), 14; T. B. Benjamin, 'Overladen with Honours: Overhauling Undergraduate Degrees Could be Made to Work in the Reform of Higher Education', *Times Higher Education Supplement* 1002 (17 January 1992), 18.

33. T. B. Benjamin, 'Public Perceptions of Higher Education', *Oxford Review of Education* 19(1) (March 1993), 47–63.

34. The W. G. Pritchard Fluids Lab, named after Bill Pritchard, Brooke's student whom he recruited to FMRI.

35. http://www.ncup.org.uk/

36. The Abdus Salam International Centre for Theoretical Physics: https://www.ictp.it/

37. https://www.academie-sciences.fr/en/Table/Membres/Liste-des-membres-depuis-la-creation-de-l-Academie-des-sciences/

38. The Bakerian Medal, one of the premier awards of the Royal Society, recognizes outstanding contributions to the physical science and the medallist is required to give a lecture. This prize lecture was established in 1775 when Henry Baker left £100 for a Fellow of the Royal Society to deliver a lecture on 'such part of natural history or experimental philosophy as the Society shall determine'.

39. Hopkins, a fellow of Peterhouse, Cambridge, was well known as a highly successful mathematics coach with the sobriquet 'senior-wrangler maker'.

40. Honorary Membership was first awarded in 1880, the founding year of the society, for lifetime service to science, research, and public service: https://www.asme.org/about-asme/honors-awards/achievement-awards/honorary-member

41. https://www.maths.ox.ac.uk/events/special-lectures/brooke-benjamin-lecture

42. https://www.siam.org/prizes-recognition/activity-group-prizes/detail/full-prize-specifications/siag-nwcs-t-brooke-benjamin-prize-in-nonlinear-waves

43. M. Berti, A. Maspero, and P. Ventura, 'Full Description of Benjamin–Feir Instability of Stokes Waves in Deep Water', arXiv:2109.11852v3 [math.AP], 26 May 2022.

NOTES ON CONTRIBUTORS

Nigel Aston is an Honorary Fellow in the School of History, Politics, and International Relations at the University of Leicester, and a Research Associate at the University of York. Educated at Durham and Christ Church, Oxford, he has written widely on British and French eighteenth-century religious, political, and intellectual history. His most recent publication was *The Anglican Episcopate 1689–1800* (University of Wales Press, 2023), co-edited with William Gibson. His next book, *Enlightened Oxford: The University in the Cultural and Political Life of Eighteenth-Century Britain and Beyond* is published in 2023 by Oxford University Press.

Sir John Ball FRS FRSE is Professor of Mathematics at Heriot-Watt University and an emeritus and visiting professor in Oxford, where he is also an Emeritus Fellow of The Queen's College. He is a former delegate of Oxford University Press. His research concerns nonlinear analysis and the calculus of variations and their applications to materials science, for example to martensitic phase transformations and liquid crystals. Among various distinctions he was awarded the 2018 King Faisal Prize for Science and the 2018 Leonardo da Vinci Award of the European Academy of Sciences. He was President of the International Mathematical Union from 2003 to 2006. He is current President of the Royal Society of Edinburgh.

June Barrow-Green is Emeritus Professor of History of Mathematics at the Open University and a visiting professor at the London School of Economics. She is Chair of the Executive Committee of the International Commission on the History of Mathematics and a past Chair of the British Society for the History of Mathematics. Her current research focuses on the history of nineteenth- and twentieth-century Western mathematics, particularly in Britain, and she has a special interest in the history of the gender gap in mathematics. She is the author of *Poincaré and the Three-Body Problem*, an editor of the *Princeton Companion to Mathematics*, and has co-authored a two-volume textbook, *The History of Mathematics: A Source-Based Approach*. Among her recent publications are studies on the role of Cambridge mathematicians during WW1, the geometric surface models of Olaus Henrici, and the contribution of Hilda Hudson to mathematical epidemiology. She is currently working on the historical representation of women in mathematics. In 2021 she was awarded the Royal Society Wilkins-Bernal-Medawar Medal for her work in the history of mathematics.

Jim Bennett is Keeper Emeritus at the Science Museum, London. He previously held curatorial posts in the Universities of Cambridge and of Oxford, where he was awarded the title professor and is currently an Emeritus Fellow of Linacre College. He is President of the Hakluyt Society. His research interests have centred on the histories of scientific instruments, astronomy, and practical mathematics. Among his publications are two articles on the subject of his chapter, Thomas Hornsby. Early in his career, while Archivist at the Royal Astronomical Society, he catalogued the Hornsby papers there.

Peter Cargill is an emeritus professor at the Blackett Laboratory, Imperial College and honorary professor at the School of Mathematics, University of St Andrews. He has published extensively over four decades on topics including the solar corona, the interplanetary medium, the Earth's magnetosphere, space weather, and more generally on shock waves in space. He is a former vice-president of the Royal Astronomical Society (RAS) and in 2013 gave the inaugural James Dungey Lecture to the RAS, named after a pioneer in solar–terrestrial relations.

Alastair Compston CBE FMedSci FRS is Professor Emeritus of Neurology in the University of Cambridge, and Fellow of Jesus College, Cambridge. He is a former president of the European Neurological Society and the Association of British Neurologists, and editor of *Brain*, a journal of neurology. His research on the clinical science of human demyelinating disease has been recognized by several international prizes. In retirement, he writes on medical history: *'All Manner of Ingenuity and Industry': A Bio-bibliography of Thomas Willis* was published by Oxford University Press in 2021.

Christopher Hollings is Departmental Lecturer in Mathematics and its History at the Oxford Mathematical Institute and Clifford Norton Senior Research Fellow in the History of Mathematics at The Queen's College, Oxford. His research interests cover a range of topics in nineteenth- and twentieth-century mathematics.

Jon Keating FRS is the current Sedleian Professor of Natural Philosophy at the University of Oxford. He gained a BA in physics from Oxford in 1985, and a PhD in theoretical physics from the University of Bristol in 1989. He works in the areas of mathematical physics relating to quantum chaos, random matrix theory, and semiclassical asymptotics, as well as in connections between mathematical physics and number theory. He was elected a Fellow of the Royal Society in 2009, and in 2010 was awarded the London Mathematical Society's Fröhlich Prize. He was President of the London Mathematical Society from 2019 to 2021.

Mark McCartney is senior lecturer in mathematics at Ulster University and a past president of the British Society for the History of Mathematics. His research interests include mathematical modelling, discrete chaos, and the history of mathematics and physics in the nineteenth and twentieth centuries.

Tom Mullin is an emeritus professor of physics at the University of Manchester and is currently a visitor at the Mathematical Institute, University of Oxford. He was the founding director of the Manchester Centre for Nonlinear Dynamics, an interdisciplinary research centre located in the Schools of Physics and Mathematics at the University of Manchester. He is a Fellow of the Royal Society of Edinburgh and the American Physical Society. His research is on experimental investigations into transition to turbulence, instabilities in elastic materials, particle motion in viscous fluids, and pattern formation in granular flows.

John Toland FRS FRSE studied for a Sussex DPhil on topological methods for nonlinear equations. In 1973 he joined the Fluid Mechanics Research Institute (FMRI) at Essex University where Brooke Benjamin was Director, and when Brooke became Sedelian Professor at Oxford he moved to Keith Stewartson's group at University College London. Then in 1982 he was appointed Professor of Mathematics at the University of Bath where, between 2002 and 2010, he was also Director of the International Centre for Mathematical Sciences in Edinburgh. In 2011 he became Director of the Isaac Newton Institute for Mathematical Sciences in Cambridge and in 2016 returned to Bath as emeritus professor. Much of his research on the rigorous mathematical theory of nonlinear waves, for which he was awarded the Royal Society Sylvester Medal, was influenced by many interactions with Brooke and his group at FMRI.

PICTURE SOURCES

CHAPTER 1: FOUR CENTURIES OF SEDLEIAN PROFESSORS

1.1 Sir Henry Savile. Source: Wikimedia Commons, Public Domain: https://commons.wikimedia.org/wiki/File:Henry_Savile.jpg.

1.2 Sedley family arms. Credit: The Master and Fellows of Trinity College, Cambridge.

1.3 St Nicholas's Church, Southfleet. Source: Brandon's *Parish Churches*, 1848.

1.4 Sedley family tomb, St Nicholas's Church, Southfleet. Credit: Christopher Hollings.

1.5 Thomas Willis. Source: Wikimedia Commons, Public Domain: https://commons.wikimedia.org/wiki/File:Thomas_Willis_ODNB.jpg.

1.6 John Keill, *Introductio ad veram physicam* (1701). Source: Eighteenth Century Collections Online, https://link.gale.com/apps/doc/CW0107139987/ECCO?u=oxford&sid=gale_marc&xid=f8513ca5&pg=1).

1.7 Baden Powell. Source: Wikimedia Commons, Public Domain: https://commons.wikimedia.org/wiki/File:Rev_Baden_Powell.jpg. Baden Powell, *The present state and future prospects of mathematical and physical studies in the University of Oxford, considered in a lecture* (1832). Source: SOLO, https://solo.bodleian.ox.ac.uk/permalink/f/89vilt/oxfaleph014415543.

1.8 Francis Jeune. Source: Wikimedia Commons, Public Domain: https://commons.wikimedia.org/wiki/File:Dr_Francis_Jeune.jpg.

1.9 The Queen's College. Credit: With the kind permission of the Provost, Fellows, and Scholars of The Queen's College Oxford.

1.10 William Thomson. Credit: With the kind permission of the Provost, Fellows, and Scholars of The Queen's College Oxford.

1.11 Upper Library, The Queen's College. Credit: Christopher Hollings.

CHAPTER 2: THOMAS WILLIS

2.1 Thomas Willis, engraved by George Vertue (1742), in Thomas Birch, *Heads of various illustrious persons of Great Britain* (1743–1751).

2.2 Image of Numbers 3 and 4 Merton Street, courtesy of Sam Hill. Image of plaque in memory of Thomas Willis, courtesy of Julian Reid.

2.3 The west prospect of St Martin's Church in the Fields, Westminster, by George Vertue (1744).

2.4 Contemporary depiction of the story of Anne Greene. Printed by T. Clowes (London, 1651). © Bodleian Libraries, the University of Oxford.

2.5 John Locke's notebook. © British Library.

2.6 Thomas Willis, *Pharmaceutice rationalis*, part 1 (1674), printed at Oxford University Press; engraving of the Sheldonian Theatre by David Loggan.

2.7 Thomas Willis, *Diatribæ duæ medico-philosophicæ* (London, 1660).

2.8 Thomas Willis, *Pharmaceutice rationalis* part 1 (Amsterdam, 1674).

2.9 Thomas Willis, *Opera Omnia* (Amsterdam, 1682).

2.10 Thomas Willis, *Cerebri anatome* (Amsterdam, 1666).

2.11 Tabula IX from Thomas Willis, *Cerebri anatome* (London, 1664).

2.12 Figura 1 (detail) from Thomas Willis, *Cerebri anatome* (London, 1664).

CHAPTER 3:   THE SEDLEIAN PROFESSORS OF THE EIGHTEENTH CENTURY

3.1 Oxford High Street, *c*.1750.

3.2 Sir Thomas Millington, mislabelled. Source: Wellcome Collection (https://www.jstor.org/stable/community.24813522).

3.3 Plan of the Oxford curriculum, *c*.1709.

3.4 Bernard Gardiner. Source: All Souls College, Oxford.

3.5 The Oxford Physic Garden, domain of the professor of botany.

CHAPTER 4:   THOMAS HORNSBY

4.1 John Bird. Image: Wikimedia Commons.

4.2 The Old Ashmolean Building. Image: author's collection.

4.3 One of two 8-ft-radius mural quadrants, Radcliffe Observatory. Image: Wikimedia Commons.

4.4 The Radcliffe Observatory. Image: author's collection.

4.5 Hornsby's transit observations. Image: Royal Astronomical Society.

4.6 Printed notice of Hornsby's Sedleian lectures. Image: RT Gunther, *Early Science in Oxford*, Volume 11 (Oxford, 1937).

CHAPTER 5:   GEORGE LEIGH COOKE

5.1 William Buckland. Credit: Wellcome Collection 544983i. (Source: Wellcome Collection, Public Domain, https://wellcomecollection.org/works/j9xh93sa).

5.2 Leigh, Cooke, and Austen family tree. Credit: Christopher Hollings using Family Echo (https://www.familyecho.com/).

5.3 Jane Austen. Source: Wikimedia Commons, Public Domain: https://commons.wikimedia.org/wiki/File:CassandraAusten-JaneAusten(c.1810)_hires.jpg).

5.4 Corpus Christi College, Oxford. Source: author's collection.

5.5 Martin Joseph Routh. Credit: With the kind permission of the Provost, Fellows, and Scholars of The Queen's College, Oxford.

5.6 A bill advertising Cooke's lectures, 1825. Source: Bodleian Library, Oxford, G.A.Oxon b.20.

5.7 George Leigh Cooke, *The three first sections and part of the seventh section of Newton's Principia: with a preface recommending a geometrical course of mathematical reading, and an introduction on the atomic constitution of matter, and the laws of motion* (Oxford, 1850). Source: SOLO, https://solo.bodleian.ox.ac.uk/permalink/44OXF_INST/35n82s/alma990129140790107026.

5.8 St Mary's Church, Cubbington. Credit: Christopher Hollings.

5.9 Parish Church of the Blessed Virgin Mary, Beckley, near Oxford. Credit: Christopher Hollings.

5.10 Cooke's tombstone in Beckley Cemetery. Credit: Christopher Hollings.

CHAPTER 6:   BARTHOLOMEW PRICE

6.1 Bartholomew Price. Credit: London Mathematical Society Tucker Collection.

6.2 Church of St James the Great, Coln St Dennis, Gloucestershire. Source: author's collection.

6.3 Pembroke College, Oxford. Source: author's collection.

6.4 Advertisements of lecture courses, *Oxford University Gazette* (14 October 1892). Source: scanned from the copy of the *Gazette* in The Queen's College Library.

6.5 Bartholomew Price. Source: Bodleian Library, OU_BODL_LP317.

6.6 Oxford University Press. Source: *The Penny Magazine of the Society for the Diffusion of Useful Knowledge* (31 January 1835), p. 48.

6.7 Bartholomew Price, *Treatise on the Differential Calculus* (1852). Source: SOLO, https://solo.bodleian.ox.ac.uk/permalink/44OXF_INST/35n82s/alma990156934400107026.

6.8 Charles Lutwidge Dodgson, a.k.a. Lewis Carroll. Source: Wikimedia Commons, Public Domain, https://commons.wikimedia.org/wiki/File:LewisCarrollSelfPhoto.jpg.

6.9 Bartholomew Price. Fig. 1.4, in *The History of Oxford University Press*, vol. II, ch. 1. https://doi.org/10.1093/acprof:oso/9780199543151.003.0002.

6.10 Bartholomew Price, *An Essay on the relation of several parts of a mathematical science to the fundamental idea therein contained*. Source: author's collection.

6.11 Astronomical clock dedicated to Price in Gloucester Cathedral. Credit: Christopher Hollings.

6.12 Price's tombstone, Holywell Cemetery, Oxford. Credit: Christopher Hollings.

CHAPTER 7: AUGUSTUS LOVE

7.1 Augustus Love, Second Wrangler. *The Graphic* (4 July 1885), p. 4.

7.2 Augustus Love, *A treatise on the mathematical theory of elasticity*, volume 1 (1892) (Balliol College, Oxford).

7.3 Augustus Love, *Lehrbuch der Elastizität* (1907) (British Library).

7.4 Augustus Love. Courtesy of the London Mathematical Society.

7.5 Rayleigh wave and Love wave. Source: Wikimedia commons. https://opentextbc.ca/geology/chapter/11-3-measuring-earthquakes/.

7.6 Twisted railway lines in the aftermath of an earthquake. Figure 5 in Piotr Kiełczyński, 'Propeties and Applications of Love Surface Waves in Seismology and Biosensors'. https://www.intechopen.com/chapters/60627.

7.7 Love's Royal Society Certificate of Election. Courtesy: The Royal Society.

7.8 Augustus Love, by C. K. Thornhill (*c*.1938).

CHAPTER 8: SYDNEY CHAPMAN

8.1 Sydney Chapman. (a) Courtesy: Royal Astronomical Society; (b) courtesy: University of Alaska.

8.2 Adapted from S. Chapman, 'Some Phenomena of the Upper Atmosphere', *Proc. Roy. Soc.* 132 (1931), 353–74.

8.3 The Bastille day flare and associated geomagnetic storm. (a, b) Reproduced courtesy of the Extreme Ultraviolet Imaging Telescope and the Large Angle and Spectrometric Coronagraph Experiment instrument teams, respectively. (c) Courtesy of the NASA Omniweb data facility].

8.4 Three possibilities of storm-associated electric currents. Adapted from Figure 3 of I. de Moortel, Isobel Falconer, and Robert Stack, 'Alfvén on Heating by Waves', A&G, 61(2) (2020), 2.34–9.

8.5 The Chapman–Ferraro model. From S. Chapman and V. C. A. Ferraro, 'A New Theory of Magnetic Storms', Nature 126 (1930) 139–40.

8.6 (a) Sketch of magnetic field lines. From J. W. Dungey, 'Interplanetary Magnetic Fields and the Auroral Zone', PRL 6, (1961), 47–8. (b, c) Ancestry unknown.

8.7 Magnetic field associated with the CME in interplanetary space, and the D_{st} index for the large geomagnetic storm associated with the July 2000 Bastille day flare. Data courtesy of NASA/Omniweb.

CHAPTER 9: GEORGE TEMPLE AND ALBERT GREEN

9.1 George Temple and his wife Dorothy. Image courtesy of Quarr Abbey.

9.2 George Temple in 1947 and in the 1980s. Images courtesy of Quarr Abbey.

9.3 Three professors watching ping-pong. Image courtesy of the Coulson family.

9.4 George Temple, *The Classic and Romantic in Natural Philosophy* (Clarendon Press, 1954).

9.6 George Temple's inscription in the copy of *100 Years of Mathematics*. Image courtesy of Quarr Abbey.

9.7 Prefects of 1930–1, Haberdashers' Aske's Hampstead School. Source: Haberdashers' Boy's School Archive.

9.8 Apparatus for measuring the elastic stability of a metal strip in. Source: A. E. Green, 'The Equilibrium and Elastic Stability of a Thin Twisted Strip', *Proc. Roy. Soc. Lond. A* 154 (1936), 452.

9.10 (a) A. E. Green and W. Zerna, *Theoretical Elasticity* (Clarendon Press, 1954). (b) A. E. Green and J. E. Adkins, *Large Elastic Deformations and Non-linear Continuum Mechanics* (Clarendon Press, 1960).

9.11 Albert Green. Image courtesy of Durham University Library and Collections.

CHAPTER 10: BROOKE BENJAMIN

10.1 Brooke Benjamin. Photo: Natalia Benjamin.

10.2 Jet formation during the high-speed collapse of a vapour cavity adjacent to a solid boundary. Source: *Philosophical Transactions of the Royal Society A* 260 (1966), plate 48.

10.3 The breakdown of a vortex core in air. Source: *Journal of Fluid Mechanics* 14 (1962), plate 1.

10.4 Russell's illustrations of his Great Wave of Translation. Source: *Report of the fourteenth meeting of the British Association for the Advancement of Science* (1845), plate 47.

10.5 Experiments performed in a large wave tank facility at the National Physical Laboratory, Feltham. Source: *Proceedings of the Royal Society A* 299 (1967), plate 1.

10.6 Cross-section of an anomalous, steady, three-cell flow of viscous liquid Source: K. A. Cliffe and T. Mullin, 'A Numerical and Experimental Study of Anomalous Modes in the Taylor Experiment', *Journal of Fluid Mechanics* 153 (1985), 243–58.

10.7 Buckling under shear of a 1-mm layer of very viscous oil. Source: *Journal of Fluid Mechanics* 195 (1988), 536.

10.8 Benjamin Brooke. Source (LHS): National Portrait Gallery. Photo (RHS): John Toland.

CHAPTER 11: AN INTERVIEW WITH JOHN BALL

11.1 John Ball, 1983. Image courtesy of George M. Bergman, Archives of the Mathematisches Forschungsinstitut Oberwolfach.

11.2 John Ball and family and King Juan Carlos of Spain, 2006. Image courtesy of John Ball.

11.3 John Ball and Kiyosi Itô, 2006. Image courtesy of John Ball.

11.4 John Ball, 2022. Images courtesy of the Royal Society of Edinburgh.

INDEX

Monthly Notices of the Royal Astronomical Society, 87–89, 171–172
Moore, Edward, 117–118
Mott, Sir Nevil, 168
Müller, Stefan, 222, 228
Mullin, Tom, 213
muscle contraction, 54

Naghdi, Paul, 197*f*, 198–200, 202
Nairne, Edward (astronomical instrument maker), 82–83
National Conference of University Professors, 214
National Mathematics Research Institute, 223
natural history, 89
The Natural History of Oxfordshire (Plot), 89
The Natural History of Staffordshire (Plot), 89
natural philosophy, 15, 89
    conception within university, 15
    eighteenth century, 61–62, 93
    George Leigh Cooke, 103
    George Temple, 186–188
    seventeenth century, 1–3
    Thomas Hornsby, 77
Navier-stokes equations, 227–228
neo-Hookian model, elasticity, 225
nervous system/neurology, 39–41, 50, 51, 52*f*, 56–57, 63–64
Neumann, Peter, 183–184
Newcastle University, 196
Newman, Francis William, 105
Newton, Sir Isaac, 16, 61–62
Newtonian philosophy, 12–13, 15, 16, 64, 93, 100
Niblett, Stephen, 67–68
nineteenth century, Sedleian professors, 16–21
nonlinear elasticity, 196–197
nonlinear partial differential equations, 199, 210–211
nonuniform gases, kinetic theory, 160–161

Ockendon, John, 200–201
O'Meara, Edmund, 59
On a new proof of the principle of virtual velocities (Price), 136
On some applications of the theory of probabilities (Price), 136
On the extension of Taylor's Theorem by the method of derivatives (Price), 136

On the principle of virtual velocities (Price), 134–136
One Hundred Years of Mathematics (Temple), 190*f*, 191
Opera omnia (Willis), 42, 46*f*
optics, 85
ordination, George Temple, 190–191
Orger, Edward Redman, 117
Oriel College, Oxford, 16–17, 131–132
orrery, 83–84, 225
Oslo conference, 151
Ovenden, Michael, 218
Oxford, eighteenth century, 62*f*
Oxford Dictionary of National Biography, 73
Oxford Experimental Philosophical Club, 10, 36–37, 42
Oxford Mathematical Instituute, 215–216
Oxford Mathematical Society, Bartholomew Price, 137–138
Oxford Movement, 105–106
Oxford Physic Garden, 69*f*
Oxford University, collegiate system, 1–3
Oxford University Gazette, 115–117, 116*f*
Oxford University Press, 118
    Bartholomew Price, 122–124, 136–137
    printing offices, 123*f*
ozone formation, 161*f*, 161–162

pacifism
    Albert Green, 195
    Sydney Chapman, 167–168
parallax, double star, 84
parallax, solar, 79–80
Parker, Eugene, 171
Pars, L. A., 192–195
Parsons, John, 16–17, 100
partial diferential equations, 199, 210, 227
particle belts, 166
Pathologiae cerebri (Willis), 46–48, 50–51, 56
patronage system, eighteenth century, 13, 62–63
Pearson, Karl, 153
Peel, Sir Robert, 101–103, 105–106
Peierls, Sir Rudolf, 168
Pembroke College, Oxford, 114*f*
    Bartholomew Price, 113, 117–122, 128–130, 136–138
    Francis Jeune, 19

William Price, 111–113
Penney, Bill, 159, 168
Pennsylvania State University, 214–215
Penrose, Roger, 219–220
personal connection, 71–72
Petty, William (surgeon), 34, 36
Pharmaceutice rationalis (Willis), 29, 41, 42, 43*f*, 45*f*, 48
philanthropy, 137–138
Philosophical Transactions of the Royal Society, 64, 153
    John Ball, 225
    Thomas Hornsby, 78–79
Philosophy Farm *see* Waddesdon manor, Buckinghamshire
physical mathematics *see* applied mathematics
Physical Society of London, 137–138
physicians, 9–13
physics, 16
Pidduck, Frederick, 149
Pingré, Alexandre Guy, 78, 80
plague, 50
A plain and easie method etc. (Willis), 50
Plot, Robert, 39, 89
pluralism, 18, 62–63, 71–72
Poisson, Siméon Denis, 146–149
Powell, Baden, 17, 18*f*, 19, 105, 119, 137
practical *vs.* theoretical aspects, mathematical teaching 19
The Present State and Future Prospects of Mathematical and Physical Studies at the University of Oxford (Powell), 17
Price, Amy Eliza (née Cole), 117–118, 140
Price, Bartholomew, 17, 19, 24, 105, 111, 112*f*, 113*f*, 130*f*, 143
    early lives and careers, 111–113
    external interests, 136–138
    final years, 138–140
    lecture, 135*f*
    Master of Pembroke College, 121*f*
    mathematical study, improvement, 128–134
    original mathematics, 134–136
    Pembroke College, 117–122
    Sedleian professor, 114–117
    tombstone, 139*f*
    university roles, 122–124
    writings, 124–128
Price, William Henry, 119